MYSTERIOUS INVADERS

THE LOCHLAINN SEABROOK COLLECTION

AMERICAN CIVIL WAR
Abraham Lincoln Was a Liberal, Jefferson Davis Was a Conservative: The Missing Key to Understanding the American Civil War
Confederacy 101: Amazing Facts You Never Knew About America's Oldest Political Tradition
Confederate Blood and Treasure: An Interview With Lochlainn Seabrook
Everything You Were Taught About African-Americans and the Civil War is Wrong, Ask a Southerner!
Everything You Were Taught About the Civil War is Wrong, Ask a Southerner!
Give This Book to a Yankee! A Southern Guide to the Civil War For Northerners
Heroes of the Southern Confederacy: The Illustrated Book of Confederate Officials, Soldiers, and Civilians
Lincoln's War: The Real Cause, the Real Winner, the Real Loser
Seabrook's Complete Battle Book: War Between the States, 1861-1865
The Great Yankee Coverup: What the North Doesn't Want You to Know About Lincoln's War!
The Hampton Roads Conference: The Southern View
The Ultimate Civil War Quiz Book: How Much Do You Really Know About America's Most Misunderstood Conflict?
Women in Gray: A Tribute to the Ladies Who Supported the Southern Confederacy

CONFEDERATE MONUMENTS
Confederate Monuments: Why Every American Should Honor Confederate Soldiers and Their Memorials

CONFEDERATE FLAG
Confederate Flag Facts: What Every American Should Know About Dixie's Southern Cross
What the Confederate Flag Means to Me: Americans Speak Out in Defense of Southern Honor, Heritage, and History

SECESSION
All We Ask Is To Be Let Alone: The Southern Secession Fact Book

RECONSTRUCTION
Twelve Years in Hell: Victorian Southerners Debunk the Myth of Reconstruction, 1865-1877

SLAVERY
Everything You Were Taught About American Slavery is Wrong, Ask a Southerner!
Slavery 101: Amazing Facts You Never Knew About America's "Peculiar Institution"
The Bittersweet Bond: Race Relations in the Old South as Described by White and Black Southerners

NATHAN BEDFORD FORREST
A Rebel Born: A Defense of Nathan Bedford Forrest -Confederate General, American Legend (winner of the 2011 Jefferson Davis Historical Gold Medal)
A Rebel Born: The Screenplay (film about N. B. Forrest)
Forrest! 99 Reasons to Love Nathan Bedford Forrest
Give 'Em Hell Boys! The Complete Military Correspondence of Nathan Bedford Forrest
I Rode With Forrest! Confederate Soldiers Who Served With the World's Greatest Cavalry Leader
Nathan Bedford Forrest and African-Americans: Yankee Myth, Confederate Fact
Nathan Bedford Forrest and the Battle of Fort Pillow: Yankee Myth, Confederate Fact
Nathan Bedford Forrest and the Ku Klux Klan: Yankee Myth, Confederate Fact
Nathan Bedford Forrest: Southern Hero, American Patriot -Honoring a Confederate Icon and the Old South
Saddle, Sword, and Gun: A Biography of Nathan Bedford Forrest For Teens
The God of War: Nathan Bedford Forrest As He Was Seen By His Contemporaries
The Quotable Nathan Bedford Forrest: Selections From the Writings and Speeches of the Confederacy's Most Brilliant Cavalryman

QUOTABLE SERIES
The Alexander H. Stephens Reader: Excerpts From the Works of a Confederate Founding Father
The Quotable Alexander H. Stephens: Selections From the Writings and Speeches of the Confederacy's First Vice President
The Quotable Jefferson Davis: Selections From the Writings and Speeches of the Confederacy's First President
The Quotable Nathan Bedford Forrest: Selections From the Writings and Speeches of the Confederacy's Most Brilliant Cavalryman
The Quotable Robert E. Lee: Selections From the Writings and Speeches of the South's Most Beloved Civil War General
The Quotable Stonewall Jackson: Selections From the Writings and Speeches of the South's Most Famous General
The Unquotable Abraham Lincoln: The President's Quotes They Don't Want You To Know!

CIVIL WAR BATTLES
Encyclopedia of the Battle of Franklin -A Comprehensive Guide to the Conflict that Changed the Civil War
Nathan Bedford Forrest and the Battle of Fort Pillow: Yankee Myth, Confederate Fact
Seabrook's Complete Battle Book: War Between the States, 1861-1865
The Battle of Franklin: Recollections of Confederate and Union Soldiers
The Battle of Nashville: Recollections of Confederate and Union Soldiers
The Battle of Spring Hill: Recollections of Confederate and Union Soldiers

CONSTITUTIONAL HISTORY
America's Three Constitutions: Complete Texts of the Articles of Confederation, Constitution of the United States of America, and Constitution of the Confederate States of America
The Articles of Confederation Explained: A Clause-by-Clause Study of America's First Constitution
The Constitution of the Confederate States of America Explained: A Clause-by-Clause Study of the South's Magna Carta

CHILDREN
Honest Jeff and Dishonest Abe: A Southern Children's Guide to the Civil War
Saddle, Sword, and Gun: A Biography of Nathan Bedford Forrest For Teens

VICTORIAN CONFEDERATE LITERATURE
I, Confederate: Why Dixie Seceded and Fought in the Words of Southern Soldiers
Rise Up and Call Them Blessed: Victorian Tributes to the Confederate Soldier, 1861-1901
Support Your Local Confederate: Wit and Humor in the Southern Confederacy
The Bittersweet Bond: Race Relations in the Old South as Described by White and Black Southerners
The God of War: Nathan Bedford Forrest As He Was Seen By His Contemporaries
The Old Rebel: Robert E. Lee As He Was Seen By His Contemporaries
Victorian Confederate Poetry: The Southern Cause in Verse, 1861-1901

ABRAHAM LINCOLN
Abraham Lincoln: The Southern View -Demythologizing America's Sixteenth President
Lincolnology: The Real Abraham Lincoln Revealed in His Own Words -A Study of Lincoln's Suppressed, Misinterpreted, and Forgotten Writings and Speeches
Lincoln's War: The Real Cause, the Real Winner, the Real Loser
The Great Impersonator! 99 Reasons to Dislike Abraham Lincoln
The Unholy Crusade: Lincoln's Legacy of Destruction in the American South
The Unquotable Abraham Lincoln: The President's Quotes They Don't Want You To Know!

NATURAL HISTORY
North America's Amazing Mammals: An Encyclopedia for the Whole Family
The Concise Book of Owls: A Guide to Nature's Most Mysterious Birds
The Concise Book of Tigers: A Guide to Nature's Most Remarkable Cats

PARANORMAL
Carnton Plantation Ghost Stories: True Tales of the Unexplained from Tennessee's Most Haunted Civil War House!
Mysterious Invaders: Twelve Famous 20th-Century Scientists Confront the UFO Phenomenon
UFOs and Aliens: The Complete Guidebook

FAMILY HISTORIES
The Blakeneys: An Etymological, Ethnological, and Genealogical Study -Uncovering the Mysterious Origins of the Blakeney Family and Name
The Caudills: An Etymological, Ethnological, and Genealogical Study -Exploring the Name and National Origins of a European-American Family
The McGavocks of Carnton Plantation: A Southern History -Celebrating One of Dixie's Most Noble Confederate Families and Their Tennessee Home

MIND, BODY, SPIRIT
Autobiography of a Non-Yogi: A Scientist's Journey From Hinduism to Christianity (Dr. Amitava Dasgupta, with Lochlainn Seabrook)
Britannia Rules: Goddess-Worship in Ancient Anglo-Celtic Society -An Academic Look at the United Kingdom's Matricentric Spiritual Past
Christ Is All and In All: Rediscovering Your Divine Nature and the Kingdom Within
Christmas Before Christianity: How the Birthday of the "Sun" Became the Birthday of the "Son"
Jesus and the Gospel of Q: Christ's Pre-Christian Teachings As Recorded in the New Testament
Jesus and the Law of Attraction: The Bible-Based Guide to Creating Perfect Health, Wealth, and Happiness Following Christ's Simple Formula
Seabrook's Bible Dictionary of Traditional and Mystical Christian Doctrines
Sea Raven Press Blank Page Journal: For Reflections, Notes, and Sketches
Secrets of Celebrity Surnames: An Onomastic Dictionary of Famous People
The Bible and the Law of Attraction: 99 Teachings of Jesus, the Apostles, and the Prophets
The Book of Kelle: An Introduction to Goddess-Worship and the Great Celtic Mother-Goddess Kelle, Original Blessed Lady of Ireland
The Goddess Dictionary of Words and Phrases: Introducing a New Core Vocabulary for the Women's Spirituality Movement
The Martian Anomalies: A Photographic Search for Intelligent Life on Mars
Victorian Hernia Cures: Nonsurgical Self-Treatment of Inguinal Hernia
Vintage Southern Cookbook: 2,000 Delicious Dishes From Dixie

WOMEN
Aphrodite's Trade: The Hidden History of Prostitution Unveiled
Princess Diana: Modern Day Moon-Goddess -A Psychoanalytical and Mythological Look at Diana Spencer's Life, Marriage, and Death (with Dr. Jane Goldberg)
Women in Gray: A Tribute to the Ladies Who Supported the Southern Confederacy

REPRINTS
A Short History of the Confederate States of America (author Jefferson Davis; editor Lochlainn Seabrook)
Prison Life of Jefferson Davis (author John J. Craven; editor Lochlainn Seabrook)
Life of Beethoven (author Ludwig Nohl; editor Lochlainn Seabrook)
The New Revelation (author Arthur Conan Doyle; editor Lochlainn Seabrook)
The Rise and Fall of the Confederate Government (author Jefferson Davis; editor Lochlainn Seabrook)

Lochlainn Seabrook does not author books for fame and glory, but for the love of writing and sharing his knowledge.

SeaRavenPress.com

Warning: SEA RAVEN PRESS BOOKS WILL EXPAND YOUR MIND!

MYSTERIOUS INVADERS

Twelve Famous 20th-Century Scientists Confront the UFO Phenomenon

CONCEIVED, COLLECTED, EDITED, ARRANGED, & DESIGNED WITH AN INTRODUCTION BY
UFO RESEARCHER, UFO EYE-WITNESS, & UFO AUTHOR

LOCHLAINN SEABROOK

JEFFERSON DAVIS HISTORICAL GOLD MEDAL WINNER

Diligently Researched and Generously Illustrated by the Author for the Elucidation of the Reader

2024

Sea Raven Press, Park County, Wyoming, USA

MYSTERIOUS INVADERS

Published by
Sea Raven Press, Cassidy Ravensdale, President
Park County, Wyoming, USA
SeaRavenPress.com

SEA RAVEN PRESS
ARTISAN-CRAFTED BOOKS & MERCH FROM THE ROCKY MOUNTAINS

Copyright © all text and illustrations Lochlainn Seabrook 2024
in accordance with U.S. and international copyright laws and regulations, as stated and protected under the Berne Union for the Protection of Literary and Artistic Property (Berne Convention), and the Universal Copyright Convention (the UCC). All rights reserved under the Pan-American and International Copyright Conventions.

PRINTING HISTORY
1st SRP paperback edition, 1st printing, January 2024 • ISBN: 978-1-955351-36-2
1st SRP hardcover edition, 1st printing, January 2024 • ISBN: 978-1-955351-37-9

ISBN: 978-1-955351-36-2 (paperback)
Library of Congress Control Number: 2024931424

This work is the copyrighted intellectual property of Lochlainn Seabrook and has been registered with the Copyright Office at the Library of Congress in Washington, D.C., USA. No part of this work (including text, covers, drawings, photos, illustrations, maps, images, diagrams, etc.), in whole or in part, may be used, reproduced, stored in a retrieval system, or transmitted, in any form or by any means now known or hereafter invented, without written permission from the publisher. The sale, duplication, hire, lending, copying, digitalization, or reproduction of this material, in any manner or form whatsoever, is also prohibited, and is a violation of federal, civil, and digital copyright law, which provides severe civil and criminal penalties for any violations.

Mysterious Invaders: Twelve Famous 20th-Century Scientists Confront the UFO Phenomenon, by Lochlainn Seabrook. Includes an introduction, illustrations, artwork, index, endnotes, appendices, and bibliography.

ARTWORK
Front and back cover design and art, book design, layout, font selection, and interior art by Lochlainn Seabrook
All images, image captions, graphic design, and graphic art copyright © Lochlainn Seabrook
All images selected, placed, manipulated, cleaned, colored, tinted, and/or created by Lochlainn Seabrook
Cover image, "Nighttime Flying Saucer Over the Mountains": Volodimir Zozulinskyi

All persons who approve of the authority and principles of Colonel Lochlainn Seabrook's literary work, and realize its benefits as a means of reeducating the world about facts left out of mainstream books, are hereby requested to avidly recommend his titles to others and to vigorously cooperate in extending their reach, scope, and influence around the globe.

The views documented in this book concerning the UFO phenomenon are those of the publisher.
WRITTEN, DESIGNED, PUBLISHED, PRINTED, & MANUFACTURED IN THE UNITED STATES OF AMERICA

REAL HISTORY MATTERS

DEDICATION

To those rare genuine scientists who approach the topic of ufology scientifically, and who are using the scientific method to help uncover the truth about UFOs, USOs, and UAPs.

EPIGRAPH

"We must rapidly escalate serious scientific attention to this extraordinarily intriguing puzzle. I believe that the scientific community has been seriously misinformed about the potential importance of UFOs."

Dr. James E. McDonald

CONTENTS

Introduction, by Lochlainn Seabrook ❧ page 11

SECTION ONE
SIX SCIENTISTS SPEAK BEFORE THE U.S. CONGRESS
Introduction to Symposium ❧ page 17
CHAPTER 1: Statement of J. Allen Hynek ❧ page 21
CHAPTER 2: Statement of James E. McDonald ❧ page 41
CHAPTER 3: Statement of Carl Sagan ❧ page 159
CHAPTER 4: Statement of Robert L. Hall ❧ page 177
CHAPTER 5: Statement of James A. Harder ❧ page 197
CHAPTER 6: Statement of Robert M. L. Baker, Jr. ❧ page 209
CHAPTER 7: Free Discussion Among the Six Scientists ❧ page 241

SECTION TWO
SUBMITTED PREPARED STATEMENTS
BY SIX ADDITIONAL SCIENTISTS
CHAPTER 8: Statement of Donald H. Menzel ❧ page 253
CHAPTER 9: Statements of Stanton T. Friedman ❧ page 267
CHAPTER 10: Statement of R. Leo Sprinkle ❧ page 283
CHAPTER 11: Statement of Garry C. Henderson ❧ page 289
CHAPTER 12: Statement of Roger N. Shepard ❧ page 295
CHAPTER 13: Statement of Frank B. Salisbury ❧ page 313

APPENDICES

Appendix A: Committee on Science & Aeronautics ❧ page 339
Appendix B: Biography of Dr. J. Allen Hynek ❧ page 341
Appendix C: Biography of Dr. James E. McDonald ❧ page 343
Appendix D: Biography of Dr. Carl Sagan ❧ page 345
Appendix E: Biography of Dr. Robert L. Hall ❧ page 347
Appendix F: Biography of Dr. James A Harder ❧ page 349
Appendix G: Biography of Dr. Robert M. L. Baker, Jr. ❧ page 351
Appendix H: Biography of Dr. Donald H. Menzel ❧ page 353
Appendix I: Biography of Stanton T. Friedman ❧ page 355
Appendix J: Biography of R. Leo Sprinkle ❧ page 357
Appendix K: Biography of Dr. Garry C. Henderson ❧ page 361
Appendix L: Biography of Roger N. Shepard ❧ page 363
Appendix M: Biography of Dr. Frank B. Salisbury ❧ page 365

Notes ❧ page 367
Bibliography ❧ page 371
Index ❧ page 373
Meet the Author-Editor ❧ page 407
Learn More ❧ page 409

INTRODUCTION

Mysterious Invaders is not my reprint of an old book, as some might assume. Like *all* of my other edited works, it is my creation of a brand new book, carefully and arduously pulled together from an often disorganized hodge-podge of sources—in this case, U.S. Congressional hearings.

The specific hearing that is the topic of this book took place in Washington, D.C., on Monday, July 29, 1968, before the U.S. House of Representatives' Committee on Science and Astronautics (Ninetieth Congress, Second Session).

Entitled the "Symposium on Unidentified Flying Objects," the committee met, "pursuant to notice," at 10:05 a.m., in room 2318, in the Rayburn House Office Building, with the Hon. J. Edward Roush (chairman of the symposium) presiding.

Six distinguished scientists were personally asked to attend, while six others were invited to submit "prepared statements." The subject at hand was "The UFO Phenomenon," or as some attendees referred to it, "The UFO Problem."

The six requested guest-scientists were:

- *Dr. J. Allen Hynek* (1910-1986): professor of astronomy at Northwestern University.
- *Dr. James E. McDonald* (1920-1971): senior physicist, University of Arizona.
- *Dr. Carl Sagan* (1934-1996): associate professor of astronomy, Cornell University.
- *Dr. Robert L. Hall* (1927-2013): head, Department of Sociology, University of Illinois
- *Dr. James A. Harder* (1926-2006): associate professor of civil engineering, University of California, Berkeley.
- *Dr. Robert M. L. Baker, Jr.* (1930-): faculty, Department of Engineering, UCLA.

The six scientists welcomed to submit prepared statements were:
- *Donald H. Menzel* (1901-1976): Smithsonian Astrophysical Observatory.
- *Stanton T. Friedman* (1934-2019): nuclear physicist, Westinghouse Astronuclear Plant.

- Dr. R. Leo Sprinkle (1930-2021): associate professor of psychology, University of Wyoming.
- Dr. Garry C. Henderson (1935-): senior research scientist, Space Sciences, Fort Worth, Texas.
- Dr. Roger N. Shepard (1929-2022): professor of psychology, Stanford University.
- Dr. Frank B. Salisbury (1926-2015): head, plant science department, Utah State University.

The views these twelve men held toward UFOs ranged from the true believer (Dr. McDonald), to the true nonbeliever (Dr. Menzel), to the noncommittal "agnostic" (Dr. Sagan). For those who have not read my books, *UFOs and Aliens: The Complete Guidebook*, and *The Martian Anomalies: A Photographic Search for Intelligent Life on Mars*, their statements will be revelatory.

In the following pages I have recorded their words, their submitted papers, and their illustrations as faithfully as possible, while correcting obvious typos, misspelled names, grammatical errors, and erroneous dates. Concurrently, I have left in our scientists' invented words, along with incorrectly bracketed words and British spellings, as these elements add character to their writings. Note that: 1) all bracketed words and sentences are mine; 2) all italicized words are mine—unless otherwise indicated; 3) all endnotes are mine—unless otherwise indicated.

THE VALUE OF CHRONICLING THIS SYMPOSIUM
Why, one might ask, preserve the statements of 12 scientists from 1968—56 years ago—all but a few who have since passed away?

It is true that this prestigious gathering of scientific luminaries is over a half century old, that there was no Internet at the time, and that the home computer had not even been invented yet. It is a fact that we had not yet landed a spacecraft on Mars, while our first Moon landing was still one year in the future. Dr. Hynek had not yet even invented the "Close Encounter" scale system we are all so familiar with.[1]

Yes, there have been countless breathtaking advancements in science since then, and if we look at this meeting purely through the supersonic lens of technological time (as opposed to the plodding of chronological time), 1968 would represent a period from several centuries ago.

Yet, despite the amazing developments we have made during this 56-year period, the field of ufology is still quite primitive. In fact, as I write these words, the U.S. government is, for the first time in its 250 year history, only now publicly—and grudgingly—admitting that UFOs exist. And this is only because it has been forced to due to the many military personnel now coming forward with eyewitness accounts, photos, and videos of UFOs.

We even have recent testimony before Congress (July 2023), by individuals—such as David Grusch, a former National Reconnaissance

Office Representative—who stoutly maintain that the government has long been in possession of both "crashed nonhuman spacecraft" and their "nonhuman biologic" occupants.

Despite all of this, the government continues to insist that it does not know the first thing about UFOs, including what they are, who makes them, who operates them, how they work, where they come from, why they are visiting our planet, or why they continually invade the air space above our military bases, airports, and nuclear power plants. Practically speaking then, the public knows as much about UFOs today, in 2024, as it did in 1968: next to nothing.[2]

With so little to go on concerning UFOs—or what the government is now referring to as UAPs ("unidentified anomalous phenomena")—the words, opinions, and theories of 20th-Century individuals are just as relevant now as they were nearly six decades ago; perhaps, as the reader shall see, even more so.

Indeed, as I was putting this book together, carefully transcribing every word by hand, I was struck by the massive amount of valuable information our twelve 20th-Century scientists had to offer; hundreds, if not thousands, of inspired ideas, hypotheses, and views, along with solid, practical recommendations, were put forth that Monday during the height of America's "love generation." And yet, not only are these sentiments and proposals seldom if ever discussed today, almost none, as far as I know, have been adopted by current mainstream scientists—much to the detriment of both ufology and humanity.

It is my hope that in resuscitating the pragmatic notions and rational concepts of these brilliant early UFO pioneers (yes, even the skeptical ones), my book will assist in opening up new doors to both the exploration of both inner and outer space, and far more significantly, our understanding of the UFO/UAP phenomenon.

It is past time to implement the ufological wisdom of those who came before us. It is not just a matter of scientific importance, government transparency, public safety, or national security anymore. Our very survival as a species could depend on it.

Lochlainn Seabrook (CE1, CE3, CE4)
Park County, Wyoming USA
January 2024

SEA RAVEN PRESS
PARK COUNTY　WYOMING USA
EST. 1995

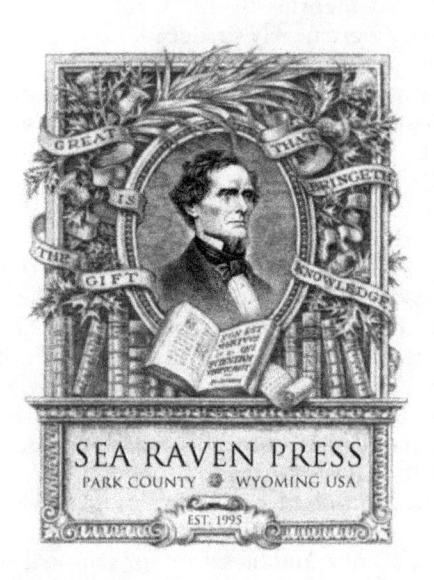

"Books invite all; they constrain none."
Hartley Burr Alexander (1873-1939)

SECTION ONE

SIX SCIENTISTS SPEAK BEFORE THE U.S. CONGRESS

INTRODUCTION TO SYMPOSIUM

SYMPOSIUM ON UNIDENTIFIED FLYING OBJECTS

MONDAY, JULY 29, 1968

HOUSE OF REPRESENTATIVES
COMMITTEE ON SCIENCE AND ASTRONAUTICS
Washington, D.C.

MR. ROUSH. The committee will be in order.

Today the House Committee on Science and Astronautics conducts a very special session, a symposium on the subject of unidentified flying objects; the name of which is a reminder to us of our ignorance on this subject and a challenge to acquire more knowledge thereof.

We approach the question of unidentified flying objects as purely a scientific problem, one of unanswered questions. Certainly the rigid and exacting discipline of science should be marshaled to explore the nature of phenomena which reliable citizens continue to report.

A significant part of the problem has been that the sightings reported have not been accompanied by so-called hardware or materials that could be investigated and analyzed. So we are left with hypotheses about the nature of UFOs. These hypotheses range from the conclusion that they are purely psychological phenomena, that is, some kind of hallucinatory phenomena; to that of some kind of natural physical phenomena; to that of advanced technological machinery manned by some kind of intelligence, that is, the extraterrestrial hypotheses.

With the range in mind, then, we have invited six outstanding scientists to address us today, men who deal with the physical, the psychological, the sociological, and the technological data relevant to the issues involved. We welcome them and look forward to their remarks. Additionally we have requested several other scientists to make their presentations in the form of papers to be added to these when published by the committee.

We take no stand on these matters. Indeed, we are here today to listen to their assessment of the nature of the problem; to any tentative conclusions or suggestions they might offer, so that our judgments and our actions might be based on reliable and expert information. We are here to listen and to learn.

Events of the last half century certainly verify the American philosopher, John Dewey's conclusion that "Every great advance in science has issued from a new audacity of imagination." With an open and inquiring attitude, then, we now turn to our speakers for the day.

They will include: Dr. J. Allen Hynek, head of the Department of Astronomy, Northwestern University; Dr. James E. McDonald, senior physicist, the Institute of Atmospheric Physics, the University of Arizona; Dr. Carl Sagan, Department of Astronomy and Center for Radiophysics and Space Research, Cornell University; Dr. Robert L. Hall, head of the Department of Sociology, University of Illinois at Chicago; Dr. James A. Harder, associate professor of civil engineering, University of California at Berkeley, and Dr. Robert M. L. Baker, Jr., Computer Sciences Corp. and Department of Engineering, UCLA.

Gentlemen, we welcome your presentations. We ask you to speak first, Dr. Hynek, followed by Dr. McDonald, and then Dr. Sagan. This afternoon Dr. Hall will commence our session, followed by Dr. Harder and then Dr. Baker. The subject matter of the presentations determines the order in which you speak. We hope at the end of the day to allow the six of you to discuss the material presented among yourselves and with the committee in a kind of roundtable discussion.

Mr. Chairman—the chairman of our full committee, Mr. George P. Miller.

CHAIRMAN MILLER. I want to join in welcoming you here. I want to point out that your presence here is not a challenge to the work that is being done by the Air Force, a particular agency that has to deal with this subject.

Unfortunately there are those who are highly critical of the Air Force, saying that the Air Force has not approached this problem properly. I want you to know that we are in no way trying to go into the field that is theirs by law, and thus we are not critical of what the Air Force is doing.

We should look at the problem from every angle, and we are herein that respect. I just want to point out we are not here to criticize the actions of the Air Force. Thank you.

MR. ROUSH. I think it is only appropriate that Dr. Hynek be introduced by our colleague, Mr. [Donald H.] Rumsfeld.

MR. RUMSFELD. Thank you, Mr. Chairman. It is a pleasure to welcome all the members of this distinguished panel, and particularly to welcome Dr. Allen Hynek, who is a son of Illinois, and presently serves in the Department of Astronomy and Director of the Lindheimer Astronomical Research Center. Dr. Hynek is a member of a number of scientific societies, and has served in the Government service as well as in the academic community. As his Congressman I am delighted he has been invited to appear on this panel, and we certainly look forward to his comments.

Thank you, Mr. Chairman.

MR. ROUSH. Dr. Hynek, the floor is yours.[3]

Instruments of astronomy.

CHAPTER ONE

STATEMENT OF J. ALLEN HYNEK

☛ Thank you.

My name is J. Allen Hynek. I am professor of astronomy at Northwestern University, Evanston, Ill., where I serve as chairman of the department of astronomy and director of the Lindheimer Astronomical Research Center. I have also served for many years, and still do, as scientific consultant to the U.S. Air Force on Unidentified Flying Objects, or UFO's. Today, however, I am speaking as a private citizen and scientist and not as a representative of the Air Force.

We are here today, I gather, to examine whether the UFO phenomenon is worthy of serious scientific attention. I hope my comments may contribute to your understanding of the problem and help lead to its eventual solution.

The UFO problem has been with us now for many years. It would be difficult to find another subject which has claimed as much attention in the world press, in the conversation of people of all walks of life, and which has captured the imagination of so many, over so long a period of time. The word UFO, or flying saucer, can be found in the languages and dictionaries of all civilized peoples, and if one were to collect all the words that have been printed in newspapers, magazines, and books in the past two decades, it would be a staggering assemblage. The bibliography of the subject recently compiled at the Library of Congress is a most impressive document, and illustrates that the UFO became a problem for the librarian even before it did for the scientist.

As we all know, the scientific world is a world of exact calculations, of quantitative data, of controlled laboratory experiments, and of seemingly well-understood laws and principles. The UFO phenomenon does not seem to fit into that world; it seems to flaunt itself before our present-day science.

The subject of UFO's has engendered an inordinate emotional reaction in certain quarters and has far more often called forth heated controversy rather than calm consideration. Most scientists have preferred to remain aloof from the fray entirely, thereby running the risk of "being down on what they were not up on," as the old adage goes.

It is unlikely that I would have become involved in the study of the UFO phenomenon had I not been officially asked to do so. I probably would have—and in fact did for a time—regarded the whole subject as rank nonsense, the product of silly seasons, and a peculiarly American craze that would run its course as all popular crazes do.

I was asked by the Air Force 20 years ago to assist them, as an astronomer, in weeding out those reports arising from misidentification of planets, stars, meteors, and other celestial objects and events. In the course of doing my "homework" I found that some 30 percent of the then current cases very probably had astronomical causes, but my curiosity was aroused by some of the patently nonastronomical reports.

These were ostensibly being explained by the consultant psychologist, but I frequently had the same feeling about the explanations offered for some of these cases that I have had when I have seen a magician saw a woman in half. How he did it was beyond my own field of competence, but I did not question his competence. Yes, I was quite sure that he did not actually saw the woman in half!

My curiosity thus once aroused led me to look into reports other than those of a purely astronomical nature, and in the course of years I have continued to do so. I have pondered over the continuing flow of strange reports from this and a great many other countries, for it is a gross mistake to think that the United States has any exclusive claim to the UFO phenomenon.

Those reports which interested me the most—and still do—were those which, apparently written in all seriousness by articulate individuals, nonetheless seemed so preposterous as to invite derisive dismissal by any scientist casually introduced to the subject. Such baffling reports, however, represent a relatively small subset of reports. I did not—and still do not—concern myself with reports which arise from obvious misidentifications by witnesses who are not aware of the many things in the sky today which have a simple, natural explanation. These have little scientific value, except perhaps to a sociologist or an ophthalmologist; it matters not whether 100 or 100,000 people fail to identify an artificial satellite or a high-altitude balloon.

The UFO reports which in my opinion have potential scientific value are those and this may serve us as a working definition of UFO's are those reports of aerial phenomena which continue to defy explanation in conventional scientific terms. Many scientists, not familiar with the really challenging UFO data, will not accept the necessity for a high order of scientific inquiry and effort to establish the validity of the data—and therefore such detailed, conscientious, and systematic inquiry has yet to be undertaken.

We cannot expect the world of science to take seriously the fare offered at airport newsstands and paperback shelves.

I have been asked by some why, as consultant to the Air Force for so many years, I did not alert the scientific world to the possible seriousness of the UFO problem years ago. The answer is simple; a scientist must try to be sure of his facts. He must not cry "wolf" unless he is reasonably sure there is a wolf.

I was painfully aware, and still am, of the amorphous nature of the UFO data, of the anecdotal nature of UFO reports, of the lack of followup and serious inquiry into reports (which would have required a large scientific staff and adequate funding), of the lack of hardware, of the lack of unimpeachable photographic evidence, and of the almost total lack of quantitative data of all those things which are part and parcel of the working environment of the scientist.

I was aware that in order to interest scientists, hard-core data were needed, and, while the store of unquestionably puzzling reports from competent witnesses continued to grow the wherewithal to obtain such hard-core data which would, once and for all, clinch the matter, was not forthcoming. Thus my scientific reticence was based on a carefully weighed decision.

In attempting analysis of the UFO problem today, I pay particular attention to reports containing large amounts of information which are made by several witnesses, if possible, who as far as I can ascertain, have unimpeachable reputations and are competent. For example, I might cite a detailed report I received from the associate director of one of the Nation's most important scientific laboratories, and his family.

Reports such as these are obviously in a different category from reports which, say, identify Venus as a hovering spaceship, and thus add to the frustrating confusion.

On the other hand, when one or more obviously reliable persons reports—as has happened many times—that a brightly illuminated object hovered a few hundred feet above their automobile, and that during the incident their car motor stopped, the headlights dimmed or went out, and the radio stopped playing, only to have these functions return to normal after the disappearance of the UFO, it is clearly another matter.

By what right can we summarily ignore their testimony and imply that they are deluded or just plain liars? Would we so treat these same people if they were testifying in court, under oath, on more mundane matters?

Or, if it is reported, as it has been in many instances over the world by reputable and competent persons, that while they were sitting quietly at home they heard the barnyard animals behaving in a greatly disturbed and atypical manner and when, upon investigating, found not only the animals in a state of panic but reported a noiseless—or sometimes humming—brightly illuminated object hovering nearby, beaming a bright red light down onto the surroundings, then clearly we should pay attention. Something very important may be going on.

Now, when in any recognized field of science an outstanding event takes place, or a new phenomenon is discovered, an account of it is quickly presented at a scientific meeting or is published in a respected appropriate journal. But this is certainly not the case with unusual UFO reports made by competent witnesses.

There appears to be a scientific taboo on even the passive tabulation of UFO reports. Clearly no serious work can be undertaken until such taboos are removed. There should be a respectable mechanism for the publication, for instance, of a paper on reported occurrences of electromagnetic phenomena in UFO encounters.

It would be foolhardy to attempt to present such a paper on UFO's to the American Physical Society or to the American Astronomical Society. The paper would be laughed down, if all that could be presented as scientific data were the anecdotal, incomplete, and nonquantitative reports available. Consequently reports of unexplainable UFO cases are likely to be found, if at all, in pulp magazines and paperbacks, of which the sole purpose of many seems to be, apart from making a fast buck for the authors, to titillate the fancy of the credulous.

Indeed, in such newsstand publications three or four UFO reports are frequently sensationalized on one page with gross disregard for accuracy and documentation; the result is that a scientist—if he reads them at all—is very likely to suffer mental nausea and to relegate the whole subject to the trash heap.

This is the first problem a scientist encounters when he takes a look at the UFO phenomenon. His publicly available source material is almost certain to consist of sensational, undocumented accounts of what may have been an actual event. Such accounts are much akin, perhaps, to the account we might expect from an aborigine encountering a helicopter for the first time, or seeing a total eclipse of the sun. There is nowhere a serious scientist can turn for what he would consider meaningful, hard-core data—as hard core and quantitative as the phenomenon itself permits at present.

Here we come to the crux of the problem of the scientist and the UFO. The ultimate problem is, of course, what are UFO's; but the immediate and crucial problem is, How do we get data for proper scientific study? The problem has been made immensely more difficult by the supposition held by most scientists, on the basis of the poor data available to them, that there couldn't possibly be anything substantial to UFO reports in the first place, and hence that there is no point to wasting time or money investigating.

This strange, but under the circumstances understandable attitude, would be akin to saying, for instance, let us not build observatories and telescopes for the study of the stars because it is obvious that those twinkling points of light up there are just illusions in the upper atmosphere and do not represent physical things.

Fortunately, centuries ago there were a few curious men who did not easily accept the notion that stars were illusory lights on a crystalline celestial sphere and judged that the study of the stars might be worthwhile though, to many, a seemingly impractical and nonsensical venture. The pursuit of that seemingly impractical and possibly unrewarding study of astronomy and related sciences, however, has given us the highly technological world we live in and the high standard of living we enjoy—a standard which would have been totally impossible in a peasant society whose eyes were never turned toward the skies.

Can we afford not to look toward the UFO skies; can we afford to overlook a potential breakthrough of great significance? And even apart from that, the public is growing impatient. The public does not want another 20 years of UFO confusion. They want to know whether there really is something to this whole UFO business—and I can tell you definitely that they are not satisfied with the answers they have been getting. The public in general may be unsophisticated in scientific matters, but they have an uncanny way of distinguishing between an honest scientific approach and the method of ridicule and persiflage.

As scientists, we may honestly wish to see whether there is any scientific paydirt in this international UFO phenomenon. But to discover this paydirt we must devote serious study to UFO's. To make serious study possible, however, requires recruiting competent scientists, engineers, and technical people, as well as psychologists and sociologists.

This in turn requires not only funds but a receptive scientific climate. Many scientists have expressed to me privately their interest in the problem and their desire to actively pursue UFO research as soon as the scientific stigma is removed. But as long as the unverified presumption is strongly entrenched that every UFO has a simple, rational everyday explanation, the required climate for a proper and definitive study will never develop.

I recall an encounter I had sometime ago with the then chief scientist at the Pentagon. He asked me just how much longer we were "going to look at this stuff." I reminded him that we hadn't really looked at it yet—that is, in the sense, say, that the FBI looks at a kidnapping, a bank robbery, or a narcotics ring.

Up to this point I have not discussed another major impediment to the acceptance of the UFO phenomenon as legitimate material for scientific study. I refer to the adoption of the UFO phenomenon by certain segments of the public for their own peculiar uses. From the very start there have been psychically unbalanced individuals and pseudoreligious cultist groups—and they persist in force today—who found in the UFO picture an opportunity to further their own fanciful cosmic and religious beliefs and who find solace and hope in the pious belief that UFO's carry kindly space brothers whose sole aim is a mission of salvation.

Such people "couldn't care less" about documentation, scientific study, and careful critical consideration. The conventions and meetings these people hold, and the literature they purvey, can only be the subject of derisive laughter and, I must stress, it is a most serious mistake for anyone to confuse this unfortunate aspect of the total UFO phenomenon with the articulate reports made by people who are unmistakably serious and make their reports out of a sense of civic duty and an abiding desire to know the cause of their experience.

It may not be amiss here to remark in passing that the "true believers" I have just referred to are rarely the ones who make UFO reports. Their beliefs do not need factual support. The reporters of the truly baffling UFO's, on the other hand, are most frequently disinterested or even skeptical people who are taken by surprise by an experience they cannot understand.

Hopefully the time is not far off when the UFO phenomenon can have an adequate and definitive hearing, and when a scholarly paper on the nature of UFO reports can be presented before scientific bodies without prejudice. Despite the scientific attitude to this subject in the past, I nevertheless decided to present a short paper on UFO's to a scientific body in 1952, following a scientific hunch that in the UFO phenomenon we were dealing with a subject of great possible importance.

In my paper (*Journal of the Optical Society of America*, 43, pp. 311-314, 1963), which I should like to have read into the record, I made reference to the many cases in 1952 and earlier which were nonastronomical in nature and did not seem to have a logical, ready explanation. (The document referred to is as follows:)

[From *Journal of the Optical Society of America*, April 1953]

UNUSUAL AERIAL PHENOMENA
J. A. HYNEK, *Ohio State University, Columbus, Ohio*
(Received December 22, 1952)

Over a period of years, diverse aerial sightings of an unusual character have been reported. On the assumption that the majority of these reports, often made in concert, come from reputable persons, and in the absence of any universal hypotheses for the phenomena which stimulated these reports, it becomes a matter of scientific obligation and responsibility to examine the reported phenomena seriously, despite their seemingly fanciful character. Accordingly, several hundred serious reports of "unidentified aerial objects" have been studied in detail in an attempt to get a pattern classification. It appears that those reported phenomena which do not admit of a ready and obvious explanation exhibit fairly well-defined patterns and that these are worthy of further study. One pattern in particular, that of a hovering nocturnal light, does not appear to be readily explainable on an astronomical basis or by mirages, balloons, or by conventional aircraft.

Perhaps the most bizarre phenomenon of our times is the continued popular interest in flying saucers. The term flying saucer, of course, dates back to the treatment by the press of the now famous triggering incident of June 24, 1947, another date which might well be said to live in infamy, when a lone pilot, Mr. [Kenneth] Arnold, reported "nine peculiar-looking aircraft" without tails, which flew in a chain-like line and "swerved in and out of the high mountain peaks [in the vicinity of Mount Rainier]." The unfortunate newspaper term, flying saucer, as you well know, captured both the press and the public imagination. One can speculate as to the turn of events, and the amount of newsprint that might have been conserved, had Mr. Arnold decided to stay on the ground that day!

Nevertheless, in the past five years, flying saucer has become a standard term in our language, with about as broad a definition as it has been the lot of any term to carry. We can define a flying saucer as *any aerial phenomenon or sighting that remains unexplained to the viewer at least long enough for him to write a report about it.* Lest anyone misunderstand what shall be meant by "flying saucers" in this paper, this definition must be emphasized.

Each flying saucer, so defined, has associated with it a probable lifetime. It wanders in the field of public inspection like an electron in a field of ions, until "captured" by an explanation which puts an end to its existence as a flying saucer.

Thus flying saucers spawned by the planet Venus have generally a short life-time. In almost no time an astronomer comes along and makes a positive identification, and another flying saucer is shattered. We can expect a host of Venus-inspired flying saucers when this planet is low in the western sky after sunset. It reaches greatest eastern elongation this year on January 31, 1953, and on March 8 attains its greatest brilliance. We can confidently predict a swarm of flying saucers from Venus!

The lifetime of a balloon-sponsored flying saucer is often longer, but before long someone like Dr. [Urner] Liddel comes along and shoots it down. And Dr. [Donald H.] Menzel has as his flying saucer ammunition a large variety of optical effects, the lethalness of which requires separate field tests.

My concern is with flying saucers of long lifetime—those which have not, as yet, been "captured" or demolished by an explanation. Let us further limit them to those that have been observed by two or more people, at least one of whom is practiced in the making of observations of some kind, that is, to pilots, control tower operators, weather observers, scientific workers, etc. Also, let us limit cases to sightings lasting a minute or more, again for obvious reasons.

The Wright Patterson Air Force Base, where the responsibility for the investigation of unidentified aerial objects has rested for the past several years, asked me in 1948 to help identify reports that had an astronomical basis. It was a relatively simple task to go through about 200 reports and pick out probable astronomical causes. Indeed, some of the most weird reports could be dismissed with clear conscience by the statement that no astronomical explanation is possible for this incident, thus leaving these unsolved cases to the psychologists.

I did wonder of course, as to how they were disposing of the nonastronomical cases. How did they explain the incident in which a pilot, co-pilot, and stewardess followed some rapidly moving dark objects which were silhouetted against the sunset sky and which disappeared presumably because of their superior speed? But my faith in the psychologists was unshaken—and when the Air Forces announced that Project Grudge had been dissolved, I assumed that my colleagues had been successful and had even solved the case in which several observers watched an object, hollow in the middle, travel at a constant slow rate, taking 15 minutes to make the journey across the sky from north to south.

After the project dissolved, Wright Field continued to take care of the slow but steady flow of reports as a part of their regular intelligence function. This spring I became curious and requested permission, through official channels, to look through the crop of reports that had accumulated since my official connection with Project Grudge had terminated. As I looked through the welter of fanciful tales, inaccurate reporting, of misobservation of natural objects, I could not help, as an astronomer, recalling another wave of stories—stories of stones that fell from heaven. Because of poor reporting and poor imagery, scientific progress in meteorites had been held back for a good century. What a difference in imagery there is between "a stone falling from the sky" and "the interception by the earth of a particle pursuing an orbit around the sun." The use of improper and inaccurate description of what actually happened kept meteorites in the category of old wives' tales and out of the niche that celestial mechanics had made ready for them a century before! In 1801, Thomas Jefferson said that he would sooner believe that two Yankee professors had lied than that stones had fallen from heaven. And the French Academy of Sciences branded stories of meteorite falls as fanciful and absurd and dismissed a bona fide meteorite whose fall had been sworn to—as an ordinary stone that had been struck by lightning. Perhaps the moral of this is: Beware the ready explanation!

Now, it is clear that stories of real flying saucers, visitors from space, and strange aircraft violating the laws of physics are as reprehensible to the scientist of today as stones that fell from heaven were to the scientist of yesteryear. But, of course, stones did not fall from heaven—that was poor reporting and a wrong slant on a perfectly natural phenomenon. And we don't have space ships that disregard physical laws. But, do we have a natural phenomenon?

The steady flow of reports, often made in concert by reliable observers, raises questions of scientific obligation and responsibility. Is there, when the welter of varied reports are shorn of, in the words of Pooh Bah, all "Corroborative detail to lend artistic verisimilitude to an otherwise bald and unconvincing narrative"—any residue that is worthy of scientific attention?

Or, if there isn't, does not obligation still exist to say so to the public—not in words of open ridicule, but seriously, to keep faith with the trust the public places in science and scientists?

The Air Forces are attempting to give all reports a fair hearing, in view of the above. They are having all reported data reduced to punch cards so that in a month or so it will be possible to compare quickly reports made by people facing west on clear Tuesday afternoons with those made on non-inversion Friday nights by pilots going south. In any event, if significant correlations between various sets of sightings exist, this method should bring them out.

In coming down to cases, to illustrate what constitutes at present the best evidence for unusual aerial phenomena, the examples submitted for examination are presented without an all-embracing explanation for them. These are presented in conformance with the school that believes that good observations and discussion of observations come before theory. It is hoped, however, that out of this discussion there may come a positive approach and, if these sightings do represent heretofore inadequately studied natural phenomena, that these examples will stimulate their serious study; if, on the other hand, no natural phenomena are involved, then an obligation exists to demonstrate explicitly how the following specific reports can be explained in terms of balloons, mirages, or conventional aircraft.

The chosen recent examples represent a definite pattern, and for each of the following there are many other similar examples in the files.

One of these patterns might be called "Nocturnal Meandering Lights." Reports falling into this category are characterized by the sighting of a bright star-like light, perhaps of −2 or -3 stellar magnitude which floats along without sound, frequently hovers, reverses its field without appearing to turn, and often abruptly speeds up. The light is most frequently described as a yellow amber or orange, changing to blue or red occasionally, and changing in brightness markedly. Sometimes the description states that the light went out as if someone had pushed a button; at other times the light is reported only as variable. A very characteristic statement by those making the reports is: "I have never seen anything like this in my whole life." The desire to identify these sightings as balloons is thwarted by the tactics observed.

As an example of a report of this kind, let us take one that came in from Florida this past July. On one night several airmen independently observed a light approach at a very slow speed, come to a halt nearly overhead, then reverse direction with no apparent turn. On two other nights, three other lights appeared in other sections of the sky, of similar appearance, but maneuvering more rapidly. They were observed for some 10 minutes by 9 airmen, including a control tower operator, an aircraft dispatcher, and two pilots from Wright Field.

In the words of one of the men, "For the next fifteen minutes we watched this light and speculated on what it might be. It was not a sharp light like a bare bulb but more like a light shining through frosted glass. No shape of any kind was discernible. It appeared to blink, but with no regularity whatever."

Also this past July at an air base in New Mexico, a similar sighting was made. Paraphrasing from sworn statements made by observers, "Our station was notified that an unconventional aircraft had been picked up with both electronic and visual contact. Our station made electronic contact with the object and two of our men and I had gone outside the building and saw it hovering under a cloud layer to the east of us. It appeared as a large light, at an uncertain distance, and was hovering at the time. A minute or so later, it moved rapidly toward the north for a short distance and stopped as suddenly as it had begun to move."

And from another statement, "Our scope operator at that time reported a strange target about thirty miles east of our station. Two of us went outside and sighted a very bright light traveling at what we estimated to be around 200 miles an hour. The light went out at least two times but did not stay out more than two or three minutes. The light seemed to have a floating effect and made no sound. At one time around seven or eight smaller lights could be seen. The object seemed to drop to about 10 or 12 thousand feet and then climbed to about 25,000 taking a northern course."

Radar observations as well as visual observations are involved in this problem. Early last month shortly before dawn colored lights were observed in the sky southeast of the radar station. At the same time and the same azimuth, unidentified targets appeared on the scope. Only a very slight temperature inversion was present, 1° at

25,000 feet. No more than two lights appeared at one time. They were observed to be moving in a rather erratic pattern and changing colors occasionally. The last thirty minutes of observation revealed the lights remaining yellow—prior to that they were red, green, and blue. They moved in no apparent formation but mostly appeared in one area and disappeared in another, when either the light went out or the objects dived behind clouds. They were starlike objects and appeared to develop long, white vapor trails, when they dived. They were motionless at times and moved rapidly at other times. This corresponded to similar movements observed on the radar scope.

One white light went out as it changed direction and continued as a black silhouette against the dawn sky. Observation was for a period of about an hour and was made by two airmen and a radar operator—all three observers were experienced aircraft control and warning operators. Objects were observed 20 to 40° above horizon. Radar gave distances of 50 to 80 miles. This implies a height of about 40 miles. There was no air traffic on radar within 100 miles.

Quoting from the observer's statement, "receiving a call concerning a strange light in the sky, I went out and scanned the sky in several directions before I saw a light. My first glimpse was a very bright blue light, but it lasted only about a minute, then it faded into a light green. It moved in a slow orbit. I was startled at first so I closed my eyes and opened them again. The light was still there. I stared at it a few minutes and now the light seemed more yellow than before. I did not think anyone would believe me, so I went inside the building and relieved the radar scope operator. I found a target at 123°, 53 miles. After that it appeared as a permanent echo. In about two minutes, it disappeared and almost immediately another pip appeared, at 134°, 73 miles. It also seemed like a permanent echo. It stayed on the scope for 1½ minutes. These pips were at no time caused by malfunction of the radar set. It was daylight when it (the object) seemed to fade both visually outdoors and electronically indoors."

And another sighting—in Northern Michigan—on July 29 of last year, a pilot chased a brilliant multicolored object close to the horizon, and due north. He flew at 21,000 feet, followed the object for over a half-hour but could not gain on it. Radar operator reported contact with the object for about thirty seconds. And ground control interceptor station reported blips too. In this case, it seems certain that our harried pilot was pursuing [the bright multi-star system] Capella! Reference to a star map will show that at his latitude, at the time of his sighting, Capella was at lower culmination, that is, at the lowest point of its swing around the pole just skirting the horizon. I have seen it at that position myself in Canada, and, can vouch for the fact that its blue, yellow, and red twinkling can be spectacular.

Unfortunately, neither Capella nor any other star can explain many other nocturnal meandering lights. But there is no question in my mind, just to make this point exceedingly clear, that there exists a relatively simple, natural explanation for them, perhaps even ordinary aircraft under special test conditions. The chief point here, is to suggest that nothing constructive is accomplished for the public at large—and therefore for science in the long run—by mere ridicule and the implication that sightings are the products of "bird-brains" and "intellectual flyweights." In short, it would appear that the flying saucer situation has always been a problem in science- public-relations, and that fine chance has consistently been missed to demonstrate on a national scale how scientists can go about analyzing a problem. A lot is said about the proper interpretation of science to the public, but the only answer they receive to a question about which they are more widely concerned than perhaps any other in this century, is ridicule. *Ridicule is not a part of the scientific method and the public should not be taught that it is* [my emphasis, L.S.].

Let me quote an additional report, to show that the original flying disks, as distinct from wandering lights, are still with us.

On the day that our pilot chased Capella, a radio from Seattle announced that flying saucers were seen heading toward Montana. At an airport in Montana several pilots gathered outside the hangars to wait and watch. A perfect set-up for suggestibility—and yet, quoting from one of the many signed statements, "Objects were seen that resembled flat disks reflecting sun's rays. One of the objects hovered from three to four minutes, while the other three circled around it like satellites. Then the stationary object moved southeast to disappear, while the three satellites moved due west and disappeared—at very high speeds!"

And from another observer: "After watching for approximately five minutes I was able to see what appeared to be a disk, white or metal in color approaching from the west. As it moved directly overhead it turned generally north at a 90° turn, then slowing down and then making several more 90° turns and proceeding east. After seeing this I knew what I was looking for and was able to pick up at least five more of these objects. *Being skeptical, I did my best to see them as either dandelion seeds or other small particles close to the surface of the earth rather than large objects at extreme distance.* However, after keeping them in sight long enough to study their appearance they definitely seemed to be very high. *I won't make an estimate of the height since I did not know their size.* All of these appeared in the west and proceeded east at what appeared to be an extremely high rate of speed."

I submit that this Air Force lieutenant was not incompetent, but rather that his manner of reporting—as far as it went—was commendable and that his report, made in good faith, is therefore entitled to a hearing without prejudice or ridicule, but also, without fanfare, hysteria, and fantastic newspaper publicity.

I cautioned against the then prevalent attitude of ridicule, pointing out that the UFO phenomenon, which had generated vast public interest, represented an unparalleled opportunity to demonstrate to the public the operation of the scientific method in attacking a problem, and that "ridicule is not a part of the scientific method and the public should not be taught that it is."

In those years and the following ones I repeatedly asked for the upgrading of the method of reporting UFO's to the Air Force. In

1960, in a hearing before Congressman [Richard] Smart and his committee I urged "immediate reaction capabilities" in the investigation of UFO reports. The recommendation was applauded but not funded.

As the scientific climate grew more receptive in giving the UFO phenomenon a scientific hearing, I published a letter in *Science* (Oct. 21, 1966), not without difficulty, in which I pointed out the following general misconceptions regarding UFO's. I should like to have that letter made a part of the record. (The letter referred to is as follows:)

UFO'S MERIT SCIENTIFIC STUDY

Twenty years after the first public furor over UFO's (called "flying saucers" then) reports of UFO's continue to accumulate. The Air Force has now decided to give increased scientific attention to the UFO phenomenon. Thus I feel under some obligation to report to my scientific colleagues, who could not be expected to keep up with so seemingly bizarre a field, the gist of my experience in "monitoring the noise level" over the years in my capacity as scientific consultant to the Air Force. In doing so, I feel somewhat like a traveler to exotic lands and faraway places, who discharges his obligation to those who stayed at home by telling them of the strange ways of the natives.

During my long period of association with the reports of strange things in the sky, I expected that each lull in the receipt of reports signaled the end of the episode, only to see the activity renew; in just the past two years it has risen to a new high. Despite the fact that the great majority of reports resulted from misidentifications of otherwise familiar things, my own concern and sense of personal responsibility have increased and caused me to urge the initiation of a meaningful scientific investigation of the residue of puzzling UFO cases by physical and social scientists. I have guardedly raised this suggestion in the literature and at various official hearings, but with little success. UFO was a term that called forth buffoonery and caustic banter; this was both a cause and an effect of the lack of scientific attention. I speak here only of the puzzling reports; there is little point to concern ourselves with reports that can be easily traced to balloons, satellites, and meteors. Neither is there any point to take account of vague oral or written reports which contain few information bits. We need only be concerned with "hard data," defined here as reports, made by several responsible witnesses, of sightings which lasted a reasonable length of time and which were reported in a coherent manner.

I have strongly urged the Air Force to ask physical and social scientists of stature to make a respectable, scholarly study of the UFO phenomenon. Now that the first firm steps have been taken toward such a study, I can set forth something of what I have learned, particularly as it relates to frequently made misstatements about UFO's. Some of these statements which lead to misconceptions are:

(1) Only UFO "buffs" report UFO's. The exact opposite is much

nearer the truth. Only a negligible handful of reports submitted to the Air Force are from the "true believers," the same who attend UFO conventions and who are members of "gee-whiz" groups. It has been my experience that quite generally the truly puzzling reports come from people who have not given much or any thought to UFO's.

(2) UFO's are reported by unreliable, unstable, and uneducated people. This is, of course, true. But *UFO's are reported in even greater numbers by reliable, stable, and educated people* [my emphasis, L.S.]. The most articulate reports come from obviously intelligent observers; dullards rarely overcome their inherent inertia toward making written reports.

(3) UFO's are never reported by scientifically trained people. This is unequivocally false. Some of the very best, most coherent reports have come from scientifically trained people. It is true that scientists are reluctant to make a public report. They also usually request anonymity which is always granted.

(4) UFO's are never seen at close range and are always reported vaguely. When we speak of the body of puzzling reports, we exclude all those which fit the above description. I have in my files several hundred reports which are fine brain teasers and could easily be made the subject of profitable discussion among physical and social scientists alike.

(5) The Air Force has no evidence that UFO's are extraterrestrial or represent advanced technology of any kind. This is a true statement but is widely interpreted to mean that there is evidence against the two hypotheses. As long as there are "unidentifieds," the question must obviously remain open. If we knew what they were, they would no longer be UFO's—they would be IFO's. Identified Flying Objects! If you know the answer beforehand, it isn't research. *No truly scientific investigation of the UFO phenomenon has ever been undertaken* [my emphasis, L.S.]. Are we making the same mistake the French Academy Sciences made when they dismissed stories of "stones that fell from the sky"? Finally, however, meteorites were made respectable in the eyes of science.

(6) UFO reports are generated by publicity. One cannot deny that there is a positive feedback, a stimulated emission of reports, when sightings are widely publicized, but it is unwarranted to assert that this is the whole cause of high incidence of UFO reports.

(7) UFO's have never been sighted on radar or photographed by meteor or satellite tracking cameras. This statement is not equivalent to saying that radar, meteor cameras, and satellite tracking stations have not picked up "oddities" on their scopes or films that have remained unidentified. It has been lightly assumed that although unidentified, the oddities were not unidentifiable as conventional objects.

For these reasons I cannot dismiss the UFO phenomenon with a shrug. The "hard data" cases contain frequent allusions to recurrent kinematic, geometric, and luminescent characteristics. I have begun to feel that there is a tendency in 20th-century science to forget that there will be a 21st-century science, and indeed, a 30th-century science, from which vantage points our knowledge of the universe

may appear quite different. We suffer perhaps, from temporal provincialism, a form of arrogance that has always irritated posterity.
— J. Allen Hynek, Dearborn Observatory, Northwestern University, Evanston, Ill.

One great misconception is that only UFO buffs report UFO's; quite the opposite is the case, as is the misconception that the most baffling reports come from unreliable, unstable, and uneducated people. Most reports of this baffling sort which I at least receive in my mail, are remarkably articulate.

Other misconceptions are that UFO's are never reported by scientifically trained people, are never seen at close range, have never been detected on radars, and have never been recorded by scientific cameras.

It is well to remind ourselves at this point of the definition of a UFO: those aerial phenomena reports which continue to defy explanation in conventional scientific terms, even after appropriate study. There is no point to be interested in anything else; lights at night which might be aircraft, balloons, meteors, or satellite reentries all these fit more readily into the category of IFO's or identified flying objects.

In other words, only truly unidentified cases should be of interest. The Air Force has its own definition of an unidentified case, and it has many hundreds in its files. The Air Force calls a sighting unidentified when a report apparently contains all pertinent data necessary to suggest a valid hypothesis concerning the cause or explanation of the report but the description of the object or its motion cannot be correlated with any known object or phenomena.

It is most logical to ask why do not the unidentified in the Air Force files call forth investigative efforts in depth and of wide scope. The answer is compound: the Air Force position is that there is no evidence that UFO's represent a threat to the national security; consequently it follows that it is not their mission to be scientifically curious about the hundreds of unidentified cases in their own files.

It may be that, properly investigated, many of the Air Force unidentifieds would turn out to be IFO's after all, but it is illogical to conclude that this would be true of all unidentified reports. As long as unidentified cases exist, thus bona fide UFO's according to definition, we don't know what they are, and these should represent a remarkable challenge to science and an open invitation to inquiry.

But so powerful and all-encompassing have the misconceptions among scientists been about the nature of UFO information that an amazing lethargy and apathy to investigation has prevailed. *This apathy is unbecoming to the ideals of science and undermines public confidence* [my emphasis, L.S.].

Now it is of interest to report that in just the past few years, probably because of the persistent flow of UFO reports from this and many other countries (one could base his whole plea for a major investigative effort solely on the reports of the years 1966 and 1967) there has been a growing but unheralded interest on the part of more and more scientists, engineers, and technicians in doing something positive about the UFO problem. To this growing body of qualified people it seems increasingly preposterous to allow another two decades of confusion to exist.

The feeling is definitely on the increase that we should either fish or cut bait, that we should mobilize in earnest adequate groups of scientists and investigators, properly funded, adopt a "we mean business" attitude, or drop the whole thing. My recommendation is to fish.

As a scientist I can form conclusions from and act upon only reliable scientific data. Such data are extremely scarce in the UFO field for reasons already pointed out: it has never been considered worthwhile to improve the data-gathering process because the whole subject has been prejudged. Even as a scientist, however, I am permitted a scientific hunch, and that hunch has told me for some time, despite the tremendous muddiness of the scientific waters in this area, the continued reporting from various parts of the world of unidentified flying objects, reports frequently made by people of high repute who would stand nothing whatever to gain from making such reports, that there is scientific paydirt in the UFO phenomenon—possibly extremely valuable paydirt—and that therefore a scientific effort on a much larger scale than any heretofore should be mounted for a frontal attack on this problem.

In saying this I do not feel that I can be labeled a flying saucer "believer"—my swamp gas record in the Michigan UFO melee should suffice to quash any such ideas—but I do feel that even though this may be an area of scientific quicksand, signals continue to point to a mystery that needs to be solved. Can we afford to overlook something that might be of great potential value to the Nation?

I am reminded of the old story of the member of Parliament who visited [Michael] Faraday's laboratory where he was at work on early experiments on electrical induction. When asked of what possible value all this might have, Faraday replied, "Sir, someday you may be able to tax it."

Apart from such inducements, I have the following recommendations to make: first, that a mechanism be set up whereby the problem posed by the reports from all over the world, but especially by those in the United States, from people of high credibility, can be adequately studied, using all methods available to modern science, and that the investigation be accorded a proper degree of scientific respectability and an absence of ridicule so that proper investigations can be carried out unhampered by matters not worthy of the ideals of scientific endeavor. I might suggest that this could be accomplished by the establishment, by the Congress, of a UFO Scientific Board of Inquiry, properly funded, for the specific purpose of an investigation in depth of the UFO phenomenon.

Secondly, I recommend that the United States seek the cooperation of the United Nations in establishing a means for the impartial and free interchange among nations of information about, and reports of, unidentified flying objects—a sort of international clearinghouse for the exchange of information on this subject. For, since the UFO phenomenon is global, it would be as inefficient to study it without enlisting the aid of other nations as it would be to study world meteorology by using weather reports from one country alone.

Now, it may be well to remind ourselves at this point, that the UFO problem may not lend itself to an immediate solution in our time. The problem may be far more complex than we imagine. Attempts to solve it may be no more productive than attempts to solve the problem of the Aurora Borealis would have been 100 years ago.

The cause of northern lights could not have been determined in the framework of the science of 1868. Scientific knowledge in those days was not sufficient to encompass the phenomenon.

Similarly, our scientific knowledge today may be grossly insufficient to encompass the problem posed by UFO's. A profound scientific obligation exists, nonetheless, to gather the best data possible for scientific posterity.

To summarize in the course of 20 years of study of UFO reports and of the interviewing of witnesses, I have been led to a conclusion quite different from the one I reached in the very first years of my work. At first I was negatively impressed with the low scientific content of UFO reports, with the lack of quantitative data, with the anecdotal nature of the reports, and especially with the lack of hardware, of unimpeachable photographs, and with the lack of instrumental recordings.

I am still aware of the paucity of truly hard-core data—but then, no effort has really been made to gather it. Nonetheless, the cumulative weight of continued reports from groups of people around the world whose competence and sanity I have no reason to doubt, reports involving close encounters with unexplainable craft, with physical effects on animals, motor vehicles, growing plants, and on the ground, has led me reluctantly to the conclusion that either there is a scientifically valuable subset of reports in the UFO phenomenon or that we have a world society containing people who are articulate, sane, and reputable in all matters save UFO reports.

Either way, I feel that there exists a phenomenon eminently worthy of study. If one asks, for what purpose, I can only answer—how does one ever know where scientific inquiry will lead. *If the sole purpose of such a study is to satisfy human curiosity, to probe the unknown, and to provide intellectual adventure, then it is in line with what science has always stood for* [my emphasis, L.S.].

Scientific inquiry has paid off, even though pioneers like Faraday, [Marie] Curie, [Otto] Hahn, [Louis] Pasteur, [Robert H.] Goddard, and many others little realized where the paths they blazed would lead. As far as UFO's are concerned, I believe we should investigate them for the simple reason that we want to know what lies behind this utterly baffling phenomenon—or even more simply, we want to find out what it's all about. Thank you. — STATEMENT OF DR. J. ALLEN HYNEK, HEAD, DEPARTMENT OF ASTRONOMY, NORTHWESTERN UNIVERSITY, EVANSTON, ILL. JULY 29, 1968.

MR. ROUSH. Thank you, Dr. Hynek. Although we have reserved the latter part of the afternoon for our roundtable discussion, the Chair is well aware the Members of Congress, because of other duties, may not find it possible to be here during that time. If any of my colleagues do have questions and can keep them brief, which I realize is impossible, I will entertain those questions at this time. But keep in mind that we have two more papers this morning, and three this afternoon.

MR. HECHLER. Mr. Chairman.

MR. ROUSH. Mr. Hechler.

MR. HECHLER. First I would like to commend you, Mr. Roush, for your initiative in setting up this symposium.

I would like to ask you, Dr. Hynek, whether you consider this scientific board of inquiry which you outlined as a sort of a one-shot thing which would make its report, or do you consider this to be a continuing body that could examine, as the Air Force has, reports and analyze them? And with this question, I would like to ask if your assumption is that the Air Force, because of its emphasis on national security, has really not measured up to a thorough scientific analysis of UFO's?

DR. HYNEK. Well, in answer to the first part of that question, sir, I would say I don't believe in a problem as complex as this the one-shot approach would be sufficient. I think there should be this board of inquiry which should be a continuing board in the same sense that we have, I presume, boards of study for world population problems, of pollution problems, of world health, and so forth.

The letter that came with the invitation to speak here, strongly stated that we would not discuss the Air Force participation in these matters, and I would like to therefore not speak to that point.

MR. ROUSH. Mr. Rumsfeld.

MR. RUMSFELD. Because of the fact it does look as though we will have a busy afternoon on the floor, I very likely will not be present for the remainder of the discussion. I would like to express the hope the other members of the panel might at some point comment on the two recommendations that Dr. Hynek has set forth in his paper. Further, I would hope that each member of the panel, during the afternoon session, might address himself to the questions of priorities.

Assuming that there is some agreement with Dr. Hynek's conclusion that this is an area worthy of additional study, then the question for Congress, of course, becomes what is the priority? This is a rather unique situation in that it is a scientific question that has reached the public prior to the time that anything beneficial can even be imagined. In many instances a scientific effort is not widely known to the public until it is successful.

Each of you are experts in one or more disciplines. I am sure there are a number of things on your shopping lists for additional funding. I would be interested to know how this effort that is proposed here might fit into your lists of priorities. Thank you, Mr. Chairman.

MR. ROUSH. Thank you, Mr. Rumsfeld. Mr. Miller.

CHAIRMAN MILLER. Doctor, you mentioned a number of things—population studies at least. A great many of these are carried out not by Government directly, but in the National Science Foundation or through the National Academy of Sciences or scientific bodies themselves.

Do you think, I merely offer this as a suggestion, perhaps the scientific community try to encourage NSF or the scientific societies dealing in this field to take the initiative in doing this, rather than to wait for Government to take the initiative?

DR. HYNEK. I know, of course, most of the bodies you have mentioned are funded by the Government anyway. Most or a great part of our scientific research today has to be so funded. Private sources are certainly not sufficient. And, therefore, I think it is rather academic, really, to worry too much about who does it. It is more a question of who is going to pay for it.

We have a rather interesting situation here, as Congressman Rumsfeld has already pointed out. This is one of those strange situations in which the cart is sort of before the horse. Generally this results in the scientific laboratories and the results of the studies of scientists finally come to the public attention, but here we have the other situation. It is the public pressure, the public wants to know actually, more than the scientists, at the moment. So you are facing public pressures, even, definitely more than scientific pressures at the moment.

CHAIRMAN MILLER. Unfortunately in some of our problems, for example the NASA problems, where the public is indifferent, the matter of waste disposal, pollution, health, and these things. They are quite indifferent to them, and it takes a lot of effort to get them interested in them sometimes.

The committee has studied this on several occasions, but we have generally had a group of the scientific community behind us to give pressure, to bring pressure, to get some of these things done.

DR. HYNEK. I think we will see, sir, in this testimony today that you will find a corps of scientists stand ready to do this. In fact, as I mentioned in my testimony, I have private information from a very large number of scientists who are interested.

CHAIRMAN MILLER. I think this one of the values of the symposium.

MR. ROUSH. Are there other questions or comments? (No response.)[4]

CHAPTER TWO

STATEMENT OF JAMES E. MCDONALD

MR. ROUSH. Our next participant is Dr. James E. McDonald. Dr. McDonald is presently with the University of Arizona. He is a senior physicist, Institute of Atmospheric Physics, the University of Arizona, and has had a long and distinguished career as a scientist.

Dr. McDonald, we are pleased to have you as one of our participants. You may proceed.

☛ DR. MCDONALD. Thank you, Mr. Roush.

I am very pleased to have this chance to make some comments and suggestions based on my own experience to the committee, and I do wish to commend the Committee on Science and Astronautics for taking this first, and I hope very significant step, to look at the problem that has puzzled many for 20 years.

As Dr. Hynek has emphasized in his remarks, it is one of the difficulties of the problem we are talking about today that the scientific community, not just in the United States but on a world basis, hastened to discount and to regard as nonsense the UFO problem. The fact that so much anecdotal data is involved has understandably discouraged many scientists from taking seriously what, in fact, I believe is *a matter of extraordinary scientific importance* [my emphasis, L.S.].

I have been studying now for about 2 years, on a rather intensive basis, the UFO problem. I have interviewed several hundred witnesses in selected cases, and I am astonished at what I have found. I had no idea that the actual UFO situation is anything like what it really appears to be.

There is a certain parallel between Dr. Hynek's slow recognition of the problem and my slow recognition of the problem. I have been curious about UFO's in a casual way for 10 or 20 years and have even checked cases in the southern Arizona area off and on rather casually, mainly encountering sincere laymen who do not recognize an aircraft strobe light, or Venus, or a bright fireball, when they see them. It is quite true that many persons misidentify natural phenomena; and my experience was mainly but not entirely limited to that sort of case.

About 2 years ago I became more than casually curious for several reasons that are not too relevant here, and began to spend much more time and very quickly changed my notions about the problem. I visited Wright-Patterson Air Force Base, saw their very impressive and surprising UFO files, the pattern of which is entirely different from what I had imagined.

At the same time, I contacted a number of private investigating UFO groups, one of the best and most constructive located here in Washington, the National Investigations Committee on Aerial Phenomena; contacted another one of the large national groups, the Aerial Phenomena Research Organization, and found again somewhat to my surprise, that these amateur groups operating on a shoestring basis, and frequently scorned by us scientists, were, in fact, doing really a rather good investigative job within their resources, and had compiled in their files, for instance in NICAP, on the order of 10,000 or 12,000 cases, many of which I have subsequently checked, and all of which imply a problem that has been lost from sight, swept under the rug, ignored, and now needs to be very rapidly brought out into the open as a problem demanding very serious and very high-caliber scientific attention.

I wish to emphasize that. We must very quickly have very good people looking into this problem, because it appears to be one of very serious concern. We are dealing here with inexplicable phenomena, baffling phenomena, that will not be clarified by any but the best scientists.

The scope of my remarks this morning, and the scope of my more detailed remarks in my prepared statement which has been submitted, deal with two broad areas:

I have been asked to summarize the results of my interviewing of witnesses in the last 2 years, what I found, the problems I have encountered and so on; and, secondly to address myself to the categories of past explanations of UFO sightings, that hinge on my own field of atmospheric physics.

Let me turn very briefly to my experience. In the past 2 years I have been able to devote a substantial part of my time to this problem. I have mainly concentrated on witnesses in UFO sightings that have already been checked by some of the independent groups; that is, I was no longer, in the last 2 years, dealing with original raw data where it was primarily misidentified phenomena, but rather, I was dealing with presifted, presorted data, leaning very heavily on groups like NICAP and APRO, and other groups in this country and other groups abroad for my leads and background material.

I have also had a chance to interview 75 or 80 witnesses in Australia, New Zealand, and Tasmania, when I was down in that area last summer. There were various kinds of atmospheric explanations that had been invoked in Australian cases. I must say that many of them are just as reasonable from the scientific point of view as many that we have heard in this country. But primarily I found in Australia that the nature of the sightings is similar to those in the United States, disk-like objects, cigar-shaped objects, objects without wings, without evident means of propulsion, frequently hovering without any sound, sometimes making sounds, hovering over cars, stopping cars, as Dr. Hynek has pointed out, causing interference with the ignition system, and the same kind of public reluctance to report was very evident.

I want to emphasize, as one of the very important misconceptions that has been fostered, that instead of dealing with witnesses who are primarily looking for notoriety, who want to tell a good story, who are all out to gain attention, it is generally quite the opposite. And this is true in Australia, too. People are quite unwilling to tell you about a UFO sighting, afraid acquaintances would think they have "gone around the bend," as Australians put it. Over and over you encounter that. People are reluctant to report what they are seeing. There is a real ridicule lid that has not been contrived by any group, it just has evolved in the way the whole problem has unfolded. This is not entirely new in science. It has occurred before.

I am sure a number here at the speakers' table are familiar with an interesting chapter in science years ago when meteorites, out of which NASA and many scientists around the world now get a very large amount of useful scientific information, were scorned and scoffed as unreal. It was regarded as nonsense that peasants were telling stories about stones falling out of the sky. The efforts of a few scientists to take a look at the problem and to get some initial data simply were ignored until a very unusual but very real event occurred in northern France, a meteorite shower. So they sent an eminent academician out to have a look at what these people were talking about, and by golly, the peasants appeared to be right. Everybody in the village, the prefect of police, the local administrators, all the peasants, had seen stones fall out of the sky, and for the first time the French Academy deigned to take a look at the problem. Meteoritics was born.

Here we now face a very similar situation in science. We have tended to ignore it because we didn't think it made sense. It definitely defies any explanation, and hence the situation has evolved where we can't get going because we aren't already going.

The scientific community as a whole won't take this problem seriously because it doesn't have scientific data. They want instrumental data.

Why don't they have instrumental data? Because the scientists don't take it seriously enough to get the scientific data. It is like the 20-year old who can't get a job because he lacks experience, and he lacks experience because he hasn't had a job. In the same way you find the scientist wishing you would give him good hard meter readings and magnetometer traces, and so on; but we don't have it yet because the collective body of scientists, including myself, have ignored UFO's.

Turning to some of the highlights of my interviewing experience, I first mention the "ridicule lid." We are not dealing with publicity seekers. We are not, and I here concur with Dr. Hynek's remarks, we are not dealing with religiosity and cultism. Those persons aren't really the least bit interested in observations. They have firm convictions entirely independent of observations. They do not cause noise that disturbs the real signal at all.

General [John A.] Samford of the Air Force put it well, 16 years ago. General Samford, then Director of Intelligence, said, and I would concur 100 percent, "Credible observers are observing relatively incredible objects." That was said 16 years ago, and it is still occurring.

I will touch in a moment or two on a sighting in Mr. [Jerry L.] Pettis' district that very well illustrates that, a sighting this year in Redlands, Calif., which I think Dr. [James A.] Harder may be able to tell still more about.

Another characteristic in interviewing the witnesses is the tendency for the UFO witness to turn first not to the hypothesis that he is looking at a spaceship, but rather it must be an ambulance out there with a blinking red light or that it is a helicopter up there. There is a conventional interpretation considered first; only then does the witness get out of the car or patrol car and realize the thing is stopped in midair and is going backwards and has six bright lights, or something like that. Only after an economical first hypothesis does the witness, in these impressive cases, go further in his hypotheses, and finally realize he is looking at something he has never seen before.

I like Dr. Hynek's phrase for this, "escalation of hypotheses." This tendency to take a simple guess first and then upgrade it is so characteristic that I emphasize it as a very important point.

Then, looking at the negative side, all of us who have checked cases are sometimes in near anguish at the typical inability of the scientifically untrained person to estimate angles, to even understand what you are asking for when you ask for an angular estimation. We are all aware of the gross errors in distances, heights, and speeds so estimated.

And I would emphasize to those who cite jury trial experience that the tendency for a group of witnesses to an accident to come in with quite different accounts, must not be overstressed here. Those witnesses don't come in from, say, a street corner accident and claim they saw a giraffe killed by a tiger. They talk about an accident. They are confused about details. There is legally confusing difference of timing and distance, and so on; but all are in agreement that it was an auto accident.

So also when you deal with multiple-witness cases in UFO sightings. There is an impressive core of consistency; everybody is talking about an object that has no wings, all of 10 people may say it was dome shaped or something like that, and then there are minor differences as to how big they thought it was, how far away, and so on. Those latter variations do pose a very real problem. It stands as a negative factor with respect to the anecdotal data, but it does not mean we are not dealing with real sightings of real objects.

Then there is the very real but not terribly serious problem of the hoaxers, fabricators, liars, and so on. You do encounter cases from time to time where you end up thinking, well, this person has some reason to have invented the whole story. Sometimes it is fairly apparent. Sometimes it takes a lot of digging to prove it.

I might say here that the independent investigative groups have done an excellent job. It takes a knowledge of human characteristics, not scientific expertise to detect lies and hoaxes.

Then there is the problem that you always have to be sure in talking with witnesses that you are not dealing with somebody already very enthusiastic about UFO's. You have to try to establish, and this is not always easy, whether he has prior knowledge of the whole UFO literature. Are you dealing with somebody who is just telling you again what he has read in a recent magazine in the barber chair?

I emphasize that my experience is that again and again you find people who were not really interested in UFO's until they saw one themselves. Then they suddenly became very, very concerned, as one more member of the public who has become a UFO witness; and in this body of citizens there are some very distressed persons who wish that the scientific community, or the Government, were doing something about this problem.

The types of objects that are being seen, and I state the word "objects" not "hazy lights," are spread over quite a range of types, a baffling range.

I want to use that word many times, because it speaks for my experience. The UFO problem is baffling. But there is a predominance of disc-shaped objects and elongated cigar-shaped objects, objects without wings, appendages, tails, and that sort of thing. Typically, wingless objects, disc- and cigar-shaped.

The same type of observations have been coming from all parts of the world, and have been for a number of years. My direct interviews with a witness in Australia speak for that global pattern.

Another characteristic that emerges is a quite fluctuatory frequency of sightings. Right now, in the past few months, there have not been very many really impressive cases that have come up; but last fall, for example, England had a wave of sightings which was unprecedented in the English experience, that led, for example, to a BBC documentary that has just been produced. It led also to a recently published study, that I got only a couple of weeks ago from the Stoke-on-Trent area in Staffordshire, 70 sightings in about a 2½-month period in this area. It happens that one of my colleagues is an English physicist from that very area. As he points out, these are no-nonsense people who are not airy-fairy types that would be on LSD, or seeing ghosts in the sky.

He is puzzled, and I am puzzled.

Well, there are many questions that are asked by skeptical scientists, skeptical members of the public; and skepticism, as Mark Twain said, is what gets you an education.

There are questions like, "Why aren't UFO's seen abroad?" "Why aren't UFO's seen by airline pilots?" "Why aren't UFO's seen by crowds of people rather than by lone individuals?" "Why aren't they tracked by radar?" "Why don't weather observers and meteorologists see UFO's?"

"Why aren't there sonic booms, or why aren't there crashed UFO's?"

Finally, a very frequently raised question, "If the UFO's are from somewhere else, if they are really devices that represent some high civilization, why no contact?" This is a question that comes up again and again, since most persons who know enough about the UFO problem to realize there must be something there, cannot, in their first view of the problem, visualize a visitation from elsewhere, surveillance, or what have you, without contact.

I want to return to that point later, but I wish to emphasize that that is a fallacious question. If we were under surveillance from some advanced technology sufficiently advanced to do what we cannot do in the sense of interstellar travel, then, as Arthur C. Clarke has put it quite well, quoted in *Time* magazine the last week, we have an odd situation. Clarke points out that any sufficiently advanced technology would be indistinguishable from magic. How well that applies to UFO sightings. You have a feeling you are dealing with some very high technology, devices of an entirely real nature which defy explanation in terms of present-day science. To say that we could anticipate the values, reasons, motivations, and so on, of any such system that has the capability of getting here from somewhere else is fallacious.

That is a homocentric fallacy of the most obvious nature, yet it is asked over and over again.

In my prepared statement I will be able to cover more of these points, of course.

The heart of the problem lies in citing cases, and I have investigated, personally, on the order of 300 cases dealing with key witnesses. I have looked as carefully as I can for all reasonable explanations.

There are many cases that fall apart when you investigate them. Then there are far too many that resist the best analysis that many of us have been able to subject them to.

Let me just cite briefly, to take a recent case rather than an old one, the instance at Redlands, and perhaps Dr. Harder can fill you in in more detail.

On February 4 of this year, at 7:20 in the evening, over a residential area in that city of population 30,000, a disc was seen. Twenty witnesses interviewed by University of Redlands' investigators, described it as having "windows" or "ports" or something of that sort. They interviewed a little over half a dozen of them and all saw something on the bottom that they described as "looking like jets."

This object was hovering at an estimated height of about 300 feet. The estimates vary, but it came out about 300 feet. The citizens had gone out in the street because dogs were barking and, because they had heard an unusual noise, and pretty soon there were people all up and down the street. It was estimated that more than 100 witnesses were involved, and 20 were directly interviewed.

Here was an object seen by many persons. It hovered, then shot up to about double the height, hovered again, and moved down across Redlands a short distance, hovered once again, and then took off rapidly to the northwest.

This case has not received any scientific attention beyond this investigation by Dr. Philip Seff and his colleagues. It has not received public notoriety. This was, in fact, only reported in a short column in the local paper and not on the wires anywhere. That happens over and over again.

Here, for example, are the reports for one month of last fall, clipping-service coverage on the things that get local coverage, but don't get on the wires, because in the present climate of the opinion, wire editors, like scientists, Congressmen, and the public at large, feel sure there is nothing to all this, and they don't put them on the wires. You have to go right to the local town to get press coverage in most cases.

The Redlands, February 1968, case illustrates that very well. Once in a while a case will get on the wires and receive national attention, but by and large, one just doesn't read about these cases in other parts of the country, because wire services don't carry them.

Let me tell you another case that answers the questions: "Why aren't there multiple witnesses?" "Why aren't they seen in cities?" "Why aren't these ever seen in the daytime?"

It is true that there is a preponderance of nighttime sights. Maybe this is merely a matter of luminosity.

It is also true that there seem to be more reports from rather remote areas, say desert areas or swampy areas, than in the middle of cities. But there are city observations. And it is also true there are more individual witness cases than sightings by large crowds. But in every instance there are striking exceptions to this.

In New York City, on November 22, 1966, a total of eight witnesses, members of the staff of the American Newspaper Publishers Association, were the witnesses in a good case. I interviewed William Leick, of that staff, the manager of the office there. I heard about it through a NICAP report. It did not appear in the papers, as I will mention. William Leick had been looking out the window, saw an object over the U.N. building. It was hovering, and as he talked to a colleague he realized there was something odd about it, so they walked out on the terrace. Soon they had six others out on the terrace. This was at 4:30 in the afternoon. It was kind of a cushion-shaped object, as he described it, and had no wings. It was rocking a little from time to time, blinked in the afternoon sun a little bit, had kind of an orange glow. All eight were watching, and after it hovered for several moments it rose vertically and then took off at high speed. There is an example of midtown sighting in New York where the witnesses are staff members of a responsible organization. Leick, himself, had been trained in intelligence, in World War II. There is no reason at all to think he and his colleagues would invent this.

They did call a New York paper, but to say they weren't the least bit interested. There was no report published in a New York paper. Next they called a local Air Force office but no one came to investigate it. It came to my attention because one of the members of the staff knew of NICAP and sent NICAP a report.

This sort of thing has happened over and over again. The ridicule lid keeps these out of sight; too many of them are occurring to delay any longer in getting at this problem with all possible scientific assistance.

A famous multiple-witness instance occurred in Farmington, N. Mex., on March 17, 1950. I interviewed seven witnesses there. A very large number of objects were involved. There were several different groups of objects, all described as disc-shaped objects. They were explained as Skyhook balloons, officially, so I checked into that.

I finally established that there was no Skyhook balloon released anywhere in the United States on or near that day. The witnesses included some of the leading citizens in the town. It was reported nationally at that time but was soon forgotten.

I have interviewed one of the witnesses in a Washington State sighting, at Longview, Wash., July 3, 1949. An air show was being held and someone spotted the UFO because there was a sky-writing aircraft overhead that some people were watching. They spotted the first of three disc-like objects that came over Longview that morning. The person whom I interviewed is a former Navy commander, Moulton B. Taylor. He was the manager of the air show, so he got on the public address system and got everybody to look at this object before it crossed the skies. It was fluttering as it went across the sky. There were pilots, engineers, police officers, and Longview residents in the audience. Many had binoculars. Taylor estimated it to be about 10 minutes of arc in diameter. Because the aircraft was still skywriting people continued to watch the sky. Two successive objects of the same type flew over in the next 20 minutes. A total of three objects came over, and they were from three different directions: one from the north, one from the northwest, and one almost from the west, quite clearly ruling out an explanation like balloons, which became the official explanation. There were no balloon stations anywhere near Longview, Wash., as a matter of fact, and the balloon explanation is quite inadequate.

Here we have a case of over a hundred witnesses to the passage of a wingless object moving at relatively high velocity. When the second and third objects went over, someone had the presence of mind to time the fluttering rate—it was 48 per minute.

Here again we have a multiple-witness case, a daytime sighting case, and one which you can't quickly write off.

If time permitted I would talk about a number of radar cases. One of the most famous is the Washington National Airport sighting. On July 19, 1952, CAA radars and Andrews Air Force radars tracked unknowns moving at variable speeds from 100 miles an hour to over 800 miles an hour, and a number of airline pilots in the air saw these, and were in some instances vectored in by the CAA radar people, and then saw luminous objects in the same area that they showed on radar up near Herndon [VA] and Martinsburg [WV].

I talked to five of these CAA people. One can still go back and check these old cases, I emphasize. I also talked to four of the airline pilots who were in the air at the time. I have gone over the quantitative aspects of the official explanation that this was ducting or trapping of the radar beams. That is quite untenable. I have gone over the radiosonde, computed the radar refractive index gradient, and it is nowhere near the ducting gradient.

Also, it is very important that at one time three different radars, two CAA and one Andrews Air Force Base radar, all got compatible echos. That is extremely significant.

And finally from a radar-propagation point of view, the angles of propagation, radar and visual, were far above any values that would permit trapping, which makes this a case which is not an explained case. It was an instance of unidentified aerial objects over our Capital, I believe.

One could go on with many cases. I want to just briefly touch two categories of atmospheric explanations that have been rather widely discussed, and close with that.

Meteorological optics is a subject that I enjoy and have looked into over the years rather carefully, and I must express for the record my very strong disagreement with Dr. Donald H. Menzel, former director of Harvard Observatory, whose two books on the subject of UFO's lean primarily on meteorological explanations. I have checked case after case of his, and his explanations are very, very far removed from what are well-known principles and quantitative aspects of meteorological optic objects. He has made statements that simply do not fit what is known about meteorological objects.

I would be prepared to talk all day on specific illustrations but time will not permit more.

Secondly, there has more recently been a suggestion made by *Aviation Week* Senior Editor Philip J. Klass, that the really interesting UFO's are atmospheric-electrical plasmas of some type similar to ball lightning, but perhaps something different, something we don't yet understand but are generated by atmospheric processes.

The first time anyone tried the ball lightning hypothesis was in Air Force Project Grudge, back in 1949. The Weather Bureau was asked to do a special study of ball lightning. I recently got a declassified copy of that, and the Air Force position at that time, and since then was that ball lightning doesn't come near to explaining these sightings. I concur in that. When you deal with multiple-witness cases involving discs with metallic luster, definite outline, seen in the daytime, completely removed from a thunderstorm, perhaps seen over center Manhattan, or perhaps in Redlands, Calif., they are not ball lightning or plasmas.

In weather completely unrelated to anything that could provide a source of energy, the continuous power source required to maintain a plasma in the face of recombination and decay of a plasma, Klass' views just do not make good sense.

It is just not reasonable to suggest that, say the BOAC Stratocruiser that was followed by six UFO's for 90 miles up in the St. Lawrence Valley in 1954 was followed by a plasma, or that these people in Redlands were looking at a plasma, or that the 20 or so objects that went over Farmington were plasmas.

One of the most characteristic features of a plasma is its very short lifetime and exceedingly great instability, as some of your members will know from your contact with fusion research problems. The difficulty of sustaining a plasma for more than microseconds is a very great difficulty. To suggest that clear weather conditions can somehow create and maintain plasmas that persist for many minutes, and fool pilots with 18,000 flight hours into thinking that they are white-and red-domed discs, to take a very famous case over Philadelphia where the pilot thought he was about 100 yards from this dome-disc, is unreasonable. It is not a scientifically well-defended viewpoint.

To conclude, then, my position is that *UFO's are entirely real* (my emphasis, L.S.) and we do not know what they are, because we have laughed them out of court. The possibility that these are extraterrestrial devices, that we are dealing with surveillance from some advanced technology, is a possibility I take very seriously.

I reach that hypothesis, as my preferred hypothesis, not by hard fact, hardware, tailfins, or reading license plates, but by having examined hundreds of cases and rejected the alternative hypothesis as capable of accounting for them.

I am afraid that this possibility has sufficiently good backing for it, despite its low *a priori* ability, that we must examine it. I think your committee, with its many concerns for the entire aerospace program, as well as our whole national scientific program, has a very special reason for examining that possibility. Should that possibility be correct, if there is even a chance of its being correct, we ought to get our best people looking at it. Instead, we are collectively laughing at this possibility.

To meet Mr. Rumsfeld's request, let me remark on Dr. Hynek's two recommendations. I strongly concur in the need for some new approach. I am sure Dr. Hynek was not suggesting there be one single UFO committee. In fact, he said, "not a one-shot approach." A pluralistic approach to the problem is needed here.

The Defense Department is already supporting some work on it. NASA definitely has a need to look at this problem. We have to pay very serious attention to the problem and get a variety of new approaches.

The other point Dr. Hynek mentioned was that we try to look at this on a worldwide basis. This is crucially important. We are dealing with a real problem here, and I insist it is a global problem. We can study it in the United States, but if we ignore what is happening in France and England—one of the greatest UFO waves that ever occurred was in France—would be a serious mistake. I strongly urge that your committee consider holding rather more extensive hearings in which a larger segment of the scientific community is given the opportunity to talk pro and con on the issue, hearings aimed at getting a new measure of scientific attention to this important problem.

Thank you.

MR. ROUSH. Thank you, Dr. McDonald, for your presentation. As we explained awhile ago, we are pressed for time. We are entertaining questions from members of the committee. Mr. Bell.

MR. [ALPHONZO] BELL. Dr. McDonald, I want to compliment you on your interesting statement. But what leads you to believe that whatever these phenomena are, they are extraterrestrial? What facts do you have?

DR. MCDONALD. May I say I wouldn't use the word "believe." I would say the "hypothesis" that these are extraterrestrial surveillance, is the hypothesis I presently regard as most likely.

As I mentioned, it is not hard facts in the sense of irrefutable proof, but dealing with case after case wherein the witnesses showed credibility I can't impugn. That impresses me. These are not at all like geophysical or astronomical phenomena; they appear to be craft-like machine-like devices. I would have to answer you in terms of case after case that I and others have investigated, to make all this clear. It is this very large body of impressive witnesses' testimony, radar tracking data on ultra-high-speed objects sometimes moving at over 5,000 miles an hour, UFO's, combined radar-visual sightings, and just too much other consistent evidence that suggests we are dealing with machine-like devices from somewhere else.

MR. BELL. Have there been pictures taken?

DR. MCDONALD. Yes; there have been pictures taken.

For instance, a photograph taken in Ohio, by an Air Force photo reconnaissance plane May 24, 1954. I recently have looked a little more closely at the data. This was explained as an undersun, but that idea is subject to quantitative observation. The angles just do not fit. There is a very important case at Edwards Air Force base with two witnesses, where they got photographs of the object. Unfortunately, in this case I have not seen the photo, but I have talked with the persons who took it. There are photographs, but not nearly as many as we would like. We would like to have lots of them. In a case in Corning, Calif., a police officer, one of five witnesses, had a loaded camera in his patrol car, 20 paces from where he watched the object, didn't even think of getting his camera. He said he was too flabbergasted to think of it. That is a part of the problem.

MR. ROUSH. Mr. Hechler.

MR. HECHLER. Have you examined any reports of communication by these objects?

DR. MCDONALD. Yes; the problem of contact is very important. There is one category of contact, not in the sense of shaking hands, but rather light response. I have a file on several of these, and I'm looking for more. For instance, in Shamokin, Pa., Kerstetter is the name of the witness, he works for a bank in Shamokin. I talked to the president of the bank as to his reliability and got very good recommendations. Last year, he and his wife and family were in a car near a mountain ridge in Shamokin, saw a thing hovering over the mountain, like the flashing lights of a theater marque. He had a flashlight. He didn't know Morse code, but it really didn't matter. He sent light flashes in various orders and he got lights back from the thing. That same thing happened in

Newton, N.H., in August of last year, where several persons saw an object coming overhead. The same thought occurred to them and they signaled with a flashlight. It wasn't Morse, it was dot dash dot, then dash dash dash, and it came back with no failure, replicated light signals. The same thing happened in West Virginia, where a pharmacist, named Sommers, did it with his headlights. When I was in Australia, I talked about some hunters out hunting kangaroo. A disk came over, one said "give them Morse"; the flash came back faithfully, and they left in a hurry. Is that contact? I don't know. Nobody got any intelligence out of it either way, if you will pardon the whimsy. It would be terrible if in fact this was surveillance and our technology was represented by the Eveready flashlight. (Laughter.)

We may be flunking our exam.

MR. ROUSH. Mr. Downing.

MR. [THOMAS N.] DOWNING. I'm interested in your testimony. On page 10 of your written statement, you say it is unfortunate no acceptable version of Reference 6 exists, though it has managed to get it into the status of limited acceptability. Why is this not available?

DR. MCDONALD. Well, that was an Air Force document. This was completed in 1949. These were classified until just a few years back. No one could get access to them, because they were under DOD classification. But the 12-year rule expired, and Dr. Leon Davidson managed to get a copy.

It is accessible in the sense that if I want to pay $90 for Xeroxes I can now get it. It is not published in the sense of being available to every library in the country. My Reference 7, which NICAP just published, is available to scientists all over the country. It is a matter of the Air Force having a policy of not publishing such items, and they were classified. I think the Moss committee and NICAP are to be highly praised to get out in the open Reference 7.

MR. DOWNING. Is there a reason why this is classified?

DR. MCDONALD. There is an understandable reason why the Air Force has had to classify this. An unidentified area object on presumption is hostile until proved otherwise. So there has been this unfortunate, but entirely understandable measuring of these two areas. The national defense mission of the Air Force has necessitated they have some part of the UFO problem inevitably, and they got it in the first instance. They have long since told us there is no hostility here, hence the scientific curiosities going unattended because it doesn't fall under the defense mission, in other words to be transferred into NASA, NSF, or something like that. That does not mean the Air Force won't continue to watch unidentified objects on the millisecond basis. But they not need worry about this other part of the problem. I think it is understandable, but needs changing.

MR. ROUSH. Mr. Pettis.

MR. PETTIS. Mr. Chairman, Doctor.

I was a little bit interested in your observations about this UFO sighting in my hometown of Redlands.

I might observe that Redlands is a rather conservative community, when people in Redlands say they saw something, they saw something. I did not happen to be in Redlands that particular date, so I did not see this.

But I would like to observe this, that having spent a great deal of my life in the air, as a pilot, professional and private pilot, I know that many pilots and professional pilots have seen phenomena that they could not explain.

These men, most of whom have talked to me, have been very reticent to talk about this publicly, because of the ridicule that they were afraid would be heaped upon them, and I'm sure that if this committee were ever to investigate this, or bring them in here, there probably would have to be a closed hearing, Mr. Chairman.

However, there is a phenomena here that isn't explained.

I think probably we ought to do a little looking into this, is my personal opinion.

MR. ROUSH. Mr. Ryan.

MR. [WILLIAM F.] RYAN. Yes, thank you, Mr. Chairman.

First I should like to commend you, Mr. Roush, for your interest in the subject matter, and the chairman of the full committee for having arranged for hearings into this problem.

I think it is important that this committee not waive its jurisdiction, but that it explore very carefully the proposals that have been made by the witnesses here, and that it have a continuing field of exploration into this whole question. I want to commend Dr. McDonald for having been persistent in presenting his views to the various members of the committee, helping to bring about these hearings. I wondered, Dr. McDonald, if you would care to evaluate the research project at the University of Colorado, and comment on that?

MR. ROUSH. Mr. Ryan, may I just say we had agreed that this was not the place to discuss that particular project, and that the purpose of the symposium was not to go into the activities of another branch of government, but rather to explore that as a scientific phenomena.

I'm sure that Dr. McDonald would be very happy to confer with you privately on this, but if you could show some restraint here, the Chair would be real grateful to you.

MR. RYAN. Well, let me rephrase my question.

In view of the fact that there has been a study conducted by a project in the Air Force, and the University of Colorado, do you believe there is anything further that should be done by any branch of the Government?

DR. MCDONALD. Emphatically, yes.

MR. RYAN. What would you recommend?

DR. MCDONALD. I think that we need to get a much broader basis of investigation of UFO's, as I did say, a few moments ago, it would be very salutary to have a group in NASA looking at this problem, and to have some NASA support of independent studies. It would be very good for the National Science Foundation to support, say, some university people interested in it. It would be good to have the Office of Naval Research et cetera involved.

We don't deal with many other important problems, space, or molecular biology or health without a pluralistic approach, a multiplicity of research programs. I don't want to touch a frayed nerve here. This problem of duplication is sometimes lamented. But by and large I think you will agree we would gain from having a lot of different people with slightly different points of view going at every problem. At the moment everything is focused through one agency, and everything now hinges on that one particular program you have asked me about, and my answer was, we very definitely need some independent programs.

I am on record elsewhere than here in my specific views on that project.

MR. RYAN. Looking back at page 14, you wrote a letter to the National Academy of Sciences, concerning this project. Have you had any reaction from the National Academy of Sciences?

DR. MCDONALD. Yes, I received a letter from Dr. [Frederick] Seitz, saying for the time being we must let the Colorado project run its course. That was the gist of the answer.

MR. ROUSH. I would appreciate it, if we dispensed with that. Let me say that the National Academy is undertaking an evaluation of the University of Colorado project, and this will be published.

MR. RYAN. I'm suggesting maybe this committee should make an investigation of the University of Colorado project.

CHAIRMAN MILLER. That is something we don't have authority to do here.

MR. RYAN. To what extent, Dr. McDonald, have sightings

been picked up by radar, and what extent of those that have been picked up been explored?

DR. MCDONALD. Well, there are many such sightings, I dare say there are thousands of military radar sightings that were for the short period unidentified. Then they identify them. But here is an impressive number of both military and civilian radar sightings that defy radar explanation in terms of unknown phenomena. Most of these deficiencies are well understood, so one can be fairly sure that many of these unidentified radar cases have no conventional explanation.

In a case where a [Northrop] P-61 [Black Widow] flew over Japan, back some years ago, made six passes at an unidentified object is was getting radar returns on, and the pilot saw it visually. Here you are dealing with an unknown. Then there was a case in Michigan where a ground radar detected an object at 600 miles an hour coming in over Saginaw Bay. The pilot got a radar return, and also saw a vast luminous object; the object turned in a very sharp 180-degree turn and went back, and eluded the [Lockheed] F-94 [Starfire]. Here you are dealing with a case where radar propagation anomalies will not explain it. There was one radar in the airplane at 20,000 feet and one radar on the ground, both showing the object. There are many cases like that which I could enlarge on.

MR. RYAN. Let me ask a further question: In the course of your investigation and your study of UFO sightings, have you found any cases where contemporaneously with the sighting of UFO's allegedly, there were any other events which took place, which might or might not be related to the UFO's?

DR. MCDONALD. Yes. Certainly there are many physical effects. For instance, in Mr. Pettis' district, several people found the fillings in their mouth hurting while this object was nearby, but there are many cases probably on record of car ignition failure. One famous case was at Levelland, Tex., in 1967. Ten vehicles were stopped within a short area, all independently in a 2-hour period, near Levelland, Tex. There was no lightning or thunder storm, and only a trace of rain. There is another which I don't know whether to bring to the committee's attention or not. The evidence is not as conclusive as the car stopping phenomenon, but there are too many instances for me to ignore. UFO's have often been seen hovering near power facilities. There are a small number but still a little too many to seem pure fortuitous chance, of system outages, coincident with the UFO sighting. One of the cases was Tamaroa, Ill. Another was a case in Shelbyville, Ky., early last year. Even the famous one, the New York blackout, involved UFO sightings. Dr.

Hynek probably would be the most appropriate man to describe the Manhattan sighting, since he interviewed several witnesses involved. I interviewed a woman in Seacliff, N.Y. She saw a disk hovering and going up and down. And then shooting away from New York just after the power failure. I went to the FPC for data, they didn't take them seriously although they had many dozens of sighting reports for that famous evening. There were reports all over New England in the midst of that blackout, and five witnesses near Syracuse, N.Y., saw a glowing object ascending within about a minute of the blackout. First they thought it was a dump burning right at the moment the lights went out. It is rather puzzling that the pulse of current that tripped the relay at the Ontario Hydro Commission plant has never been identified, but initially the tentative suspicion was centered on the Clay Substation of the Niagara Mohawk network right there in the Syracuse area, where, unidentified aerial phenomenon has been seen by some of the witnesses.

This extends down to the limit of single houses losing their power when a UFO is near. The hypothesis in the case of car stopping is that there might be high magnetic fields, d.c. fields, which saturate the core and thus prevent the pulses going through the system to the other side. Just how a UFO could trigger an outage on a large power network is however not yet clear. But this is a disturbing series of coincidences that I think warrant much more attention than they have so far received.

MR. RYAN. As far as you know, has any agency investigated the New York blackout in relation to UFO?

DR. MCDONALD. None at all. When I spoke to the FPC people, I was dissatisfied with the amount of information I could gain. I am saying there is a puzzling and slightly disturbing coincidence here. I'm not going on record as saying, yes, these are clear-cut cause and effect relations. I'm saying it ought to be looked at. There is no one looking at this relation between UFO's and outages.

MR. ROUSH. Our time is really running short, Mr. Ryan.

MR. RYAN. One final question. Do you think it is imperative that the Federal Power Commission, or Federal Communications Commission, investigate the relation if any between the sightings and the blackout?

DR. MCDONALD. My position would call for a somewhat weaker adjective. I'd say extremely desirable.

MR. ROUSH. Thank you.

Thank you, Dr. McDonald.[5]

PREPARED STATEMENT ON UNIDENTIFIED FLYING OBJECTS

James E. McDonald, Senior Physicist, Institute of Atmospheric Physics, and Professor, Department of Meteorology, The University of Arizona, Tucson, Arizona

INTRODUCTION

I should like first to commend the House Committee on Science and Astronautics for recognizing the need for a closer look at scientific aspects of the long-standing puzzle of the Unidentified Flying Objects (UFOs). From time to time in the history of science, situations have arisen in which a problem of ultimately enormous importance went begging for adequate attention simply because that problem appeared to involve phenomena so far outside the current bounds of scientific knowledge that it was not even regarded as a legitimate subject of serious scientific concern. That is precisely the situation in which the UFO problem now lies. One of the principal results of my own recent intensive study of the UFO enigma is this: I have become convinced that the scientific community, not only in this country but throughout the world, has been casually ignoring as nonsense a matter of extraordinary scientific importance. The attention of your Committee can, and I hope will, aid greatly in correcting this situation. As you will note in the following, my own present opinion, based on two years of careful study, is that *UFOs are probably extraterrestrial devices engaged in something that might very tentatively be termed "surveillance"* [my emphasis, L.S.].

If the extraterrestrial hypothesis is proved correct (and I emphasize that the present evidence only points in that direction but cannot be said to constitute irrefutable proof), then clearly UFOs will become a top-priority scientific problem. I believe you might agree that, even if there were a slight chance of the correctness of that hypothesis, the UFOs would demand the most careful attention. In fact, that chance seems to some of us a long way from trivial. We share the view of Vice Adm. Roscoe H. Hillenkoetter, former CIA Director, who said eight years ago, "It is imperative that we learn where the UFOs come from and what their purpose is (Ref. 1)." Since your Committee is concerned not only with broad aspects of our national scientific program but also with the prosecution of our entire space program, and since that space program has been tied in for some years now with the dramatic goal of a search for life in the universe, I submit that the topic of today's Symposium is eminently deserving of your attention. Indeed, I have to state, for the record, that I believe no other problem within your jurisdiction is of comparable scientific and national importance. Those are strong words, and I intend them to be.

In addition to your Committee responsibilities with respect to science and the aerospace programs, there is another still broader basis upon which it is highly appropriate that you now take up the UFO problem: Twenty years of public interest, public puzzlement, and even some public disquiet demand that we all push toward early clarification of this unparalleled scientific mystery. I hope that our session here today will prove a significant turning point, orienting new scientific efforts towards illumination of this scientific problem that has been with us for over 20 years.

SCOPE & BACKGROUND OF PRESENT COMMENTS

It has been suggested that I review for you my experiences in interviewing UFO witnesses here and abroad and that I discuss ways in which my professional experience in the field of atmospheric physics and meteorology illuminates past and present attempts at accounting for UFO phenomena. To understand the basis of my comments, it may be helpful to note briefly the nature of my own studies on UFOS.

I have had a moderate interest in the UFO problem for twenty years, much as have a scattering of other scientists. In southern Arizona, during the period 1956-66, I interviewed, on a generally rather random basis, witnesses in such local sightings as happened to come to my attention via press or personal communications. This experience taught me much about lay misinterpretations of observations of aircraft, planets, meteors, balloons, flares, and the like. The frequency with which laymen misconstrue phenomena associated with fireballs (meteors brighter than magnitude -5), led me to devote special study to meteor physics; other topics in my own field of atmospheric physics also drew my closer attention as a result of their bearing on various categories of UFO reports. This period of rather casual UFO-witness interviewing on a local basis proved mainly educational; yet on a few occasions I encountered witnesses of seemingly high credibility whose reports lay well outside any evident meteorological, astronomical, or other conventional bounds. Because I was quite unaware, before 1966, that those cases were, in fact, paralleled by astonishing numbers of comparable cases elsewhere in the U.S. and the rest of the world, they left me only moderately puzzled and mildly bothered, since I came upon relatively few impressive cases within the environs of Tucson in those dozen years of discursive study. I was aware of the work of non-official national investigative groups like NICAP (National Investigations Committee on Aerial Phenomena) and APRO (Aerial Phenomena Research Organization); but lacking basis for detailed personal evaluation of their investigative methods, I simply did not take their publications very seriously. I was under other misimpressions, I found later, as to the nature of the official UFO program, but I shall not enlarge on this before this Committee. (I cite all of this here because I regard it relevant to an appreciation, by the Committee, of the way in which at least one scientist has developed his present strong concern for the UFO problem, after a prior period of some years of only mild interest. Despite having interviewed a total of perhaps 150-200 Tucson- area witnesses prior to 1966 (75 of them in a single inconclusive case in 1958), I was far from overwhelmed with the importance of the UFO problem.

A particular sighting incident in Tucson in early 1966, followed by the widely publicized March, 1966, Michigan sightings (I, too, felt that the "swamp gas" explanation was quite absurd once I checked a few relevant points), led me finally to take certain steps to devote the coming summer vacation months to a much closer look at the UFO problem. Within only a few weeks in May and June of 1966, after taking a close look at the files and modes of operation of both private and official (i.e., *Project Bluebook*) UFO investigative programs, after seeing for the first time press-clipping files of (to me) astonishing bulk, covering innumerable intriguing cases I had never before heard of, and (above all) after the beginning of what became a long period of personal interviewing of key witnesses in important UFO cases, I rapidly altered my conception of the scientific importance of the UFO question. By mid-1966, I had already begun what became months of effort to arouse new interest and to generate new UFO investigative programs in various science agencies of the Federal government and in various scientific organizations. Now, two years later, with very much more background upon which to base an opinion, I find myself increasingly more concerned with what has happened during the past twenty years' neglect, by almost the entire scientific community, of a problem that appears to be one of extremely high order of scientific importance.

THE UNCONVENTIONAL NATURE
OF THE UFO PROBLEM

To both laymen and scientists, the impressive progress that science has made towards understanding our total environment prompts doubt that there could be machine-like objects of entirely unconventional nature moving through our atmosphere, hovering over automobiles, power installations, cities, and the like, yet all the while going unnoticed by our body scientific. Such suggestions are hard to take seriously, and I assure you that, until I had taken a close look at the evidence, I did not take them seriously. We have managed to so let our preconceptions block serious consideration of the possibility that *some form of alien technology is operating within our midst* [my emphasis, L.S.] that we have succeeded in simply ignoring the facts. And we scientists have ignored the pleas of groups like NICAP and APRO, who have for years been stressing the remarkable nature of the UFO evidence. Abroad, science has reacted in precisely this same manner, ignoring as nonsensical the report-material gathered by private groups operating outside the main channels of science. I understand this neglect all too well; I was just one more of those scientists who almost ignored those facts, just one more of those scientists who was rather sure that such a situation really could not exist, one more citizen rather sure that official statements must be basically meaningful on the non-existence of any substantial evidence for the reality of UFOS.

The UFO problem is so unconventional, involves such improbable events, such inexplicable phenomenology, so defies ready explanation in terms of present-day scientific knowledge, has such a curiously elusive quality in many respects, that it is not surprising (given certain features in the past twenty years' handling of the problem) that scientists have not taken it very seriously. We scientists are, as a group, not too well-oriented towards taking up problems that lie, not just on the frontiers of our scientific knowledge, but far across some gulf whose very breadth cannot be properly estimated. These parenthetical remarks are made here to convey, in introductory manner, viewpoints that will probably prove to be correct when many more scientists begin to scrutinize this unprecedented and neglected problem. The UFO problem is, if anything, a highly unconventional problem. Hence, before reviewing my own investigations in detail, and before examining various proposed explanations lying within atmospheric physics, it may be well to take note of some of the principal hypotheses that have been proposed, at one time or another, to account for UFOs.

SOME ALTERNATIVE HYPOTHESES

In seeking explanations for UFO reports, I like to weigh witness-accounts in terms of eight principal UFO hypotheses:

1. Hoaxes, fabrications, and frauds.
2. Hallucination, mass hysteria, rumor phenomena.
3. Lay misinterpretations of well-known physical phenomena (meteorological, astronomical, optical, aeronautical, etc.).
4. Semi-secret advanced technology (new test vehicles, satellites, novel weapons, flares, re-entry phenomena, etc.).
5. Poorly understood physical phenomena (rare atmospheric-electrical or atmospheric-electrical effects, unusual meteoric phenomena, natural or artificial plasmoids, etc.).
6. Poorly understood psychological phenomena.
7. Extraterrestrial devices of some surveillance nature.
8. Spaceships bringing messengers of terrestrial salvation and occult truth.

Because I have discussed elsewhere all of these hypotheses in some detail (Ref. 2), I shall here only very briefly comment on certain points. Hoaxes and fabrications do crop up, though in percentually far smaller numbers than many UFO scoffers seem to think. Some of the independent groups like APRO and NICAP have done good work in exposing certain of these. Although there has been a good deal of armchair-psychologizing about unstable UFO witnesses, with easy charges of hallucination and hysteria, such charges seem to have almost no bearing in the hundreds of cases I have now personally investigated. Misinterpreted natural phenomena (Hypothesis 3) do explain many sincerely-submitted UFO reports; but, as I shall elaborate below, efforts to explain away almost the entirety of all UFO incidents in such terms have been based on quite unacceptable reasoning. Almost no one any longer seriously proposes that the truly puzzling UFO reports of close-range sighting of what appear to be machines of some sort are chance sightings of secret test devices (ours or theirs); the reasons weighing against Hypothesis 4 are both obvious and numerous. That some still-not-understood physical phenomena of perhaps astronomical or meteorological nature can account for the UFO observations that have prompted some to speak in terms of extraterrestrial devices would hold some weight if it were true that we dealt therein only with reports of hazy, glowing masses comparable to, say, ball lightning or if we dealt only with fast-moving luminous bodies racing across the sky in meteoric fashion. Not so, as I shall enlarge upon below.

Jumping to Hypothesis 6, it seems to receive little support from the many psychologists with whom I have managed to have discussions on this possibility; I do not omit it from consideration, but, as my own witness-interviewing has proceeded, I regard it with decreasing favor. As for Hypothesis 8, it can only be remarked that, in all of the extensive literature published in support thereof, practically none of it has enough ring of authenticity to warrant serious attention. A bizarre "literature" of pseudo-scientific discussion of communications between benign extraterrestrials bent on saving the better elements of humanity from some dire fate implicit in nuclear weapons testing or other forms of environmental contamination is certainly obtrusive on any paperback stand. That "literature" has been one of the prime factors in discouraging serious scientists from looking into the UFO matter to the extent that might have led them to recognize quickly enough that cultism and wishful thinking have essentially nothing to do with the core of the UFO problem. Again, one must here criticize a good deal of armchair-researching (done chiefly via the daily newspapers that enjoy feature-writing the antics of the more extreme of such groups). A disturbing number of prominent scientists have jumped all too easily, to the conclusion that only the nuts see UFOs.

The seventh hypothesis, that UFOs may be some form of extraterrestrial devices, origin and objective still unknown, is a hypothesis that has been seriously proposed by many investigators of the UFO problem. Although there seems to be some evidence that this hypothesis was first seriously considered within official investigative channels in 1948 (a year after the June 24, 1947, sighting over Mt. Rainier that brought the UFO problem before the general public), the first open defense of that Hypothesis 7 to be based on any substantial volume of evidence was made by [Donald E.] Keyhoe (Ref. 3) in about 1950. His subsequent writings, based on far more evidence than was available to him in 1950, have presented further arguments favoring an extraterrestrial origin of UFOs. Before I began an intensive examination of the UFO problem in 1966, I was disposed to strong doubt that the numerous cases discussed at length in Keyhoe's rather dramatically written and dramatically-titled books (Ref. 4) could be real cases from real witnesses of any appreciable credibility. I had the same reaction to a 1956 book (Ref. 5) written by [Edward J.] Ruppelt, an engineer in charge of the official investigations in the important 1951-3 period. Ruppelt did not go as far as Keyhoe in suggesting the extraterrestrial UFO hypothesis, but he left his readers little room for doubt that he leaned toward that hypothesis. I elaborate these two writers' viewpoints because, within the past month, I have had an opportunity to examine in detail a large amount of formerly classified official file material which substantiates to an almost alarming degree the authenticity and hence the scientific import of the case-material upon which Keyhoe and Ruppelt drew for much of their discussions of UFO history in the 1947-53 period (Refs. 6 and 7).

One of these sources has just been published by NICAP (Ref. 7), and constitutes, in my opinion, an exceedingly valuable addition to the growing UFO literature. The defense of the extraterrestrial hypothesis by Keyhoe, and later many others (still not within what are conventionally regarded as scientific circles), has had little impact on the scientific community, which based its write-off of the UFO problem on press accounts and official assurances that careful investigations were turning up nothing that suggested phenomena beyond present scientific explanation. Hypothesis No. 7 has thus received short shrift from science to date.

As one scientist who has gone to some effort to try to examine the facts, I say that this has been an egregious, if basically unwitting, scientific error—an error that must be rectified with minimum further delay. On the basis of the evidence I have examined, and on the basis of my own weighing of alternative hypotheses (including some not listed above), I now regard Hypothesis 7 as the one most likely to prove correct. My scientific instincts lead me to hedge that prediction just to the extent of suggesting that if the UFOs are not of extramundane origin, then I suspect that they will prove to be something very much more bizarre, something of perhaps even greater scientific interest than extraterrestrial devices.

SOME REMARKS ON INTERVIEWING EXPERIENCE AND TYPES OF UFO CASES ENCOUNTERED

1. Sources of cases dealt with

Prior to 1966, I had interviewed about 150–200 persons reporting UFOs; since 1966, I have interviewed about 200-250 more. The basis of my post-1966 interviewing has been quite different from the earlier period of interviewing of local witnesses, whose sightings I heard about essentially by chance. Almost all of my post-1966 interviews have been with witnesses in cases already investigated by one or more of the private UFO investigatory groups such as NICAP or APRO, or by the official investigative agency (*Project Bluebook*). Thus, after 1966, I was not dealing with a body of witnesses reporting Venus, fireballs, and aircraft strobe lights, because such cases are so easily recognizable that the groups whose prior checks I was taking advantage of had already culled out and rejected most of such irrelevant material. Many of the cases I checked were older cases, some over 20 years old. It was primarily the background work of the many independent investigatory groups here and in other parts of the world (especially the Australian area where I had an opportunity to interview about 80 witnesses) that made possible my dealing with that type of once-sifted data that yields up scientifically interesting information so quickly.

I wish to put on record my indebtedness to these "dedicated amateurs," to use the astronomer's genial term; their contribution to the ultimate clarification of the UFO problem will become recognized as having been of basic importance, notwithstanding the scorn with which scientists have, on more than one occasion, dismissed their efforts. Although I cite only the larger of these groups (NICAP about 12,000 members, APRO about 8,000), there are many smaller groups here and abroad that have done a most commendable job on almost no resources. (Needless to add, there are other small groups whose concern is only with sensational and speculative aspects.)

2. Some relevant witness-characteristics

By frequently discussing my own interviewing experience with members of those non-official UFO groups whose past work has been so indispensable to my own studies, I have learned that most of my own reactions to the UFO witness interview problem are shared by those investigators. The recurrent problem of securing unequivocal descriptions, the almost excruciating difficulty in securing meaningful estimates of angular size, angular elevation, and angular displacements from laymen, the inevitable variance of witness-descriptions of a shared observation, and other difficulties of non-instrumental observing are familiar to all who have investigated UFO reports. But so also are the impressions of widespread concern among UFO witnesses to avoid (rather than to seek) publicity over their sightings. The strong disinclination to make an open report of an observation of something the witness realizes is far outside the bounds of accepted experience crops up again and again.

In my interviewing in 1947 sightings, done as a crosscheck on case material used in a very valuable recent publication by Bloecher (Ref. 8), I came to realize clearly for the first time that this reluctance was not something instilled by post-1947 scoffing at UFOs, but is part of a broadly disseminated attitude to discount the anomalous and the inexplicable, to be unwilling even to report what one has seen with his own eyes if it is well outside normal experience as currently accepted. I have heard fellow-scientists express dismay at the unscientific credulity with which the general public jumps to the conclusion that UFOs are spaceships. Those scientists have certainly not interviewed many UFO witnesses; for almost precisely the opposite attitude is overwhelmingly the characteristic response.

In my Australian interviewing, I found the same uneasy feeling about openly reporting an observation of a well-defined UFO sighting, lest acquaintances think one "has gone round the bend." Investigators in still other parts of the world where modern scientific values dominate world-views have told me of encountering just this same witness-reluctance. The charge that UFO witnesses, as a group, are hyperexcitable types is entirely incorrect.

I would agree with the way Maj. Gen. John A. Samford, then Director of Air Force Intelligence, put it in a 1952 Pentagon press conference: "Credible observers have sighted relatively incredible objects." Not only is the charge of notoriety-seeking wrong, not only is the charge of hyperexcitability quite inappropriate to the witnesses I have interviewed, but so also is the easy charge that they see an unusual aerial phenomenon and directly leap to some kind of "spaceship hypothesis."

My experience in interviewing witnesses in the selected sample I have examined since 1966 is that the witness first attempts to fit the anomalous observation into some entirely conventional category. "I thought it must be an airplane." Or, "At first, I thought it was an auto-wrecker with its red light blinking." Or, "I thought it was a meteor—until it stopped dead in midair," etc. Hynek has a very happy phrase for this very typical pattern of witness-response: he terms it "escalation of explanation," to denote the often rapid succession of increasingly more involved attempts to account for and to assimilate what is passing before the witness' eyes, almost invariably starting with an everyday interpretation, not with a spaceship hypothesis. Indeed, I probably react in a way characteristic of all UFO investigators; in those comparatively rare cases where the witness discloses that he immediately interpreted what he sighted as an extraterrestrial device, I back away from what is likely to be a most unprofitable interview.

I repeat: such instances are really quite rare; most of the general population has soaked up a degree of scientific conventionalism that reflects the net result of decades, if not centuries of scientific shaping of our views. I might interject that the segment of the population drawn to Hypothesis 8 above might be quick to jump to a spaceship interpretation on seeing something unusual in the sky, but, on the whole, those persons convinced of Hypothesis 8 are quite uninterested in observations, per se. Their conviction is firm without bothering about such things as observational matters. At least that is what I have sensed from such exposure as I have had to those who support Hypothesis 8 fervently.

3. Credibility of witnesses

Evaluating credibility of witnesses is, of course, an ever-present problem at the present stage of UFO studies. Again, from discussions with other investigators, I have concluded that common sense and previous everyday experience with prevaricators and unreliable persons lead each serious UFO investigator to evolve a set of criteria that do not differ much from those used in jury instructions in our courts (e.g., Federal Jury Instructions). It seems tedious to enlarge here on those obvious matters. One can be fooled, of course; but it would be rash indeed to suggest that the thousands of UFO reports now on record are simply a testimony to confabulation, as will be better argued by some of the cases to be recounted below.

4. Observational reliability of witnesses

Separate from credibility in the sense of trustworthiness and honesty is the question of the human being as a sensing system. Clearly, it is indispensable to be aware of psychophysical factors limiting visual discrimination, time estimation, distance estimation, angular estimation, etc. In dealing with the total sample of all observations which laymen initially label as UFOs, such factors play a large role in sorting out dubious cases. In the type of UFO reports that are of primary significance at present, close-range sightings of objects of large size moving at low velocities, or at rest, and in sight for many seconds rather than fractions of a second, all of these perceptual problems diminish in significance, though they can never be overlooked.

A frequent objection to serious consideration of UFO reports, made by skeptics who have done no first-hand case investigations, is based on the widely discrepant accounts known to be presented by trial-witnesses who have all been present at some incident. To be sure, the same kind of discrepancies emerge in multiple-witness UFO incidents. People differ as to directions, relative times, sizes, etc. But I believe it is not unfair to remark, as the basic rebuttal to this attack on UFO accounts, that a group of witnesses who see a street-corner automobile collision do not come to court and proceed, in turn, to describe the event as a rhinoceros ramming a baby carriage, or as an airplane exploding on impact with a nearby building.

There are, it needs to be soberly remembered, quite reasonable bounds upon the variance of witness testimonies in such cases. Thus, when one finds a half-dozen persons all saying that they were a few hundred feet from a domed disk with no resemblance to any known aircraft, that it took off without a sound, and was gone from sight in five seconds, the almost inevitable variations in descriptions of distances, shape, secondary features, noises, and times cannot be allowed to discount, per se, the basically significant nature of their collective account. I have talked with a few scientists, especially some psychologists, whose puristic insistence on the miserable observing equipment with which the human species is cursed almost makes me wonder how they dare cross a busy traffic intersection. Some balance in evaluating witness perceptual limitations is surely called for in all of these situations. With that balance must go a healthy skepticism as to most of the finer details, unless agreed upon by several independent witnesses. There is no blinking that anecdotal data are less than ideal; but sometimes you have to go with what you've got. To make a beginning at UFO study has required scrutiny of such anecdotal data; the urgent need is to get on to something much better.

5. *Problem of witness' prior knowledge of UFO phenomena*

In interviewing UFO witnesses, it is important to try to ascertain whether the witness was, prior to his reported sighting, familiar or unfamiliar with books and writings on UFOs. Although a strong degree of familiarity with the literature of UFOs does not negate witness testimony, it dictates caution. Anyone who has done a lot of interviewing at the local level, involving previously unsifted cases, will be familiar with occasional instances where the witness exhibited such an obvious enthusiasm for the UFO problem that prudence demanded rejection of his account. However, in my own experience, a much more common reaction to questions concerning pre-sighting interest in UFO matters is some comment to the effect that the witness not only knew little about UFOs beyond what he'd happened to read in newspapers, but he was strongly disinclined to take the whole business seriously. The repetitiveness and yet the spontaneity with which witnesses of seeming high credibility make statements similar to, "I didn't believe there was anything to all the talk about UFOs until I actually saw this thing," is a notable feature of the interview-experience of all of the investigators with whom I have talked.

Obviously, an intending prevaricator might seek to deceive his interrogator by inventing such an assertion; but I can only say that suspicion of being so duped has not been aroused more than once or twice in all of the hundreds of witnesses I have interviewed. On the other hand, I suppose that, in several dozen instances, I have lost interest in a case because of a witness openly stressing his own prior and subsequent interest in the extraterrestrial hypothesis.

Occasionally one encounters witnesses for whom the chance of prior knowledge is so low as to be almost amusing. An Anglican missionary in New Guinea, Rev. N. E. G. Cruttwell (Ref. 9), who has done much interviewing of UFO witnesses in his area, has described testimony of natives who come down into the mission area from their highland home territory only when they are wallaby-hunting, natives who could not read UFO reports in any language of the world, yet who come around, in their descriptions of what they have seen, to the communications-shortcut of picking up a bowl or dish from a nearby table to suggest the shape they are seeking to describe in native tongue. Little chance of bias gained from reading magazines in a barber-chair in such instances.

6. *Types of UFO accounts of present interest*

The scope of the present statement precludes anything approaching an exhaustive listing of categories of UFO phenomena: much of what might be made clear at great length will have to be compressed into my remark that *the scientific world at large is in for a shock when it becomes aware of the astonishing nature of the UFO phenomenon and its bewildering complexity* [my emphasis, L.S.]. I make that terse comment well aware that it invites easy ridicule; but intellectual honesty demands that I make clear that my two years' study convinces me that in the UFO problem lie scientific and technological questions that will challenge the ability of the world's outstanding scientists to explain—as soon as they start examining the facts.

a) *Lights in the night sky.*—("DLs" as they are called by the NICAP staff, on the basis that the profusion of reports of "damnable lights" meandering or hovering or racing across the night sky in unexplainable manner are one of the most common, yet one of the least useful and significant categories of UFO reports.) Ultimately, I think their significance could become scientifically very substantial when instrumental observing techniques are in wide use to monitor UFO movements. But there are many ways that observers can be misled by lights in the night sky, so I shall discuss below only such few cases as are of extremely unconventional nature and where the protocols of the observations are unusually strong.

b) *Close-range sightings of wingless discs and cigar-shaped objects.* This category is far more interesting. Many are daytime sightings, many have been made by witnesses of quite high credibility. Structural details such as "ports"and "legs" (to use the terms the witnesses have adopted to suggest most closely what they think they have seen) are described in many instances. Lack of wings and lack of evident means of propulsion clearly rule out conventional aircraft and helicopters. Many are soundless, many move at such speeds and with such accelerations that they defy understanding in terms of present technology. It is to be understood that I speak here only of reports from what I regard as credible observers.

c) *Close-range nighttime sightings of glowing, hovering objects, often with blinking or pulsating discrete lights.*—In these instances, distinct shape is not seen, evidently in many cases because of the brilliance of the lights. Less significant than those of the preceding category, these nonetheless cannot be accounted for in terms of any known vehicles. Frequently they are reported hovering over vehicles on the ground or following them. Sometimes they are reported hovering over structures, factories, power installations, and the like. Soundlessness is typical. Estimated sizes vary widely, over a range that I do not believe can be accounted for simply in terms of the known unreliability of distance and size estimates when one views an unknown object.

d) *Radar-tracked objects, sometimes seen visually simultaneously by observers on the ground or in the air.*—In many of these cases, the clues to the non-conventional nature of the radar target is high speed (estimated at thousands of miles per hour in certain instances); in others, it is alternate motion and hovering; in still others, it has been the unconventional vertical motions that make the radar observations significant. Clearly, most important are those instances in which there was close agreement between the visual and the radar unknown. There are far more such cases than either scientists or public would guess.

Those four categories do not exhaust the list by any means. But they constitute four commonly encountered categories that are of interest here. Examples will be found below.

7. *Commonly encountered questions*

As Mark Twain said, "Faith is a great thing, but it's doubt that gets you an education." There are many questions that one encounters again and again from persons who have done no personal case-checking and who maintain a healthy skepticism about UFOs. Why don't pilots report these things if they are buzzing around in our skies? Why aren't they tracked on radar? Why don't our satellite and astronomical tracking systems get photos of UFOs? Why are they always seen in out of- the-way rural areas but never over large cities? Why don't large groups of people ever simultaneously see UFOs, instead of lone individuals? Why don't astronomers see them? Shouldn't UFOs occasionally crash and leave clear-cut physical evidence of their reality? Or shouldn't they at least leave some residual physical evidence in those alleged instances where the objects have landed? Shouldn't they affect radios and produce other electromagnetic effects at times? If UFOs are a product of some high civilization, wouldn't one expect something of the nature of inquisitive behavior, since innate curiosity must be a common denominator of anything we would call "intelligence"? Why haven't they contacted us if they're from somewhere else in the universe and have been here for at least two decades? Is there any evidence of hostility or hazard? Are UFOs seen only in this country? Why didn't we see them before 1947, if they come from remote sources? And so on.

In the following sections, I shall show how some of these questions do have quite satisfactory answers, and how some of them still defy adequate rebuttal. I shall use mostly cases that I have personally investigated, but, in a few instances (clearly indicated), I shall draw upon cases which I have not directly checked but for which I regard the case-credentials as very strong.

8. Useful source materials on UFO's

Hoping that Committee staff personnel will be pursuing these matters further, I remark next on some of the more significant items in the UFO literature. All of these have been helpful in my own studies.

One of the outstanding UFO references (though little-known in scientific circles) is *The UFO Evidence*, edited by R. H. Hall and published by NICAP (Ref. 10). It summarizes about 750 UFO cases in the NICAP files up to about 1964. I have cross-checked a sufficiently large sample of cases from this reference to have confidence in its generally very high reliability. A sequel volume, now in editorial preparation at NICAP, will cover the 1964-68 period. Reference 8, by Bloecher, is one of the few sources of extensive documentation (here primarily from national newspaper sources) of the large cluster of sightings in a period of just a few weeks in the summer of 1947; its study is essential to appreciation of the opening phases of the publicly recognized UFO problem. Reference 7 is another now-accessible source of extremely significant UFO documentation; it is unfortunate that no generally accessible version of Reference 6 exists, though the Moss Subcommittee, through pleas of Dr. Leon Davidson, has managed to get it into a status of at least limited accessibility. I am indebted to Davidson for a recent opportunity to study it for details I missed when I saw it two years ago at Bluebook headquarters.

The 1956 book by Ruppelt (Ref. 5) is a source whose authenticity I have learned, through much personal cross-checking, is far higher than I surmised when I first read it a dozen years ago. It was for years difficult for me to believe that the case-material which he summarized could come from real cases. References 5 and 6, plus other sources, do, however, now attest to Ruppelt's generally high reliability. Similarly Keyhoe's books (Refs. 3 and 4) emerge as sources of UFO case material whose reliability far exceeds my own first estimates thereof. As a scientist, I would have been much more comfortable about Keyhoe's books had they been shorn of extensive direct quotes and suspenseful dramatizations; but I must stress that much checking on my part has convinced me that Keyhoe's reportorial accuracy was almost uniformly high. Scientists will tend to be put off by some of his scientific commentary, as well as by his style; but on UFO case material, his reliability must be recognized as impressive. (Perhaps it is well to insert here the general proviso that none of these sources, including myself, can be expected to be characterized by 100 per cent accuracy in a problem as intrinsically messy as the UFO problem;

here I am trying to draw attention to sources whose reliability appears to be in the 90 +% range.)

A useful collection of 160 UFO cases drawn from a wide variety of sources has been published by Olsen (Ref. 11), 32 of which he obtained directly from the official files of *Project Bluebook*, a feature of particular interest. A book devoted to a single short period of numerous UFO observations within a small geographic area, centering around an important sighting near Exeter, N.H., is Fuller's *Incident at Exeter* (Ref. 12). Having checked personally on a number of features of the main Sept. 3, 1965, sighting, and having checked indirectly on other aspects, I would describe Reference 12 as one of the significant source-items on UFOS.

Several books by the Lorenzens, organizers of APRO, the oldest continuing UFO investigating group in this country, contain valuable UFO reference material (Ref. 13). Through their writing, and especially through the APRO Bulletin, they have transmitted from South American sources numerous unusual sightings from that country. I have had almost no opportunity to cross-check those sightings, but am satisfied that some quite reliable sources are being drawn upon.

An extremely unusual category of cases, those involving reports of humanoid occupants of landed UFOs, has been explored to a greater extent by APRO than by NICAP. Like NICAP, I have tended to skirt such cases on tactical grounds; the reports are bizarre, and the circumstances of all such sightings are automatically charged in a psychological sense not found in other types of close-range sightings of mere machine-like devices. Since I shall not take up below this occupant problem, let me add the comment that I do regard the total number of such seemingly reliable reports (well over a hundred came just from central France in the outstanding 1954 sighting wave in that country), far too great to brush aside. Expert psychological opinion is badly needed in assessing such reports (expert but not close-minded opinion).

For the record, I should have to state that my interviewing results dispose me toward acceptance of the existence of humanoid occupants in some UFOs. I would not argue with those who say that this might be the single most important element of the entire UFO puzzle; I would only say that most of my efforts over the past two years, being aimed at arousing a new degree of scientific interest among my colleagues in the physical sciences, have led me to play down even the little that I do know about occupant sightings. One or two early attempts to touch upon that point within the time-limits of a one-hour colloquium taught me that one loses more than he gains in speaking briefly about UFO occupants. (Occupant sightings must be carefully distinguished from elaborate "contact-claims" with the Space Brothers; I hold no brief at all for the latter in terms of my present knowledge and interviewing experience. But occupants there seem to be, and contact of a limited sort may well have occurred, according to certain of the reports. I do not regard myself as very well-informed on this point, and will say little more on this below.)

It is, of course, somewhat more difficult to assess the reliability of foreign UFO references. Michel (Ref. 13) has assembled a day-by-day account of the remarkable French UFO wave of the fall of 1954, translated into English by the staff of CSI (Civilian Saucer Intelligence) of New York City, a now-inactive but once very productive independent group. I have spoken with persons having first-hand knowledge of the French 1954 episode, and they attest to its astonishing nature. *Life* and *The New Yorker* published full contemporary accounts at the time of the 1954 European wave. An earlier book by Michel (Ref. 14), also available in English, deals with a broader temporal and geographic range of European UFO sightings.

A just-published account of about 70 UFO sightings that occurred within a relatively small area around Stoke-on-Trent, England, in the summer and fall of 1967 (Ref. 15) presents an unusual cross-section of sightings that appear to be well-documented. A number of foreign UFO journals are helpful sources of the steady flow of UFO reports from other parts of the world, but a cataloging will not be attempted here. Information on some of these, as well as on smaller American groups, can be found in the two important books by [Jacques] Vallee (Refs. 16 and 17). Information on pre-1947 UFO-type sightings form the subject of a recent study by Lore and Denault (Ref. 18). I shall return to this phase of the UFO problem below; I regard it as being of potentially very great significance, though there is need for far more scholarly and scientific research before much of it can be safely interpreted.

Another source of sightings of which many may ultimately be found to fall within the presently understood category of UFO sightings is the writings of Charles Fort (Ref. 19). His curious books are often drawn upon for material on old sightings, but not often duly acknowledged for the mine of information they comprise. I am afraid that it has not been fashionable to take Fort seriously; it certainly took me some time to recognize that, mixed into his voluminous writings, is much that remains untapped for its scientific import. I cannot imagine any escalated program of research on the UFO program that would not have a subgroup studying Fortean reports documented from 19th century sources.

To close this brief compilation of useful UFO references, two recent commentaries (not primarily source-references) of merit may be cited, books by Stanton (Ref. 20) and by Young (Ref. 21).

Next, I examine a number of specific UFO cases that shed light on many of the recurrent questions of skeptical slant often raised against serious consideration of the UFO problem.

WHY DON'T PILOTS SEE UFO'S?

This question may come in just that form from persons with essentially no knowledge of UFO history. From others who do know that there have been "a few" pilot-sightings, it comes in some altered form, such as, "Why don't airline and military pilots see UFO's all the time if they are in our atmosphere?" By way of partial answer, consider the following cases. (To facilitate internal reference, I shall number sequentially all cases hereafter treated in detail.)

1. Case 1. Boise, Idaho, July 4, 1947

Only about a week after the now-famous Mt. Rainier sighting by private pilot Kenneth Arnold, a United Air Lines [Douglas] DC-3 [airliner] crew sighted two separate formations of wingless discs, shortly after takeoff from Boise (Refs. 8, 10, 22, 23). I located and interviewed the pilot, Capt. Emil J. Smith, now with United's New York office. He confirmed the reliability of previously published accounts. United Flight105 had left Boise at 9:04 p.m. About eight minutes out, en route to Seattle, roughly over Emmet, Idaho, Co-pilot Stevens, who spotted the first of two groups of objects, turned on his landing lights under the initial impression the objects were aircraft. But, studying them against the twilight sky, Smith and Stevens soon realized that neither wings nor tails were visible on the five objects ahead. After calling a stewardess, in order to get a third confirming witness, they watched the formation a bit longer, called Ontario, Oregon, CAA to try to get ground-confirmation, and then saw the formation spurt ahead and disappear at high speed off to the west.

Smith emphasized to me that there were no cloud phenomena to confuse them here and that they observed these objects long enough to be quite certain that they were no conventional aircraft. They appeared "flat on the bottom, rounded on top," he told me, and he added that there seemed to be perceptible "roughness" of some sort on top, though he could not refine that description.

Almost immediately after they lost sight of the first five, a second formation of four (three in line and a fourth off to the side) moved in ahead of their position, again travelling westward but at a somewhat higher altitude than the DC-3's 8000 ft. These passed quickly out of sight to the west at speeds which they felt were far beyond then-known speeds.

Smith emphasized that they were never certain of sizes and distances, but that they had the general impression that these disc-like craft were appreciably larger than ordinary aircraft. Smith emphasized that he had not taken seriously the previous week's news accounts that coined the since persistent term, "flying saucer." But, after seeing this total of nine unconventional, high-speed wingless craft on the evening of 7/4/47, he became much more interested in the matter. Nevertheless, in talking with me, he stressed that he would not speculate on their real nature or origin. I have spoken with United Air Lines personnel who have known Smith for years and vouch for his complete reliability.

Discussion.—The 7/4/47 United Air Lines sighting is of historic interest because it was obviously given much more credence than any of the other 85 UFO reports published in press accounts on July 4, 1947 (see Ref. 8). By no means the most impressive UFO sighting by an airliner crew, nevertheless, it is a significant one. It occurred in clear weather, spanned a total time estimated at 10-12 minutes, was a multiple-witness case including two experienced observers familiar with airborne devices, and was made over a 1000-ft. altitude range (climb-out) that, taken together with the fact that the nine objects were seen well above the horizon, entirely rules out optical phenomena as a ready explanation. It is officially listed as a Unidentified.

2. Case 2. Montgomery, Alabama, July 24, 1948

Another one of the famous airline sightings of earlier years is the Chiles-Whitted Eastern Airlines case (Refs. 3, 5, 6, 10, 23, 24, 25, 26). An Eastern [Douglas] DC-3 [airliner], en route from Houston to Atlanta, was flying at an altitude of about 5000 ft., near Montgomery at 2:45 a.m. The pilot, Capt. Clarence S. Chiles, and the co-pilot, John B. Whitted, both of whom now fly jets for Eastern, were experienced fliers (for example, Chiles then had 8500 hours in the air, and both had wartime military flying duty behind them). I interviewed both Chiles and Whitted earlier this year to cross-check the many points of interests in this case. Space precludes a full account of all relevant details.

Chiles pointed out to me that they first saw the object coming out of a distant squall-line area which they were just reconnoitering. At first, they thought it was a jet, whose exhaust was somehow accounting for the advancing glow that had first caught their eyes. Coming almost directly at them at nearly their flight altitude, it passed off their starboard wing at a distance on which the two men could not closely agree: one felt it was under 1000 ft., the other put it at several times that. But both agreed, then and in my 1968 interview, that the object was some kind of vehicle. They saw no wings or empennage, but both were struck by a pair of rows of windows or some apparent openings from which there came a bright glow "like burning magnesium." The object had a pointed "nose," and from the nose to the rear along its underside there was a bluish glow. Out of the rear end came an orange-red exhaust or wake that extended back by about the same distance as the object's length. The two men agreed that its size approximated that of a [Boeing] B-29 [Superfortress], though perhaps twice as thick. Their uncertainty as to true distance, of course, renders this only a rough impression.

There is uncertainty in the record, and in their respective recollections, as to whether their DC-3 was rocked by something like a wake. Perception of such an effect would have been masked by Chiles' spontaneous reaction of turning the DC-3 off to the left as the object came in on their right. Both saw it pass aft of them and do an abrupt pull-up; but only Whitted, on the right side, saw the terminal phase in which the object disappeared after a short but fast vertical ascent. By "disappeared," Whitted made clear to me that he meant just that; earlier interrogations evidently construed this to mean "disappeared aloft" or into the broken cloud deck that lay above them. Whitted said that was not so; the object vanished instantaneously after its sharp pull-up. (This is not an isolated instance of abrupt disappearance. Obviously I cannot account for such cases.)

Discussion. This case has been the subject of much comment over the years, and rightly so. Menzel (Ref. 24) first proposed that this was a "mirage," but gave no basis for such an unreasonable interpretation. The large azimuth-change of the pilots' line of sight, the lack of any obvious light source to provide a basis for the rather detailed structure of what was seen, the sharp pull-up, and the high flight altitude involved all argue quite strongly against such a casual disposition of the case. In his second book, Menzel (Ref. 25) shifts to the explanation that they had obviously seen a meteor. A horizontally-moving fireball under a cloud-deck, at 5000 ft., exhibiting two rows of lights construed by experienced pilots as ports, and finally executing a most non-ballistic 90-degree sharp pull-up, is a strange fireball indeed. Menzel's 1963 explanation is even more objectionable, in that he implies, via a page of side-discussion, that the Eastern pilots had seen a fireball from the Delta Aquarid meteor stream. As I have pointed out elsewhere (Ref. 2), the radiant of that stream was well over 90° away from the origin point of the unknown object. Also, bright fireballs are, with only rare exceptions, not typical of meteor streams. The official explanation was shifted recently from "Unidentified" to "Meteor," following publication of Menzel's 1963 discussion (see Ref. 20, p. 88).

Wingless, cigar-shaped or "rocket-shaped" objects, some emitting glowing wakes, have been reported by other witnesses. Thus, Air Force Capt. Jack Puckett, flying near 4000 ft. over Tampa in a [Douglas] C-47 [Skytrain] on August 1, 1946 (Ref. 10, p. 23), described seeing "a long, cylindrical shape approximately twice the size of a B-29 with luminous portholes," from the aft end of which there came a stream of fire as it flew near his aircraft. Puckett states that he, his copilot, Lt. H. F. Glass, and the flight engineer also saw it as it came in to within an estimated 1000 yards before veering off. Another somewhat similar airborne sighting, made in January 22, 1956 by TWA Flight Engineer Robert Mueller at night over New Orleans, is on record (Ref. 27). Still another similar sighting is the AAL case cited below (Sperry case). Again, over Truk Island, in the Pacific, a Feb. 6, 1953, mid-day sighting by a weather officer involved a bullet-shaped object without wings or tail (Ref. 7, Rept. No. 10). Finally, within an hour's time of the Chiles-Whitted sighting, Air Force ground personnel at Robins AFB, Georgia, saw a rocket-like object shoot overhead in a westerly direction (Refs.3, 5, 10, 6). In none of these instances does a meteorological or astronomical explanation suffice to explain the sightings.

3. Case 3. Sioux City, Iowa, January 20, 1951

Another of the many airline-crew sightings of highly unconventional aerial devices that I have personally checked was, like Cases 1 and 2, widely reported in the national press (for a day or two, and then forgotten like the rest). A check of weather data confirms that the night of 1/20/51 was clear and cold at Sioux City at the time that a Mid-Continent Airlines DC-3, piloted by Lawrence W. Vinther, was about to take off for Omaha and Kansas City, at 8:20 p.m. CST. In the CAA control tower, John M. Williams had been noting an oddly maneuvering light high in a westerly direction. Suddenly the light abruptly accelerated, in a manner clearly precluding either meteoric or aircraft origin, so Williams alerted Vinther and his co-pilot, James F. Bachmeier. The incident has been discussed many times (Ref. 4, 5, 10, and 28), but to check details of these reports, I searched for and finally located all three of the above-named men. Vinther and Bachmeier are now Braniff pilots, Williams is with the FAA in Sacramento. From them I confirmed the principal features of previous accounts and learned additional information too lengthy to recapitulate in full here.

The essential point to be emphasized is that, shortly after Vinther got his DC-3 airborne, under Williams' instructions to investigate the oddly-behaving light, the object executed a sudden dive and flew over the DC-3 at an estimated 200 ft. vertical clearance, passing aft and downward. Then a surprising maneuver unfolded. As Vinther described it to me, and as described in contemporary accounts, the object suddenly reversed course almost 180°, without slowing down or slewing, and was momentarily flying formation with their DC-3, off its port wing. (Vinther's dry comment to me was: "This is something we don't see airplanes do.") Vinther and Bachmeier agreed that the object was very big, perhaps somewhat larger than a B-29, they suggested to newspapermen who interviewed them the following day. Moonlight gave them a good silhouetted view of the object, which they described as having the form of a fuselage and unswept wing, but not a sign of any empennage, nor any sign of engine-pods, propellers, or jets. Prior to its dive, it had been seen only as a light; while pacing their DC-3, the men saw no luminosity, though during the dive they saw a light on its underside. After about five seconds, the unknown object began to descend below them and flew under their plane. They put the DC-3 into a steep bank to try to keep it in view as it began this maneuver; and as it crossed under them, they lost it, not to regain sight of it subsequently.

There is much more detail, not all mutually consistent as to maneuvers and directions, in the full accounts I obtained from Vinther, Bachmeier, and Williams. The dive, pacing, and fly-under maneuvers were made quickly and at such a distance from the field that Williams did not see them clearly, though he did see the object leave the vicinity of the DC-3. An Air Force colonel and his aide were among the passengers, and the aide caught a glimpse of the unknown object, but I have been unable to locate him for further cross-check.

Discussion.—The erratic maneuvers exhibited by the unknown object while under observation from the control tower would, by themselves, make this a better-than-average case. But the fact that those maneuvers prompted a tower operator to alert a departing aircrew to investigate, only to have the object dive upon and pace the aircraft after a non-inertial course-reversal, makes this an unusually interesting UFO. Its configuration, about which Vinther and Bachmeier were quite positive in their remarks to me (they repeatedly emphasized the bright moonlight, which checks with the near-full moon on 1/20/51 and the sky-cover data I obtained from the Sioux City Weather Bureau), combines with other features of the sighting to make it a most significant case.

The reported shape (tailless, engineless, unswept aircraft of large size) does not match that of any other UFO that I am aware of; but my exposure to the bewildering range of reported configurations now on record makes this point less difficult to assimilate. This case is officially carried as Unidentified, and, in a 1955 publication (Ref. 29), was one of 12 Unidentifieds singled out for special comment. A contemporary account (Ref. 28), taking note of a then recent pronouncement that virtually all UFOs are explainable in terms of misidentified Skyhook balloons, carried a lead-caption, "The Office of Naval Research claims that cosmic ray balloons explain all saucer reports. If so, what did this pilot see?" Certainly it would not be readily explained away as a balloon, a meteor, a sundog, or ball lighting. Rather, it seems to be just one more of thousands of Unidentified Flying Objects for which we have no present explanations because we have laughed such reports out of scientific court. Bachmeier stated to me that, at the time, he felt it had to be some kind of secret device, but, in the ensuing 17 years, we have not heard of any aircraft that can execute instantaneous course-reversal. Vinther's comment to me on a final question I asked as to what he thinks, in general, about the many airline-pilot sightings of unidentified objects over the past 20 years, was: "We're not all having hallucinations."

4. Case 4. Minneapolis, Minn., October 11, 1951

There are far more private pilots than airline pilots, so it is not surprising that there are more UFO sightings from the former than the latter. An engineer and former Air Force [Lockheed] P-38 [Lightning] pilot, Joseph J. Kaliszewski, flying for the General Mills Skyhook balloon program on balloon-tracking missions, saw highly unconventional objects on two successive days in October, 1951 (Refs. 5, 7, 10). Both were reported through company channels to the official investigative agency (Bluebook), whose report (Ref. 7) describes the witnesses as "very reliable" and as "experienced high altitude balloon observers."

On October 10, at about 10:10 a.m., Kaliszewski and Jack Donaghue were at 6000 ft. in their light plane, climbing toward their target balloon, when Kaliszewski spotted "a strange object crossing the skies from East to West, a great deal higher and behind our balloon (which was near 20,000 ft. at that time)." When I interviewed Kaliszewski, he confirmed that this object "had a peculiar glow to it, crossing behind and above our balloon from east to west very rapidly, first coming in at a slight dive, leveling off for about a minute and slowing down, then into a sharp left turn and climbing at an angle of 50 to 60° into the southeast with a terrific acceleration." The two observers had the object in view for an estimated two minutes, during which it crossed a span of some 45° of the sky. No vapor trail was seen, and Kaliszewski was emphatic in asserting that it was not a balloon, jet, or conventional aircraft.

The following morning, near 0630, Kaliszewski was flying on another balloon mission with Richard Reilly and, while airborne north of Minneapolis, the two of them noticed an odd object. Quoting from the account submitted to the official agency (Ref. 7, Rept. No. 2):

> "The object was moving from east to west at a high rate and very high. We tried keeping the ship on a constant course and using the reinforcing member of the windshield as a point. The object moved past this member at about 50 degrees per second. This object was peculiar in that it had what can be described as a halo around it with a dark undersurface. It crossed rapidly and then slowed down and started to climb in lazy circles slowly. The pattern it made was like a falling oak leaf inverted. It went through these gyrations for a couple minutes and then with a very rapid acceleration disappeared to the east. This object Dick and I watched for approximately five minutes."

Shortly after, still another unknown object shot straight across the sky from west to east, but not before Kaliszewski succeeded in radioing theodolite observers at the University of Minnesota Airport. Two observers there (Douglas Smith, Richard Dorian) got fleeting glimpses of what appeared to them to be a cigar-shaped object viewed through the theodolite, but could not keep it in view due to its fast angular motion. In my conversations with Kaliszewski about these sightings, I gained the impression of talking with a careful observer, in full accord with impressions held by three other independent sources, including Air Force investigators.

Discussion. The October 10 sighting is officially categorized as "Aircraft," the October 11 main sighting as "Unidentified." When

I mentioned this to Kaliszewski, he was unable to understand how any distinction could be so drawn between the two sightings, both of which he felt matched no known aeronautical device. Clearly, objects performing such intricate maneuvers are not meteors, nor can they be fitted to any known meteorological explanations of which I am aware. Instead, these objects seem best described as devices well beyond the state of 1951 (or 1968) technology.

5. *Case 5. Willow Grove, Pa., May 21, 1966*
Skipping over many other pilot observations to a more recent one which I have personally checked, I call attention to a close-range airborne sighting of a domed-disc, seen under midday conditions by two observers. One of them, William C. Powell, of Radnor, Pa., is a pilot with 18,000 logged flight hours. He and a passenger, Miss Muriel McClave, were flying in Powell's Luscombe [Aircraft] in the Philadelphia area on the afternoon of 5/21/66 when an object that had been first spotted as it apparently followed an outbound flight of Navy jets from Willow Grove NAS made a sharp (non-banking) turn and headed for Powell's plane on a near-collision course. As the object passed close by, at a distance that Powell put at roughly 100 yards, they both got a good look at the object. It was circular in plan form and had no wings or visible means of propulsion, both witnesses emphasized to me in interviews. The upper domed portion they described as "porcelain-white," while the lower discoid portion was bright red ("dayglow red" Powell put it).

It was slightly below their altitude as it passed on their right, and Powell pointed out that it was entirely solid, for it obscured the distant horizon areas. His brief comment about its solidity and reality was, "It was just like looking at a Cadillac." He estimated its airspeed as perhaps 200 mph, and it moved in a steady, non-fluttering manner. He estimated its diameter at perhaps 20 feet. Miss McClave thought it might have been nearer 40 feet across. Each put the thickness-to-diameter ratio as about one-half. After it passed their starboard wing, Powell could see it only by looking back over his shoulder through a small aft window, but Miss McClave had it in full view when suddenly, she stated to me, it disappeared instantaneously, and they saw no more of it.

Discussion.—Powell flies executive transports for a large Eastern firm, after years of military and airline duty. I have discussed the case with one of his superiors, who speaks without qualification for Powell's trustworthiness. At a UFO panel discussion held on April 22, 1967 at the annual meeting of the American Society of

Newspaper Editors, Powell was asked to summarize his sighting. His account is in the proceedings of that session (Ref. 30). I know of no natural phenomenon that could come close to explaining this sighting. The visibility was about 15 miles, they were flying in the clear at 4500 ft., and the object passed nearby. A pilot with 18,000 hours flight experience is not capable of precise midair distance and speed estimates, but his survival has probably hinged on not commonly making errors of much over a factor or two. Given the account and accepting its reliability, it seems necessary to say that here was one more case of what Gen. Samford described as "credible observers seeing relatively incredible objects."

I felt that Powell's summary of his sighting at the ASNE meeting was particularly relevant because, in addition to my being on the panel there, Dr. D. H. Menzel and Mr. Philip J. Klass, both strong exponents of meteorological-type UFO theories, were present to hear his account. I cannot see how one could explain this incident in terms of meteorological optics nor in terms of ball lighting plasmoids. Here again, we appear to be dealing with a meaningful observation of some vehicle or craft of nonterrestrial origin. Its reported instantaneous disappearance defies (as does the same phenomenon reported by J. B. Whitted and numerous other UFO witnesses) ready explanation in terms of present-day scientific knowledge. Powell reported his sighting at Willow Grove NAS, but it engendered no interest.

6. *Case 6. Eastern Quebec, June 29, 1954*

A case in which I have not been able to directly interview any witnesses, but about which a great deal is on record, through contemporary press accounts, through the pilot's subsequent report, and through recent interviews by BBC staff members, occurred near Seven Islands, Quebec, just after sunset on 6/29/54. A BOAC [377 Monarch] Stratocruiser, bound from New York to London with 51 passengers, was followed for 18 minutes (about 80 miles of airpath) by one large object and six smaller objects that flew curious "formations" about it. The pilot of the Stratocruiser was Capt. James Howard, a highly respected BOAC flight officer still flying with BOAC. At the time, he had had 7500 flight hours. About 20 witnesses, including both passengers and crew, gave statements as to the unprecedented nature of these objects (Refs. 4, 10, and Associated Press wire stories datelined June 30,1954).

Discussion.—The flight was at 19,000 ft. in an area of generally fair weather, with good visibility, attested by Howard and by weather maps for that day. No obvious optical or electrical explanation seems capable of accounting for this long-duration sighting. The objects were dark, not glowing, and their position relative to the sunset point precludes sundogs as an explanation. Mirage phenomena could not account for the eighty-mile persistence, nor for the type of systematic shape-changes described by the witnesses, nor for the geometrically regular formations taken up by the satellite objects as they shifted positions from time to time. Just before an [North American] F-86 [Sabre] arrived from Goose AFB at Howard's request, First Officer Boyd and Navigator George Allen, who were watching the objects at that moment, said the small objects seemed to merge into the larger object. Then the large object receded rapidly towards the northwest and was out of sight in a matter of seconds. Such a maneuver of a number of satellite objects seeming to merge with or to enter a larger object has been reported in other UFO incidents around the world.

7. *Case 7. Goshen, Ind., April 27, 1950*
Another early airline sighting that seemed worth personally cross-checking involved the crew and passengers of a TWA DC-3 on the evening of 4/27/50 (Refs. 4, 5, 10, 23). I have interviewed both the pilot, Capt. Robert Adickes, and the copilot, Capt. Robert F. Manning, and confirmed all of the principal features first reported in detail in a magazine account by Keyhoe (Ref. 31). The DC-3 was at about 2000 ft., headed for Chicago, when, at about 8:25 p.m., Manning spotted a glowing red object aft of the starboard wing, well to their rear. Manning sent to me a copy of notes that he had made later that night at his Chicago hotel. Quoting from the notes:

> "It was similar in appearance to a rising blood red moon, and appeared to be closing with us at a relatively slow rate of convergence. I watched its approach for about two minutes, trying to determine what it might be. I then attracted Adickes' attention to the object asking what he thought it was. He rang for our hostess, Gloria Henshaw, and pointed it out to her. At that time the object was at a relative bearing of about 100 degrees and slightly lower than we were. It was seemingly holding its position relative to us, about one-half mile away."

Manning's account then notes that Capt. Adickes sent the stewardess back to alert the passengers (see Keyhoe's account, Ref. 31), and then banked the DC-3 to starboard to try to close on the unknown object. Manning continues in his 4/27/50 notes:

"As we turned, the object seemed to veer away from us in a direction just west of north, toward the airport area of South Bend. It seemed to descend as it increased its velocity, and within a few minutes was lost to our sight..."

Discussion.—Although, in my interview, I found some differences in the recollected shape of the object, as remembered by the two TWA pilots, both were positive it was no aircraft, both emphasized its red glow, and both were impressed by its high speed departure. Manning remarked to me that he'd never seen anything else like it before or since; and he conceded, in response to my query, that the decreased number of airline reports on UFOS in recent years probably stems chiefly from pilot reluctance to report. Both he and Adickes, like most other pilots I have asked, indicated they were unaware of any airline regulations precluding reporting, however. I mentioned to Adickes that there is indirect indication in one reference (Ref. 5) that the official explanation for this sighting was "blast-furnace reflections off clouds." He indicated this was absolutely out of the question. It is to be noted that here, as in many other pilot sightings, an upper bound, even if rough, is imposed on the range to the unknown by virtue of a downward-slanting line of sight. In such instances, meteor-explanations are almost automatically excluded. The Goshen case has no evident meteorological, astronomical, or optical explanation.

8. Case 8, Newport News, Va., July 14, 1952
Another case in which experienced pilots viewed UFOs below them, and hence had helpful background-cues to distance and size, occurred near 8:12 p.m. EST, July 14, 1952. A Pan American DC-4, en route from New York to Miami, was at 8000 ft over Chesapeake Bay, northeast of Newport News, when its cockpit crew witnessed glowing, disc-shaped objects approaching them at a lower altitude (estimated at perhaps 2000 ft). First Officer Wm. B. Nash, at the controls for Capt. Koepke (who was not on the flight deck during the sighting) and Second Officer Wm. H. Fortenberry saw six amber-glowing objects come in at high velocity and execute a peculiar flipping maneuver during an acute-angle direction-change. Almost immediately after the first six reversed course, two other apparently identical discs shot in under the DC-4, joining the other six. I am omitting here certain other maneuver details of significance, since these are on record in many accounts (4, 5, 10, 11, 25). Although I have not interviewed Nash (now in Germany with PAA, and Fortenberry is deceased), I

believe that there has never been any dispute as to the observed facts. Nash has stated to T. M. Olsen (author of Ref. 11) that one of the most accurate accounts of the facts has been given by Menzel (Ref. 25), adding that Menzel's explanation seems entirely out of the question to him. A half-dozen witnesses on the ground also saw unknowns at that time, according to official investigators. The objects had definite edges, and glowed "like hot coals," except when they blinked out, as they did in unison just after the first six were joined by the latter two. When the lights came back on, Nash and Fortenberry saw them climbing westward, eight in line, north of Newport News. The objects climbed above the altitude of the DC-4 and then blinked out in random order and were seen no more.

Discussion.—Menzel explains this famous sighting as resulting from a searchlight playing on thin haze layers, an almost entirely *ad hoc* assumption, and one that will not account for the amber color, nor for the distinct edges, nor for the final climb-out of the objects. The rapid motion, abrupt course-reversal, and the change from negative to positive angles of elevation of the line of sight to the unknowns seem to preclude any meteorological-optical explanation, and there is, of course, no possibility of explaining cases like this in terms of ball lightning, meteors, balloons, or many of the other frequently adduced phenomena. Nash has stated that he feels these were "intelligently operated craft." This case is officially "Unidentified."

9. Many other pilot sightings, both recent and old, could readily be cited. Not only civilian pilots but dozens of military pilots have sighted wholly unconventional objects defying ready explanation (see esp. Ref. 10 and Ref. 7 for many such instances). Thus, the answer to the question, "Why don't pilots see UFOs?" is: "They do."

WHY ARE UFO'S ONLY SEEN BY LONE INDIVIDUALS? WHY NO MULTIPLE-WITNESS SIGHTINGS?

It is true that there are more single-witness UFO reports than multiple-witness cases. But, to indicate that by no means all interesting UFO reports entail lone witnesses, consider the following examples:

1. Case 9. Farmington, N.M., March 17, 1950
In the course of checking this famous case that made short-lived press headlines in 1950, I interviewed seven Farmington witnesses

out of a total that was contemporarily estimated at "hundreds" to "over a thousand." (Refs. 5, 25) It became clear from my interviewing that the streets were full of residents looking up at the strange aerial display that day. It was not only a multiple-witness case, but also a multiple-object case. My checking was done seventeen years after the fact, so the somewhat confused recollective impressions I gained are not surprising. But that unidentified aerial objects moved in numbers over Farmington on 3/17/50 seems clear.

One witness with whom I spoke, Clayton J. Boddy, estimated that he had observed a total of 20 to 30 disc-shaped objects, including one red one substantially larger than the others, moving at high velocity across the Farmington sky on the late morning of 3/17/50. John Eaton, a Farmington realtor, described being called out of a barbershop when the excitement began and seeing a high, fast object suddenly joined by many objects that darted after it. Eaton sent me a copy of an account he had jotted down shortly after the incident. A former Navy pilot, Eaton put their height at perhaps 15,000 ft. "The object that has me puzzled was the one we saw that was definitely red. It was seen by several and stated by all to be red and travelling northeast at a terrific speed." Eaton also spoke of the way the smaller objects would "turn and appear to be flat, then turn and appear to be round," a description matching an oscillating disc-shaped object. No one described seeing any wings or tails, and the emphasis upon the darting, "bee-like" motion was in several of the accounts I obtained from witnesses. I obtained more details, but the above must suffice here for a brief summary.

Discussion.—This once-headlined, but now almost forgotten multiple-witness case has been explained as resulting from the breakup of a Skyhook balloon (Ref.25). Skyhooks do shatter at the very low temperatures of the upper troposphere, and occasionally break into a number of smaller pieces. But to suggest that such fragments of transparent plastic at altitudes of the order of 40-50,000 ft. could be detected by the naked eye, and to intimate that these distant objects of low angular velocity could confuse dozens of persons into describing fast-moving disc-shaped objects (including a large red object) is simply not reasonable. However, to check further on this, I contacted first Holloman AFB and then the Office of Naval Research, who jointly hold records on all Alamogordo Skyhook releases. No Skyhooks or other experimental balloons had been released from the Holloman area or any other part of the country on or near the date of this incident. A suggestion that the witnesses were seeing only cotton-wisps was not only unreasonable, given the witness accounts, but was in fact tracked down by a local journalist to comments casually made by a law enforcement officer and overheard by another reporter. From my examination of this case, I see no ready explanation for the numerous disc-shaped objects moving in unconventional manner and seen by large numbers of Farmington residents on 3/17/50.

2. Case 10. Longview, Wash., July 3, 1949

Many of the UFO cases I am citing are drawn intentionally from earlier years, in order to illustrate that the evidence for the existence of a quite real and scientifically significant phenomenon has been with us for a disturbing number of years.

I discuss next a case on which I hold copies of material from the official investigative files, copies that state that this incident was "observed by 150 other people at an Air Show," in addition to the reporting witness, Moulton B. Taylor. I have interviewed Mr. Taylor and have obtained strong recommendations of his reliability from a former superior officer, Adm. D. S. Fahrney, under whom Taylor served in Navy guided missiles work prior to the incident. Taylor is an aeronautical engineer, and was airport manager at Longview, in charge of an air show that was to be held on the afternoon of 7/3/49, the day of the incident in question.

A skywriting Stearman [probably the Boeing-Stearman Kaydet Model 75 biplane] was at 10,000 ft. at 10:40 a.m., laying down "Air Show Today," and hence holding the attention of a number of the personnel already at the airport, when the first of three unidentified objects flew over at high altitude. Alerted by one of the persons who first spotted the object coming from the northwest, Taylor got on the public address system and announced to all persons at hand that they should look up to see the odd object. Many had binoculars, and among the over 150 persons present were police officers, city officials and a number of Longview's leading citizens, Taylor emphasized. The object was observed by a number of experienced pilots; and, according to official file summaries, all agreed that the object was shaped much like a discus. It seemed to have metallic luster and oscillated periodically as it crossed the sky from northwest to southeast until lost in mill-smoke. Taylor described the motion as a "sculling or falling-leaf motion rather than a movement through the axis of the disc." Its angular size he estimated as about that of a pinhead at arm's length, or about that of a DC-3 at 30,000 ft., both of which come out to be near 10 minutes of arc (one-third of moon's diameter). The crowd's attention to events in the sky did not lapse when the first object was lost from view, and, about nine minutes later, someone spotted a second object, whereupon the event was again announced via the public address system. Still a third object was brought to the attention of the crowd in the same manner at 11:25. The second object came out of the north, the third came from almost due west. In the third case, someone thought of timing the oscillation frequency (all three exhibited the same unconventional oscillation, with sun-glint perceptible in certain of the instances of tipping, Taylor mentioned). The oscillation frequency was clocked at 48 per minute.

In the official report are height estimates and some disparate comments on color, etc., from several other witnesses, as well as remarks on other sightings in the same area on the same day. Full details cannot be recounted here, for reasons of space limitation. Taylor, in his statement submitted to official investigators, said:

> "My experience in radio control of pilotless aircraft and guided missiles for the Navy at NAMU during the war, and over 20 years of aircraft study, does not permit my identification of the objects which were seen. They definitely were not balloons, birds, common aircraft, parachutes, stars, meteors, paper, clouds, or other common objects. They moved in a regular motion either straight or in curved lines. They were all at approximately the same altitude, but moved on different courses as indicated on the sketch. The oscillations were clearly visible and timed on the 3rd sighting."

Discussion.—The official explanation for this case is "Balloons." I obtained information on upper winds over that part of Washington on that day (700 and 500 mb charts), and the flow aloft between 10,000 and 20,000 ft was from the southwest. The objects, all reported as about the same angular size, came from three distinctly different directions, all within a period of less than an hour. This immediately casts very strong doubt on the balloon hypothesis, as does the flipping motion, the sun-glint, and, above all, the fact that no pilot balloon stations were located close upwind of Longview. Furthermore, a typical pilot balloon of about 1 meter diameter could be no higher than about 2500 ft altitude to subtend as large an angle as 10 minutes of arc. Taylor's report (official files) gave transit times of 2-3 minutes for the unknowns to cross the Longview sky, and, during such a time interval, the normal ascent rate of a pilot balloon would carry it up by 1200-1800 ft. To then fit the angular-size requirements would clearly require that the balloons have been released at some nearby location, which fails to match known pibal-station locations at that time. Furthermore, surface winds were from the west, and winds a short distance above the ground were southwesterly, as indicated by pulpmill smoke-drift described in Taylor's report. This, plus the previously cited upper-flow directions, contradict the balloon hypothesis for all three directions of arrival, particularly those coming from north and northwest. To hypothesize that these were, say, Skyhook balloons coming from three different (unknown) sites, at three different high altitudes, but all so arranged that the apparent balloon diameter came out at about the same 10 minutes of arc each time is scarcely reasonable. In all, I can only regard the balloon explanation as untenable.

Disc-shaped objects have been sighted in dozens of instances, including Arnold's 6/24/47 Mt. Rainier sighting. In many, though not all, the odd flipping or fluttering motion has been described by witnesses (Refs. 8, 10). What the dynamical significance of this might be is unclear. We know no more about this in 1968 than we knew in 1947, because such observations have been ignored as nonsense—or misidentified balloons.

3. Case 11. Salt Lake City, Utah, Oct. 3, 1961

A midday sighting of a lens-shaped object involving one airborne witness and seven witnesses on the ground became headline news in Salt Lake City (Ref. 32). Accounts of the incident have been summarized elsewhere (Refs. 2, 10, 13, 25).A private pilot, Mr. Waldo J. Harris, was taking off on Runway 160 at Utah Central Airport at almost exactly noon on 10/2/61 when he noted what he at first idly viewed as a distant airplane. He noted it again in the same area just after becoming airborne, once more after gaining some altitude, and then became somewhat puzzled that it had not exhibited any appreciable change of position. About then it seemed to tilt, glinting in the noonday sun, and exhibiting a shape unlike any aircraft.

To get a better view, Harris climbed towards the southeast and found himself at its altitude when he was somewhat above 6000 ft. By then it appeared as a biconvex metallic gray object, decidedly different from conventional aircraft, so he radioed back to the airport, where eventually seven persons were taking turns viewing it with binoculars. I have interviewed not only Harris, but also Jay W. Galbraith, operator of the airport, who, with his wife, watched the object, and Robert G. Butler, another of those at the airport.

As Harris attempted to close in, he got to a minimal distance that he thought might have been approximately two or three miles from the object, when it abruptly rose vertically by about 1000 ft, a maneuver confirmed by the ground witnesses. They indicated to me that it took only a second or perhaps less to ascend. Just before the abrupt rise, Harris had been viewing the object on an essentially dead-level line of sight, with distant Mt. Nebo behind it, a significant feature of the case, as will be brought out in a moment.

Before Harris could close his distance much more, the object began moving off to the southeast at a speed well above his light-plane top speed. It was soon an estimated ten miles or so away, but Harris continued his attempt to close. However, after seeming to hover a short time in its new location, it began rising and moving westward, at an extremely rapid speed, and passed out of sight aloft to the southwest in only a few seconds. Some, but not all of the ground witnesses, observed this final fast climb-out, I was told. Military jets were called, but the object had gone before they arrived.

Both Harris and the ground observers using binoculars attested to lack of wings or tail, and to the biconvex side view. Harris said he had the impression its surface resembled "sand-blasted aluminum," but his closest view was about 2-3 miles away, and its estimated size was put at about 50-60 ft. diameter (and only a tenth as thick), so the impression of surface texture must be regarded as uncertain.

All witnesses confirmed that the object "wobbled" during its hovering. Jay Galbraith said that, when Harris' Mooney Mark 20A was only a speck, they could see the disc rather easily by naked eye, suggesting that its size may have been substantially larger than Harris' estimated 50 ft. Galbraith's recollection of its final departure was that it climbed at a very steep angle, perhaps within about 20° of the vertical, he thought. Butler also recalled the final departure and stressed that it was a surprisingly steep climb-out, quite beyond any known jet speed. All remarked on 10/2/61 being a beautifully clear day.

Discussion.—Once again we deal with observed performance characteristics far beyond anything of which we have present knowledge: a wingless device that can hover, shoot straight up, and move fast enough to pass out of sight in a matter of a few seconds does not correspond to any known terrestrial craft. The official explanation was originally that Harris saw Venus. From astronomical data, one finds that Venus was in the Utah sky at noon in early October, but lay in the southwest, whereas everyone's line of sight to the object lay to the southeast. Furthermore, Harris' statement that at one stage he viewed the disc against a distant mountain would contradict such an explanation. Finally, it is well known to astronomers that Venus, even at peak brilliance, is not very easily spotted in daytime, whereas he had no difficulty relocating it repeatedly as he flew.

Menzel (Ref. 25) proposed that it was merely a sundog that Harris and the others were observing, and this was subsequently adopted as the official explanation. But sundogs (parhelia), for well-known reasons, occur at elevation angles equal to or slightly greater than the sun, which lay about 40° above the southern horizon at noon in Salt Lake that day. Such a solar position would imply that a sundog might have lain to the southeast (22° to the left of the sun), but at an elevation angle that completely fails to match Harris' dead-level viewing (against a distant mountain, to further embarrass the sundog hypothesis).

Finally, to check the witness' statements about cloud-free skies, I checked with the Salt Lake City Weather Bureau office, and their logs showed completely clear skies and 40 miles visibility. Sundogs cannot occur without ice crystal clouds present. The only weather balloon released that morning was sent up at 10:00 a.m.; but in any event, one would have to write off almost all of the observed details to propose that this incident was a misinterpretation of a weather balloon. As I see it the 10/2/61 Salt Lake City sighting is just one more of the hundreds of very well-observed cases of machine-like craft exhibiting "flight performance" far beyond the state and present-day technology.

4. Case 12 Larson AFB, Moses Lake, Washington, January 8, 1953

NICAP's recent publication of long-inaccessible official report-summaries (Ref. 7) makes readily available to interested scientists a large number of fascinating UFO reports. Many are in the multiple-witness category, for example, the dawn (0715 PST) sighting at Larson AFB where

> "over sixty varied military and civilian sources observed one green disc-shaped object. The observations continued for fifteen minutes during which time the object moved in a southwesterly direction while bobbing vertically and going sideways. There was no sound. An [Republic] F-84 [Thunderjet] aircraft was scrambled but a thirty minute search of the area produced negative intercept results."

The official summary also notes that the

> "winds were generally from 240° below an overcast at 13,000 ft. Thus the object would appear to move against the wind since it must have been below the clouds. There was no air traffic reported in the area."

No radar sites in the area had unusual returns or activity, according to the same report.

Discussion.—This green disc, moving against the wind below an overcast and seen by over sixty witnesses, is an official Unidentified.

5. *Case 13. Savannah River AEC Plant, Summer, 1952*

A rather illuminating multiple-witness case was called to my attention by John A. Anderson, now at Sandia Base, [near Albuquerque] New Mexico, but in 1952 working as a young engineer in the Savannah River AEC facility near Aiken, S.C. After a considerable amount of cross-checking on the part of both Anderson and myself, the date was inferred to be late July, 1952, probably 7/19/52. The circumstance giving a clue to the date was that, at about 10:00 a.m. on the day in question, Anderson, along with what he estimated at perhaps a hundred other engineers, scientists and technicians from his group were outside watching a "required attendance" skit presented from a truck-trailer and commemorating the 150th anniversary of the founding of the DuPont company, July 18, 1802. Anderson indicated that some less than absorbed in the skit first spotted the unidentified object in the clear skies overhead, and soon most eyes had left the skit to watch more technically intriguing events overhead.

A greenish glowing object of no discernible shape, and of angular size estimated by Anderson to be not over a fifth of full-moon diameter, was darting back and forth erratically at very high speed. Anderson had the impression it was at great altitude, but conceded that perhaps nothing but the complete lack of sound yielded that impression. It was in view for about two minutes, moving at all times. He stressed its "phenomenal maneuverability"; it repeatedly changed direction abruptly in sharp-angle manner, he stressed. The observation was terminated when the object disappeared over the horizon "at apparently tremendous velocity."

Discussion.—Anderson said that the event was discussed among his group afterwards, and all agreed it could not possibly have been a conventional aircraft. He remarked that no one even thought of suggesting the unreasonable notion that it was an hallucination or illusion. Despite searching local papers for some days thereafter, not a word of this sighting was published, and no further information or comment on it came from within the very security-conscious AEC plant. He was unaware of any official report.

Months after hearing of this from Anderson, in one of my

numerous rereadings of Ruppelt's book (Ref. 5), I came across a single sentence in which Ruppelt, referring to the high concentration of reports in the Southeast around September of 1952, states that: "Many of the reports came from people in the vicinity of the then new super-hush-hush AEC facility at Savannah River, Georgia." Whether one of those reports to the official investigative agency came from within Anderson's group or other Savannah River personnel on the 7/52 incident is unknown. If not, then we may have here a case where dozens of technically trained personnel witnessed an entirely unexplainable aerial performance, yet reported nothing. Anderson knew of no report, and was unaware of any assembling of witness-information within his group, so the evidence points in the direction that this event may have gone unreported. If, as Anderson is inclined to think, this event was on July 19, 1952, it occurred only about twelve hours before the famous Washington National Airport radar-visual sightings; but this date remains uncertain.

6. *Case 14. Trinidad, Colo., March 23, 1966*
A daytime sighting by at least a dozen persons, in several parts of town, occurred near 5:00 p.m. on 3/23/66 in Trinidad, Colo. Following up a report in the APRO Bulletin on this interesting case, I eventually interviewed ten witnesses (seven children of average age near 12, and five adults). This case came just a few days after the famous "swamp gas" UFO incidents in southern Michigan, which made headline news all over the country.
As APRO noted in its account, the Trinidad case seems in several respects a distinctly better case, yet went essentially unnoted outside of Trinidad. (Press reporting of UFO sightings leaves very much to be desired; I concur in the cited APRO comment. However, press shortcomings in the UFO area are only secondary factors in the long failure to get this matter out into the open.) The witness-variance that skeptics like to cite is fairly well illustrated in the results of my ten interviews. I wish space permitted a full exposition of what each witness told me, for it would not only attest to that well-known variance but would also illustrate the point made earlier, namely, that despite those bothersome differences in details, there nevertheless comes through a consistent core of information on observations of something that was of scientific interest.
Mrs. Frank R. Hoch paid no attention when her son first tried to call her out to see something in the sky. Knowing it was kite season, dinner preparations took precedence, and she told the

10-year-old boy to go ride his bike. The second time he was more insistent, and she went outside to look. Two objects, domed on the top but nearly flat on the bottom, shaped like a cup upside down, having no rim or "sombrero brim," she said, were moving slowly westward from Fisher's Peak, which lies just south of Trinidad. Her son, Dean, told her he had seen three such objects when he tried to get her to come out earlier. (Mr. Louis DiPaolo, a Trinidad postman whom I interviewed, had also seen three objects.) Interestingly, when Mrs. Hoch saw the objects, one was between her and the ridge, the other just above the low ridgeline. The ridge is about a half-mile from the Hoch residence. A photo of the ridge, with roughly-scaled objects sketched on it, suggests an angular diameter of perhaps a degree (object size of order 100 ft.), in disagreement with her earlier angular estimates. It was clear that Mrs. Hoch was, as are most, unfamiliar with angular-size estimating. The objects, Mrs. Hoch said, moved up and down in bobbing manner as they progressed slowly westward along the ridgeline. Occasionally they tilted, glinting in the late afternoon sun as if metallic. No sound was mentioned by any witness except one young boy whose attention was drawn to the object by a "ricocheting sound," as he put it. DiPaolo's observations were made with 7x35 binoculars; he also described the objects as metallic in appearance and shaped like a saucer upside down. His attention had been called to it by neighborhood boys playing outside.

Mrs. Amelia Berry, in another part of Trinidad, evidently saw the objects somewhat earlier, when they were farther east, circling near Fisher's Peak, but she was uncertain of the precise time. She saw only two, and remarked that they seemed to "glitter," and she described them as "saucer shaped," "oblong and narrow." Mrs. J. R. Duran, horseback-riding with a 12-year-old son on the opposite (north) side of town also saw two objects, "flat on the bottom, and domed on top, silvery," when her son called them to her attention. She described them as "floating along slowly, bobbing up and down, somewhat to the west of Fisher's Peak." She, like the other witnesses, was positive that these were not airplanes.

No one described anything like wings or tail. A number of witnesses were so close that, had this been an unconventional helicopter, its engine-noise would have been unmistakable.

Discussion.—Notwithstanding differences in the witness accounts (more of which would emerge from a more complete recounting), the common features of the observers' descriptions would seem to rule out known types of aircraft, astronomical, meteorological, and other explanations.

7. Case 15, Redlands, Calif., February 4, 1968

A still more recent multiple-witness case of great interest was well-documented by three University of Redlands professors shortly after it occurred on the evening of 2/4/68. APRO plans a fairly detailed summary-report. Dr. Philip Seff kindly sent me a copy of the witness-testimony he and his colleagues secured in interviewing about twenty out of an estimated hundred-plus witnesses to this low-altitude sighting in a residential area of Redlands. Because I understand that Dr. Harder will be giving a fairly detailed report of this case to your Committee, I shall give only a much-abbreviated version.

At 7:20 p.m., many persons went outdoors to investigate either (a) the unusual barking of neighborhood dogs, or (b) a disturbing and unusual sound. Soon many persons up and down several streets were observing an object round in planiform, estimated at perhaps 50-60 feet in diameter, moving slowly towards the east-northeast at an altitude put by most witnesses as perhaps 300 feet. Glowing ports or panels lay around its upper perimeter and "jet-like" orange-red flames or something resembling flames emanated from a number of sources on the undersurface. A number of odd psychological effects were remarked by various witnesses, and the animal-reactions were a notable feature of this case. The object at one point rose abruptly by some hundreds of feet before continuing its somewhat "jerky" motion to the east. It then hovered a short time and moved off with acceleration to the northwest.

Discussion.—The Redlands University trio inquired concerning radar detection, but were informed that the nearest radar was at March AFB, Riverside, and the beam clearing intervening ridges could not detect so low a target over Redlands. An interesting aspect of press coverage of UFOs, a very characteristic aspect, is illustrated here. The local Redlands-area papers carried only short pieces on the event; beyond that no press coverage occurred, as far as I have been able to ascertain. Evidently even the state wires did not carry it.

(I think this fact deserves very strong emphasis. One has to see national clipping-service coverage, drawing upon many small-town papers, to gain even a dim glimpse of the astonishing number of UFO reports that occur steadily, but go unreported on state and national wires so that none but very diligent UFO investigators have any appreciation of the true frequency of UFO sightings. This is no "press clampdown," no censorship; wire editors simply "know" that there's nothing to all this nonsense about UFOS. A local story will be run simply for its local interest, but that interest falls off steeply with radial distance from the observation site.)

Thus, we must confront a situation, developed over 20 years, in which over a hundred citizens in a city of about 30,000 population can see an utterly unconventional aerial machine just overhead and, almost by the time the dogs have stopped barking, press and officialdom are uninterested.

Dr. Seff told me just last week that he had encountered a Redlands University coed who had seen the object (he hadn't interviewed her previously), and she seemed still terrified by the incident. I believe that your Committee must recognize an unfilled scientific obligation to get to the bottom of such matters.

8. Many other multiple-witness cases could be cited, some from my own interviewing experience, far more from other sources within this country and abroad. An October 28, 1954 sighting in Rome was estimated to have been viewed by thousands of people, one of whom was U.S. Ambassador Clare Booth Luce (Ref. 10) with her embassy staff. Mrs. Luce said it had the shape of a silver dollar and crossed the skies in about 30 seconds. A now-famous group of sightings of June 26/27, 1959, near Boiani [or Boianai], New Guinea, was observed by several dozen witnesses, the principal one of whom I interviewed in Melbourne, in 1967, Rev. Wm. B. Gill. Bloecher (Ref. 8) describes a number of mid-1947 incidents where the witness-totals ranged from dozens up to well over a hundred persons. Hall (Ref. 10) cites more recent instances. Many other sources could be cited to show that the intimation that UFOs are never seen except by lone individuals driving along some remote back road (a frequent setting, to be sure!) does not accord with the actual facts. Multiple-witness UFO cases are impressively numerous.

WHY AREN'T UFO'S EVER SEEN IN CITIES?
WHY JUST IN OUT-OF-THE-WAY PLACES?

One cannot study the UFO problem long without being struck by the preponderance of reports that come from somewhat remote areas, non-urban areas. Similarly, one cannot escape the conclusion that more UFOs are reported at night than in day. For the latter, luminosity and its obvious effect on probability of chance visual detection may go far towards explaining the diurnal variation of UFO sightings (though I suspect that most students of the problem would conclude that there is a real excess of nighttime occurrences for quite unknown reasons). Why, some ask with respect to the geographical distribution, don't the UFOs, if real and if extraterrestrial, spend most of their time looking over our cities? That's what we'd do, if we got to Mars and found huge urban complexes, some skeptics insist.

It is surprising to find scientists who do not see through the transparency of that homocentric fallacy. If it were true that we were under surveillance from some advanced civilization of extraterrestrial origin, the pattern of the observations, the motivation of the surveillance, and the degree of interest in one versus another aspect of our planet could be almost incomprehensible to us. Aboriginal natives under anthropological observation must find almost incomprehensible the motives behind the strange things that the field-teams do, the odd things in which they are interested. But the cultural and the intellectual gulf that would separate us from any intelligent beings commanding a technology so advanced that they could cross interplanetary or interstellar distances to inspect us would be a gulf vastly greater than that which separates a Harvard field-anthropologist from a New Guinea native. And, for this reason, I think one must concede that, within the argumentation carried out under tentative consideration of an extraterrestrial hypothesis for UFOs, incomprehensibility must be expected as almost inevitable. Hence there is more whimsy than good reasoning in queries such as, "Why don't they land on the White House lawn and shake hands with the President?" Nevertheless, the evidence affords a fairly definite answer to the skeptics' question, "Why aren't they ever seen over or in cities?" They are.

1. *Case 16. New York City, November 22, 1966*

A report in a 1967 issue of the NICAP *UFO Investigator* (Ref. 33) reads as follows:

> "A UFO over the United Nations in New York City was reportedly seen on November 22, 1966. Witnesses included at least eight employees of the American Newspaper Publishers Association, who watched from their offices on the 17th floor of 750 Third Avenue at 4:20 p.m. on a bright, sunny day. The UFO was a rectangular, cushion-shaped object (which) came southward over the East River, then hovered over the UN Building... It fluttered and bobbed like a ship on agitated water."

Witnesses mentioned were D. R. McVay, assistant general manager of ANPA and Mr. W. H. Leick, manager of the ANPA's Publications Department. I telephoned the ANPA offices and spoke at some length with Mr. Leick about the sighting. He confirmed that eight or nine persons went out on the 17th floor terrace, watching the object hover over the UN Building (as nearly as they could estimate) for a number of minutes as it rocked and reflected the sun's rays with a golden glint before rising and moving off eastward at high speed. I asked Leick if they reported it to any official channels, and he said that A. A. LaSalle called a New York office of the Air Force and was assured that an officer would be in the next day to interview them. But no one ever came. Leick added that they also phoned a New York newspaper "which shall go unnamed," but "they weren't interested." It got to NICAP almost by accident, and NICAP sent up their standard witness-questionnaires which Leick said they all filled out.

Discussion.—When an incident such as this is cited to the skeptic who asks,"Why no UFOs near cities?," I find that his almost invariable retort is something like: "If that had really happened, why wouldn't hundreds to thousands of persons have reported it?" There are, I believe, two factors that explain the latter situation.

First, consider the tiny fraction of persons on any city street whose vision is directed upwards at any given moment. In absence of loud noises aloft, most urbanites don't spend any large amount of time scanning the skies. In addition to infrequency of sky-scanning, another urban obstacle to UFO detection is typically restricted vision of the full dome of the sky; buildings or trees cut down the field of view in a way not so typical of the view afforded the farmer, the forest ranger, or a person driving in open country. Finally, in UFO studies, it is always necessary to draw sharp distinction between a "sighting" and a "report." The first becomes the second only if a witness takes the step of notifying a newspaper, a law-enforcement office, a university, or some official agency. It is abundantly clear, from the experience of UFO investigations in many parts of the world, that psychological factors centering around unwillingness to be ridiculed deter most witnesses from filing any official report on a very unusual event. Again and again one learns of a UFO sighting quite indirectly, from someone who knows someone who once mentioned that he'd seen something rather unusual. On following such leads, one frequently comes upon extremely significant sightings that were withheld from official reporting channels because of the "ridicule lid," as I like to term it, that imposes a filter screening out a large number of good sightings at their source.

Returning to the 11/22/66 New York City report, I must say that, between the information NICAP secured from the witnesses and my own direct conversations with Leick, I accept this as a quite real sighting, made by reliable observers under viewing circumstances that would seem to rule out obvious conventional explanations. When the object left its hovering location, it rose straight upward rapidly, before heading east, Leick said. Although he and his colleagues may well have erred in their slant-range estimate which put it over the UN Building, their description of its shape and its maneuvers would appear to rule out helicopters, aircraft, balloons, etc.

2. Case 17. Hollywood, Calif., February 5-6, 1960
A still more striking instance in which entirely unconventional objects were observed by many city-dwellers, where low-altitude objects hovered and exhibited baffling phenomena, is a central Hollywood case that was rather carefully checked by LANS, the Los Angeles NICAP Subcommittee (Ref. 34). The two incidents occurred just after 11:00 p.m. on two successive nights, Friday 2/5/60 and Saturday 2/6/60, over or near the intersection of Sunset Blvd. and La Brea Ave., i.e., in the heart of downtown Hollywood.

I have gone over the site area with one of the principal investigators of these incidents, Mrs. Idabel Epperson of LANS, have examined press accounts (Ref. 35) that dealt (very superficially) with the event, and have studied correspondence between the LANS investigators and official agencies concerning this case. The phenomenology is far too complex to report in full detail here; even the 21-page single-spaced LANS report was only a digest of results of all the NICAP witness-interviewing carried out to substantiate the events.

The LANS report summarizes object-descriptions given by eight witnesses Friday night and eighteen witnesses Saturday night, several of them police officers. Cars were stopped bumper-to-bumper, according to employees of several businesses on the Sunset-La Brea intersection in the midst of the main events, with people gaping at the object overhead. Persons on hotel and apartment rooftops were out looking at the bright "cherry-red, circular light" that figured in both incidents. On the two successive nights, the red object first appeared at about 11:15 p.m., and on both nights it stopped and hovered motionless for periods of about 10 minutes at a time. The angular estimates of the size of the red light varied, but seemed to suggest a value of one-fourth to one-third of the lunar diameter, say 5-10 minutes of arc. Almost all agreed that the light was sharp-edged rather than hazy or fuzzy. The usual witness-variances are exhibited in the total of about two dozen persons interviewed, e.g., some thought the light pulsated, others recalled it as steady, etc., but the common features, consistent throughout almost all the testimony, bespeak a quite unusual phenomenon.

On Friday night, the red light was first seen directly overhead at Sunset and La Brea. Two service-station attendants at that intersection, Jerry Darr and Charles Walker, described to LANS interviewers how, ". . . hundreds of people saw it—everybody was looking" as the light hovered for at least five minutes over a busy drive-in there. Ken Meyer, another service station attendant a third of a mile to the north, estimated it hovered for about 10 minutes. Harold Sherman, his wife, and two others watched it in the later phases (also described by the above-cited witnesses) as it resumed motion very slowly eastward. After proceeding east for a distance that witnesses roughly estimated at a block or two, it veered southeastward and passed out of sight. (It is not clear whether it was occulted by buildings for some witnesses, or diminished in intensity, or actually passed off into the distance.) No sound was heard over street- noise background.

The following night, an object which appeared to be the same, to those several witnesses who saw both events, again showed up overhead, this time first seen about one block farther east than on Friday night. Triangulation based on estimates of angular elevations as seen from various locations was used to approximate the height above ground. LANS concluded that, when first seen, it lay about 500-600 ft. above the intersection of Sunset and Sycamore. A number of witnesses observed it hovering motionless in that position for about 10 minutes. Then a loud explosion and brilliant bluish-white flash was emitted by the object, the noise described by all witnesses as unlike any sonic boom or ordinary explosion they had ever heard. The sound alerted witnesses as far away as Curson and Hollywood Blvd., i.e., Tom Burns and two friends who asked LANS interviewers not to use their names.

Condensing very greatly here the descriptions given to the interviewers by independent witnesses who viewed the "explosion" from various locations scattered over a circle of about a 1-mile radius yields a summary-description as follows: What had, just before the explosion, looked much "like a big red Christmas ball hanging there in the sky," was suddenly the source of a flash that extended downward and to the west, lighting up the ground all around one interviewed (Soe Rosi) on La Brea Ave. A "mushroom-shaped cloud," with coloration that impressed all who saw it, emerged upward and soon dissipated. Concurrently, as the red light extinguished, an object described by most, but not all, witnesses as long and tubular shot upwards. Angular estimates implied an object a number of tens of feet long, 70 ft. from Harold Sherman's rough estimates.

Clearly, it is difficult to explain how an object of such size could have materialized from a light at 500 ft. elevation and subtending an angle of only 10 minutes of arc, unless it had been there all along, unseen because of the brilliance of the red light beneath it. Or perhaps the angular-size estimates are in error.

Some witnesses followed only the tubular ascending object, others saw only something that "spiraled downwards" beneath the explosion source. No witness seemed certain of what it was that came down; some spoke of "glowing embers"; no one gave indication of following it to ground.

Glossing over other details bearing on this "explosion" at an estimated 5-600 ft. above Sunset and Sycamore, witnesses next became aware that the just extinguished red light had evidently reappeared in a new location, about a block to the west. Police officers Ray Lopez and Daniel Jaffee, of LAPD, located at the corner of Sunset and La Brea, heard the explosion and looked up, seeing the light in its new location "directly overhead," as did many others at that intersection who then watched the red light hovering in its new location for about 8 minutes.

(Space precludes my giving all pertinent information on time estimates as set out in the 21-page LANS summary. For example, a good time-fix on the explosion came from the fact that E. W. Cass, a contractor living almost a mile west, was just winding his alarm clock, looking at it, when flare-like illumination "lit up the whole bedroom," just at the indicated time of 11:30. He went out, watched the hovering red light in its new location, and added further details I shall omit here. Others took their time clues from the fact that 11:30 commercials had just come on TV when they heard the peculiar explosion and hastened outside to check, etc.)

The red light, now over Sunset and La Brea, was roughly triangulated at about 1000 ft. up, a figure in accord with several witness comments that, when it reappeared some 4-5 seconds after the "explosion," it lay not only somewhat west of its first location, but noticeably higher. After hovering there for a time inferred to be eight minutes, it began slowly drifting eastward, much as on the previous night when much less spectacular events had occurred. Larry Moquin, one witness who had taken rather careful alignment fixes using rooflines as an aid, remarked that, at this stage, La Brea and Sunset was filled with watchers: "Everybody was standing outside their cars looking up—cars were backed up in the streets—and everyone was asking each other, 'What is it?'"

After moving slowly but steadily (observers mentioned absence of bobbing, weaving, or irregularity in its motion) for about a block east, to its first location it turned sharply towards the north-northeast, accelerated, and climbed steeply, not stopping again until at a very high altitude well to the north. From crude triangulation, LANS investigators inferred a new hovering altitude of over 25,000 ft, but it is clear from the data involved that this estimate is extremely rough.

Discussion.—Although I have done no personal witness-interviewing to date in the 2/60 Hollywood case, I can vouch for the diligence and reliability with which the LANS group pursues its case-studies. The large number of interviews secured and the degree of consistency found therein seem to argue that some extremely unusual devices maneuvered over Hollywood on the two nights in question. Unless one simply rejects most of the salient features of the reports, it is quite clear that no meteorological or astronomical explanation is at all reasonable. Nor does any conventional aircraft match the reports.

The question that arises almost immediately is that of a practical joke, a hoax. However, the resources required to fabricate some device yielding the complex behavior (stop motionless, move against wind, explosively emit secondary devices, and finally, in the 2/6 event, climb to rather high altitude) would scarcely be available to college pranksters. The phenomena go so far beyond the gas-balloon level of hoaxing that one must have some much more elaborate hoax hypothesis to account for the reported events. Balloons must drift with the winds, and the LANS group secured the local upper-wind data for both nights, and there is no match between the reported motions and the winds in the surface-1000-ft. layer. And, in any event, the alternation between hovering and moving, plus the distinct direction-shifts without change of apparent altitude, cannot be squared with balloon-drift. This would mean that some highly controlled device was involved, capable (in the 2/6 incident) of hovering in an almost precisely stationary position relative to the ground (Moquin sighted carefully, using structural objects to secure a fix when the red light lay right over La Brea and Sunset, and perceived no motion for many minutes). Yet the Weather Bureau was reporting 5 mph winds from the southwest at 1000 ft. (triangulated altitude when hovering there). Only if one hypothesized that this was an expensively elaborate experiment in psychological warfare could one account for financial resources needed to build a device capable of simulating some of these phenomena. Such a hypothesis seems quite unreasonable in the 100-megaton age where ever-present realities of weaponry pose more psychological strains than Disney-like pyrotechnics. In fact, UFO sightings with equally peculiar phenomenology are so much apart of the total record that this Hollywood incident is not as unparalleled as it might first seem.

In Hobard, Tasmania, I interviewed an electrical engineer who, along with a fellow engineer also employed by the Tasmanian Hydroelectric Commission, observed phenomena occurring in broad daylight over and near the River Derwent at Risdon that have the same "absurd" nature that one meets in the Hollywood case. The wife of a Texas rancher described to me an incident she witnessed in Juarez, Mexico, with about the same absurdity-quotient.

We simply do not understand what we are dealing with in these UFO phenomena; my present opinion is that we must simply concede that, in the Hollywood case, we are confronted with decidedly odd UFO phenomena, in a decidedly urban locale. There appears to have been no official investigation of these striking events (Ref. 35), and local newspapers gave it only the briefest attention. In the New York City case cited above, the particulars were phoned to a large New York paper, but the paper was not interested, and no account was reported. Similarly in the 2/4/68 Redlands case, the local papers felt it warranted only an extremely brief article. This pattern is repeated over and over again; newspapermen have been led to believe that UFOs are really no more than occasional feature-story material. On rare occasions, for reasons not too clear to students of the UFO problem, some one case like the Michigan incident of 1966 will command national headlines for a day or two and then be consigned to journalistic limbo. This, in company with scientific rejection of the problem, plus official positions on the matter have combined to keep the public almost entirely unaware of the real situation with respect to frequency and nature of UFO incidents.

For emphasis, let me repeat that I do not see design in that, nothing I construe as any well-planned attempt to keep us all uninformed for some sinister or protective reason. The longer I reflect on the history of the past handling of the UFO problem, the more I can see how one thing led to another until we have reached the intolerable present situation that so urgently calls for change.

3. *Case 18. Baytown, Texas, July 18, 1966*

Baytown, Texas, on Galveston Bay, has a population near 30,000. Several persons evidently saw an interesting object there at about 9:00 a.m. on 7/18/66. My original source on this case was an article that appeared in the 10/8/66 *Houston Post* from NICAP files. The article, by *Post* reporter Jimmie Woods, represents one of those rare UFO feature stories in which fact is well blended with human interest, as I found when I subsequently interviewed one of the principal witnesses, W. T. Jackson, at whose service station he and assistant Kelly Dikeman made the sighting. Both were inside the station when Jackson spotted the object hovering motionless about 100 yards away. (The *Post* said 1000 yards, but Jackson pointed out that Woods interviewed him while he was waiting on customers at the station and the reporter didn't get all of it correct.) Jackson explained to me that the object "lay right over the Dairy Queen." He described it as a white object that "looked like two saucers turned together with a row of square windows in between," and he thought it might have been 50 feet in diameter. He called Dikeman over, and they both looked at it for a few seconds and then simultaneously started for the door to get a better look. Almost at that moment it started moving westward. Dikeman was at the door before Jackson and had the last view of it as it passed over a water tower, beyond buildings and a refinery and was gone, "faster than any airplane." Jackson described it as pure white, and definitely not spinning, since he saw clearly the features that he termed "windows." Jackson kept the incident to himself for a time; when it got out, two nurses who were unwilling to give him their names because "they didn't want to be laughed at" stopped at his station and told him they had seen it from another part of Baytown.

Discussion.—"Swamp gas" explanations were then still featured in press discussions of UFOs, and Jackson volunteered the comment that there are no swamps nearby and that it was "too high for any gas formations" he knew of. "It damned sure wasn't no fireball," Jackson told the *Post* reporter, and also commented, "Feller, when you set there and count the windows it ain't no damn reflection." I received similar salty commentary on various hypotheses when I spoke with Jackson. No sound was heard, yet, as Jackson put it, "if it had been any kind of jet, we'd have been deafened." As in many other cases, a distinctly machine-like configuration, definite outlines, secondary "structural" features here termed "windows," add up to a description that does not suggest any misinterpreted natural phenomenon. That it hovered within a city of moderate size with only a total of two declared and two other undeclared witnesses is not entirely difficult to understand when one has interviewed large numbers of witnesses for whom the likelihood of ridicule was an almost sufficient deterrent to open reporting.

4. Case 19. Portland, Oregon, July 4, 1947

In the course of cross-checking a sampling of the 1947 cases that went into Bloecher's study (Ref. 8), the numerous daytime sightings in central Portland on 7/4/47 seemed worth checking, especially because many of the reports came from police and harbor patrolmen. Here again, we deal with a case for which there are so many relevant details available that space precludes an adequate summary (see Ref. 8).

I spoke with Sheriff's Deputy Fred Krives who, along with several other deputies, had seen some of the many objects over Portland from the Court House across the Columbia River in Vancouver, Wash. Krives recalled that over half a dozen deputies were outside looking at what they estimated to be about 20 disc-shaped objects in several subgroups racing across the sky at an estimated height of perhaps 1000 ft., heading to the southwest. Both from contemporary press accounts and my own checks, it became evident that more than one formation of discs flew over Portland that day. Harbor Patrol Capt. K. A. Prehn, whom I located by telephone, told me that he had been called outside by another officer who spotted objects moving overhead towards the south. Their speed seemed comparable to that of aircraft, their outlines were quite sharp, and they looked metallic as they flashed in the sun. They occasionally wobbled, and their path seemed to be slightly irregular.

Other officers with whom I spoke sighted discs from other parts of the Portland area; one of them, Officer Walter Lissy, emphasized that he recalled them as zig-zagging along at "terrific speed." Another officer, Earl Patterson, told me of seeing a single object that "made sudden 90-degree turns with no difficulty." I also obtained letter accounts from others in the Portland area who saw disc-like objects that day. Here was an early instance of unidentified objects maneuvering in full daylight over a major city.

Discussion.—The July 4, 1947, sightings (for which Bloecher gathered press accounts for more than 80 from various parts of the U.S.) were made the subject of a good deal of press ridicule, as Bloecher's study makes clear. However, after interviewing a number of the witnesses to the Portland sighting concerning their recollections of what they saw that day, I see no basis at all for rejecting these sightings. The official explanation for the Portland observations is "Radar Chaff," based evidently (Ref. 6) on a report that some aircraft had made a chaff-drop in that area sometime on that day. "Chaff" is metal-foil cut into short strips, typically a few inches in length, ejected from military aircraft to jam radar. The strips float down through the air, intercepting and returning the radar pulses. To suggest that numerous police officers would confuse strips of foil, so small as to be invisible beyond a few hundred yards, with maneuvering disc-like objects seems unreasonable. I doubt that anyone who had talked directly to these officers could have seriously proposed such an explanation.

Herein lies a difficulty: In an overwhelming majority of cases, official explanations have been conceived without any direct witness-interviewing on the part of those responsible for conceiving the explanations.

5. Perhaps, for present purposes, the foregoing cases will suffice to indicate that there have been significant UFO incidents in cities. Many other examples could easily be cited. Elsewhere (Ref. 2) I have discussed my interviews with witnesses in a case at Beverly, Mass., on the evening of April 22, 1966, where three adult women and subsequently a total of more than half a dozen adults (including two police officers) observed three round lighted objects hovering near a school building in the middle of Beverly. At one early stage of the sightings, one of the discs moved rapidly over the three women, hovering above one of them at an altitude of only a few tens of feet and terrifying the hapless women until she bolted. This case was quite thoroughly checked by Mr. Raymond E. Fowler, one of NICAP's most able investigators, who has studied numerous other UFO incidents in the New England area. I interviewed witnesses in a most interesting sighting in Omaha in January 1966, where a stubby cigar-shaped object had been seen by a number of persons on the northwest side of the city. Urban UFO cases in other parts of the world are also a matter of at least journalistic if not yet scientific record. To sum up, though non-urban reports are definitely more numerous, urban reports do indeed exist.

WHY DON'T ASTRONOMERS EVER SEE UFO'S?

I have had this question put to me by many persons, including a number of astronomers. Once I was speaking to a group from an important laboratory of astronomy when the director asked why astronomers never see them. In the room, among his staff, were two astronomers who had seen unconventional objects while doing observing but who had asked that the information they had given me about their sightings be kept confidential. I understand such strictures, but some of them make things a bit difficult. This phenomenon of professional persons seeing unidentified objects and then being extremely loath to admit it is far more common than one might guess. After hearing of an evidently very significant sighting by a prominent physical scientist who was hiking in some western mountains when he spotted a metallic-looking disc, examined it with binoculars, and saw it shoot up into the air (according to my second-hand report from a professional colleague). I tried for months to secure a direct report of it from him; he was unwilling to discuss it openly with me. NICAP has had reports from prominent executives in large technical corporations who insisted that, just because of their positions, their names not be used publicly. Similar instances could be cited almost *ad infinitum*. The very types of witnesses whose testimony would carry greatest credence often prove to be the most reluctant to admit their sightings; they seem to feel they have the most to lose.

Within a day of this writing, I spoke to a veteran airlines pilot about a sighting in which he was involved about a decade ago. After the official "explanation" was publicized, he decided he'd never report another one. I predict that social psychologists are going to have a field day, in a few years, studying the "pluralistic ignorance" that led so many persons to conceal so many sightings for so long.

Returning, however, to the question of why astronomers never see UFOs, a relevant quantitative consideration needs to be cited at once. According to a recent count, the membership of the American Astronomical Society is about 1800; by contrast, our country has about 350,000 law-enforcement officers. With almost 200 times as many police, sheriffs' deputies, state troopers, etc., as there are professional astronomers, it is no surprise that many more UFO reports come from the law-enforcement officers than from the astronomers. Furthermore, the notion that astronomers spend most of their time scanning the skies is quite incorrect; the average patrolman almost certainly does more random looking about than the average professional astronomer. Despite these considerations, there are on record many sightings from astronomers, particularly the amateurs, who far outnumber the professionals. A few examples will be considered.

1. Case 20. Las Cruces, N.M., August 20, 1949
    A good account of the sighting by Dr. Clyde Tombaugh,

discoverer of the planet Pluto, is given by Menzel (Ref. 25). From my own discussions with Dr. Tombaugh, I confirmed the main outlines of this incident.

At about 10:00 p.m. on 8/20/49, he, his wife, and his mother-in-law were in the yard of his Las Cruces home, admiring what Tombaugh described as a sky of rare transparency, when Tombaugh, looking almost directly towards zenith, spotted an array of pale yellow lights moving rapidly across the sky towards the southeast. He called them to the attention of the two others, who saw them just before they disappeared halfway to the horizon. The entire array subtended an angle which Tombaugh put at about one degree, and it took only a few seconds to cross 50 or 60 degrees of sky. The array comprised six "window-like" rectangles of light, formed into a symmetric pattern; they moved too fast for aircraft, too slowly for a meteor, and made no sound. Menzel quotes Tombaugh as saying, "I have never seen anything like it before or since, and I have spent a lot of time where the night sky could be seen well."

Discussion.—Dr. Menzel explains this phenomenon as resulting from reflection of lights from the ground, possibly "the lighted windows of a house" reflected by an inversion or haze layer aloft. The movement he explains as resulting from a ripple on the haze layer. Such an "explanation" is not merely difficult to understand; it is incredible. For an "inversion layer" to produce such a near-normal reflection of window lights would demand a discontinuity of refractive index so enormously large compared with anything known to occur in our atmosphere as to make it utterly out of the question. However, it has been just such casual *ad hoc* explanations as this by which Menzel has, in his writings, used meteorological optics to rationalize case after case with no attention to crucial quantitative details. It is a simple matter to show that even inversions of intensity many orders of magnitude larger than have ever been observed yield reflectivities (at the kind of near-normal incidence involved in Tombaugh's sighting) that are only a tiny fraction of one per cent (Ref. 36). In fact, I see no way of accounting for the Tombaugh observation in terms of known meteorological or astronomical phenomena.

2. *Case 21. Ft. Sumner, New Mexico, July 10, 1947*

A midday sighting by a University of New Mexico meteoriticist, Dr. Lincoln La Paz, and members of his family was summarized by *Life* magazine years ago (Ref. 37) without identifying La Paz's name. Bloecher (Ref. 8) gives more details and notes that this is officially an Unidentified.

At 4:47 p.m. MST on 7/10/47, four members of the La Paz family nearly simultaneously noted "a curious bright object almost motionless" low on the western horizon, near a cloudbank. The object was described as ellipsoidal, whitish, and having sharply-outlined edges. It wobbled a bit as it hovered stationary just above the horizon, then moved upwards, passed behind clouds and re-emerged farther north in a time interval which La Paz estimated to be so short as to call for speeds in excess of conventional aircraft speeds. It passed in front of dark clouds and seemed self-luminous by contrast. It finally disappeared amongst the clouds. La Paz estimated it to be perhaps 20 miles away, judging from the clouds involved; and he put its length at perhaps 100-200 ft.

Discussion.—This observation is attributed by Menzel (Ref. 24, p. 29) to "some sort of horizontal mirage, perhaps one of a very brilliant cloud shining like silver in the sunlight—a cloud that was itself invisible because of the darker clouds in the foreground." As nearly as I am able to understand that explanation, it seems to be based on the notion that mirage-refraction can neatly superimpose the image of some distant object (here his "brilliant cloud") upon some nearer object in the middle distance (here his "darker clouds"). That is a fallacious notion. If any optical distortions did here bring into view some distant bright cloud, it would not be possible to receive along immediately adjacent optical paths an image of the intermediate clouds. Furthermore, the extremely unstable lapse rates typical of the southwestern desert areas under afternoon conditions produce inferior mirages, not superior mirages of the looming type here invoked by Menzel. Rapid displacements, vertically and horizontally, are not typical of mirage phenomena. Hence Menzel's' explanations cannot be accepted for this sighting.

3. *Case 22. Harborside, Me., July 3, 1947*

An observation by an amateur astronomer, John F. Cole, reported to official investigative offices near the beginning of the period of general public awareness of the UFO problem, involves an erratically maneuvering cluster of about 10 objects, seen near 2:30 p.m. EDT on 7/3/47 on the eastern shore of Penobscot Bay. Hearing a roar overhead, Cole looked up to see the objects milling about like a moving swarm of bees as they traveled northwestward at a seemingly high speed, as nearly as he could judge size and distance. The objects were light-colored, and no wings could be discerned on most, although two appeared to have some sort of darker projections somewhat resembling wings. In 10-15 seconds they passed out of sight.

Discussion.—This is one of several dozen cases admitted to the Unidentified category in one of the earliest official reports on UFOS (Ref. 6). I have tried, unsuccessfully, to locate J. F. Cole. An account of the case is given by Bloecher (Ref. 8). It might be remarked that "swarming bee" UFO observations have cropped up repeatedly over the years, and from all over the world.

4. Case 23. Ogra, Latvia, July 26, 1965

An astronomer whom I know recently toured a number of observatories in the USSR, and brought back the word that a majority of Russian astronomers have paid little attention to Russian UFO reports (details of which are quite similar to American UFO reports, my colleagues established), a frequently-cited reason being that the American astronomer, Menzel, had given adequate optical explanations of all such sightings. I must agree with Dr. Felix Zigel who, writing on the UFO problem in *Soviet Life* (Ref. 38), remarked that Menzel's explanation in terms of atmospheric optics "does not hold water." It would, for example, be straining meteorological optics to try to account in such terms for a sighting by three Latvian astronomers whose report Zigel cites in his article.

At 9:35 p.m. on 7/26/65 while studying noctilucent clouds, R. Vitolniek and two colleagues visually observed a starlike object drifting slowly westward. Under 8-power binocular magnification, the light exhibited finite angular diameter, so a telescope was used to examine it. In the telescope, it appeared as a composite of four smaller objects. There was a central sphere around which, "at a distance of two diameters, were three spheres resembling the one in the center." The outer spheres slowly rotated around the central sphere as the array gradually moved across the sky, diminishing in size as if leaving the Earth. After about 20 minutes' observation, the astronomers noted the outer spheres moving away from the central object, and by about 10:00 p.m., the entire group had moved so far away that they were no longer visible.

Discussion.—I have no first-hand information on this report, of course. The group of objects was seen at an angular elevation of about 60 degrees, far too high to invoke any mirage-effects or other familiar refractive anomalies. Furthermore, the composite nature of the array scarcely suggests an optical distortion of the telescope, a possibility also rendered improbable from the observed angular velocity and apparent recessional motion.

5. *Case 24. Kislovodsk, Caucasus, August 8, 1967*
Zigel, who is affiliated with the Moscow Aviation Institute, reports in the same article (Ref. 38), a sighting at 8:40 p.m., 8/8/67, made by astronomer Anatoli Sazanov and colleagues working at the Mountain Astrophysical Station of the USSR Academy of Sciences, near Kislovodsk. Sazanov and ten other staff members watched an "asymmetric crescent, with its convex side turned in the direction of its movement" moving eastward across the northern sky at an angular elevation of about 20 degrees. Just ahead of it, and moving at the same angular speed was a point of light comparable to a star of the first magnitude. The crescent-like object was reddish-yellow, had an angular breadth of about two-thirds that of the moon, and left vapor-like trails aft of the ends of the crescent horns. As it receded, it diminished in size and thus "instantly disappeared."

Discussion.—If we may accept as reliable the principal features of the sighting, how might we account for it? The "faintly luminous ribbons" trailing from the horns suggest a high-flying jet, of course; but the asymmetry and the reddish-yellow coloration fail to fit that notion. Also, it was an object of rather large angular size, about 20 minutes of arc, so that an aircraft of wingspan, say, 150 feet would have been only about five miles away whence engine-noise would have been audible under the quiet conditions of a mountain observatory. More significant, if it had been an aircraft at a slant range of five miles, and at 20 degrees elevation, its altitude would have been only about 9000 ft. above the observatory. For the latitude and date, the sun was about ten degrees below the western horizon, so direct sun-illumination on the aircraft at 9000 ft above observatory level would be out of the question. Hence the luminosity goes unexplained. Clearly, satellites and meteors can be ruled out. The astronomers' observation cannot be readily explained in any conventional terms. Zigel remarks that the object was also seen in the town of Kislovodsk, and that another reddish crescent was observed in the same area on the evening of July 17, 1967.

6. *Case 25. Flagstaff, Ariz., May 20, 1950*
Near noon on 5/20/50, Dr. Seymour Hess observed an object from the grounds of the Lowell Observatory. Although Hess' principal field of interest has been meteorology, we may here consider him an astronomer-by-association, since he was at Lowell doing work on planetary atmospheres, on leave from Florida State University. Spotting an unusual small object moving from SE to

NW, he had time to send his son after binoculars, which he used in the later portions of his observation. He said it looked somewhat disc-shaped, or perhaps somewhat like a tipped parachute. It had no wings or visible means of propulsion. Dr. Hess indicated to me that he probably had it in sight a total of about three minutes, during which it passed directly between him and a cloud, before disappearing (into a cloud Hess feels, though this point was not certain). From meteorological data bearing on the cloud-base height, Hess deduced that the cloud bases lay 12,000 ft. above terrain (vs. Weather Bureau visual estimate of 6000 ft. above terrain). The zenith angle was about 45 degrees, so the slant range would have been 17,000 ft. or 8,000 ft., depending on which cloud height is accepted. For its 3 minutes estimated angular diameter (dime at 50 ft. Hess estimated), the diameter would then come out of the order of 10 to 15 feet. His subjective impression was that it was possibly smaller than that.

Discussion.—The possibility that this might have been a balloon or some other freely drifting device comes to mind. However, Hess noted carefully that the clouds were drifting from SW to NE, i.e., at right angles to the object's motion. He estimated its speed to be in the neighborhood of 100 to 200 mph, yet no engine noises of any kind were audible. It appeared dark against the bright cloud background, but bright when it was seen against blue sky. No obvious explanation in conventional terms seems to fit this sighting.

7. Many other sightings by both professional and amateur astronomers could be listed. Vallee (Ref. 17) discusses in detail a November 8, 1957 observation by J. L. Chapuis of Toulouse Observatory in France of what appeared through a small telescope to be a yellowish, elliptical body, with distinct outlines, leaving a short trail behind it. It was seen by other observers in three separate locations, executed maneuvers entirely excluding meteoric origin, and was regarded as an unexplainable phenomenon by all of the witnesses. Hall (Ref. 10) lists nine examples of astronomer sightings of unidentified objects, several of which are quite striking. Ruppelt (Ref. 5) remarks that an astronomer working under contract to the official UFO investigatory program interviewed 45 American astronomers during the summer of 1952, of whom five (11 per cent) had seen what they regarded to be UFOS. Although the sample is small, that percentage is well above the population percentage who say they have seen UFOs, which suggests that perhaps astronomers may sight more UFOs than they report as such. Indeed, with the recent publication of Ref. 7, further

interesting information on that 1952 poll is now at hand. The contract astronomer wrote at that time (Ref. 7, Rept. 8),

> "... certainly another contributing factor to their desire not to talk about these things is their overwhelming fear of publicity. One headline in the nation's papers to the effect that 'Astronomer Sees Flying Saucer' would be enough to brand the astronomer as questionable among his colleagues."

Unfortunately, *we scientists are by no means as open-minded and fearlessly independent as we are sometimes pictured* [my emphasis, L.S.]. It is often quite difficult to persuade a scientist to let his confidential report of a UFO sighting become a fully open UFO report; and my own experience suggests that perhaps astronomers, as a group, are just a bit more sensitive on this score than other scientists. At any event, perhaps the above-cited cases will suggest that some astronomers have seen unidentified flying objects.

METEOROLOGISTS & WEATHER OBSERVERS LOOK AT THE SKIES FREQUENTLY. WHY DON'T THEY SEE UFOS?

1. *Case 26. Richmond, Va., April 1947*

To begin an answer to that rhetorical question, we might consider an observation made by a weather observer at the Richmond, Va., U.S. Weather Bureau station, about two months before the first national publicity concerning UFOs.

Walter A. Minczewski, whom I located at the same Weather Bureau office where he made the sighting in 1947, was making a pilot balloon observation, when he spotted a silvery object that entered the field of his theodolite (which was trained on the balloon he had released). In the account that Minczewski sent me, he stated that "the bottom was flat and the top was slightly dome-shaped": and when he tried to see it with naked eye, he could not spot it. (Typical pilot balloon theodolites have magnifications of about 20 to 25, and angular fields that are usually about a degree across.) It was a "clear bright morning" when he spotted the object, and it lay to his NNE at an elevation of about 45 degrees. Whether Minczewski really saw the upper surface or formed his mental impressions without realizing that the theodolite may have inverted the image is now unclear, and my questioning did not settle that point.

Discussion.—A report of this sighting is in the official files, a circumstance which greatly surprised Minczewski, since he had discussed it only with his fellow workers. In the ensuing two decades, he has never again seen anything like it. Clearly, the probability of an object crossing the small angular field of a meteorological theodolite is quite low, if only chance were involved here. He tried to track it but lost it, due to its high angular velocity, after about five or six seconds, he recalled. No obvious conventional explanation suggests itself for this early sighting.

2. *Case 27. Yuma, Ariz., February 4, 1953*

Weather Bureau observer S. H. Brown was tracking a pilot balloon at 6000 ft. over Yuma at 1:50 p.m. MST on 2/4/53 when first one and then a second unidentified object moved across his theodolite field, somewhat as in the preceding case. I obtained an account of this sighting from V. B. Cotten, Meteorologist-in-Charge at the Yuma station. The full account is too long for recapitulation here. Both objects appeared to be of the order of a minute of arc in diameter and appeared "almost round, a solid dull pure white color, with a thin white mist completely edging each object." The first object moved into the optical field and curved upwards to the west, with the theodolite oriented to about 53 degrees elevation, 157 degrees azimuth. About 20 seconds later, a second object entered the field and moved in and out of the field erratically two times, to rejoin the first object. Brown was able to track the pair thereafter, as they jointly changed both azimuth and elevation. Because he had a stopwatch at hand for the balloon observation (which he did not complete because of following the unknown objects), he was able to determine that he followed the pair of objects for five minutes (1350 to 1355), until he lost sight of them against a cirrus cloud deck to the SSW. At the termination of the observation, his instrument was pointed to 29 degrees elevation, 204 degrees azimuth.

Discussion.—This case is carried as Unidentified in the official files (see Ref.7 for official summary). At times these objects lay near the sun's position in the sky, which might suggest forward-angle scattering of sunlight by airborne particles. However, initially, the objects were detected at angular distance of about 40 degrees from the solar position, which would not yield appreciable low angle scattering. Furthermore, if these were airborne scatterers, they would almost certainly be separated by random turbulence within as long a period as five minutes, yet the observer's report indicates that they maneuvered together within angular separations of the order of the roughly one-degree field of such theodolites. The fact that the second object did go out of the field only to return to the vicinity of the first object strains the airborne-particle hypothesis. Thus the official categorization of Unidentified seems reasonable here.

3. Case 28. Upington, Cape Province, [South Africa] December 7, 1954
R. H. Kleyweg, Officer-in-Charge of the Upington Meteorological Station, had just released a balloon for upper-wind measurement and was shielding his eyes from the sun trying to spot the balloon to get his theodolite on it. Seeing an object east of the sun, moving slowly to the west, he thought it was his balloon and got the theodolite on it, only to find that it was white, whereas he had released a red balloon. An account in the *Natal Mercury*, January 28, 1955, quoted Kleyweg as saying that it seemed "like a half-circle with the sun reflecting off the sloping top." He had no difficulty following it for about three minutes, but then it began to accelerate and, after another minute, he was unable to track fast enough to keep it in optical view (Ref. 10).

Discussion.—Kleyweg was quoted in the cited press source as saying, "I have followed thousands of meteorological balloons. This object was no balloon." A South African student doing graduate work in my Department, Petrus DuToit, has confirmed this sighting, having had an account of it directly from Kleyweg. An accelerating airborne half-circular object with sloping top seems best categorized as an unidentified flying object.

4. Case 29. Arrey, New Mexico, April 24, 1949
Charles B. Moore, Jr., was with four enlisted Navy personnel making a pilot balloon observation preparatory to release of a Skyhook balloon at the White Sands Proving Ground in the middle of the morning of 4/24/49. The balloon was airborne and was under observation by one of the men when Moore became aware that a white object which he took to be the balloon was in a part of the sky well away from where the theodolite operator had this instrument trained. As Moore has explained directly to me in discussing this famous case, he thought the operator had lost the balloon. Moore took over, swung the 25-power scope onto the "balloon" he had spotted, and found that it was in fact an ellipsoidal white object moving at a rapid angular velocity towards the NE. With stopwatch and recording forms at hand, it was possible for the team of five men to secure some semi-quantitative data on this sighting; Moore disengaged the vernier drives to track manually, and followed the object as it sped from the southwest into the northeast skies. At its closest approach, it was moving at about 5 degrees/sec. Just before Moore lost it in the distance to the

northeast, its angular elevation began to increase, as if it were climbing, a quite significant point. The object had a horizontal length about two to three times greater than its vertical thickness. Moore never got a sufficiently clear view to identify any finer details if any were present. Another balloon was immediately released to check the slim possibility that a high-speed jet from SW to NE might have carried some airborne object across the sky; but the winds were blowing more or less at right angles to the object's path to the 93,000 ft. level, and were rather weak (Ref. 10). The angular diameter of the object was estimated at about a minute of arc (which in the 25-power theodolite would appear to Moore as about three-fourths the apparent size of the moon).

Discussion.—Moore's sighting is carried as Unidentified in official files. Menzel (Ref. 24) says of it:

> "This incident, kept in the classified files for more than two years, presents no serious difficulty to the person who understands the optics of the earth's atmosphere. The air can, under special conditions, produce formations similar to lenses. And, just as a burning glass can project the sun into a point of light, so can these lenses of air—imperfect though they are—form an image. What Moore saw was an out-of-focus and badly astigmatic image of the balloon above."

It would be interesting to hear Menzel present a quantitative defense of that astonishing disposition of this interesting sighting. Here five witnesses, with aid of a tracking device giving better than rough angular-coordinate information on the movements of an unknown object, observe the object move through an arc of over 90 degrees that took it into a part of the sky about that same large angular distance from the real balloon's location, and Menzel adduces a "lens of air" to explain it away. Astronomers find atmospheric scintillation a very serious observational problem because stellar images are often erratically shifted by tens of seconds of arc from their mean position as a result of atmospheric turbulence effects.

In the 5/24/49 Moore sighting, Menzel is proposing that the atmosphere carried a refracted image of the balloon northeastward at a steady rate of excursion that finally totalled several thousand times the magnitude of refractive angular image-displacements known to occur with bad seeing. I feel obliged to repeat an observation I have made before: If the transmission properties of the Earth's atmosphere were as anomalous as Menzel assumes in his handling of UFO observations, he and his colleagues would be out of business. The official categorization of Unidentified for the

Moore sighting seems inescapable. It might be added that, over the years, there have been very many UFO observations of significant nature from the vicinity of White Sands Proving Ground, many involving instrumental tracking, many made by experienced observers. A long and impressive list of them could easily be compiled, yet all have been slowly lost from official cognizance by a process that is characteristically at the heart of response to the UFO problem.

5. Case 30. Admiralty Bay, Antarctica, March 16, 1961
This listing of UFO sightings by meteorologists could be extended very considerably by drawing on my file of such cases. To cite just one more that also indicates the global scale of the UFO phenomena, a very unusual luminous unidentified aerial object seen by a meteorologist and others aboard the U.S.S. *Glacier* at about 6:15 p.m. on 3/16/61 in the Antarctic will be mentioned.

I have quite recently received, through French UFO investigator René Fouéré, a rather detailed summary of this sighting by Brazilian meteorologist Rubens J. Villela, whose earlier account I had seen but paid little attention to (Ref. 10). The point I had missed, prior to reading Villela's detailed description of the circumstances of the sighting, was the very important feature of a low cloud overcast present at about 1500 ft. above the sea. With three shipmates on the flying bridge, Villela suddenly saw "a multicolored luminous object crossing the sky," an object which for a moment they took to be an unusual meteor.

> "It was egg-shaped, colored mainly reddish at first, and traveled slowly from NE to SW at about 50 degrees above the horizon, on a straight horizontal trajectory. From its frontal part, several multicolored, perfectly straight 'rays' extended backwards, diverging outwards at an angle; the colors of these rays changed continually, predominantly green, red, and blue. Most striking of all, it left a long trail of orange color in the form of a perfectly straight tube which gave the distinct impression of being hollow, faintly comparable to a neon light," Villela stated in his summary.

Then,

> "Suddenly the object divided in two. It was not an explosion, it was a controlled division in two equal parts, one behind the other, each egg-shaped as before and each radiating outwards its V-shaped lateral rays. Then the object shone with a slightly stronger light, changing color to blue and white, and disappeared completely. That's it—just disappeared, abruptly."

His account emphasizes that the boundaries of the object (s) were definite and sharp, not diffuse. Villela's account indicates that a total of six persons were above-decks and saw this striking phenomenon. It is to be emphasized that, in the estimated 10 seconds that this lasted, the object was moving below a cloud deck that lay only about 1500 feet above the sea, so that, for the reported elevation angle of about 50 degrees, the slant range from observers to object was perhaps of the order of 2000 ft. Villela had the subjective impression that the egg-shaped initial form was about as big as a small airplane.

Discussion.—In a recent book aimed at showing that a majority of the most interesting UFOs are an atmospheric-electrical plasma related to ball lightning, Philip J. Klass (Ref. 39) cites the preceding case as a good example of the sort of observation which he feels he can encompass in his "plasma-UFO" hypothesis. To the extent that he treats only the breakup into two parts, he has some observational basis for trying to interpret this as something akin to ball lightning. But almost at that point the similarity ends as far as meteorologically recognized characteristics of ball lightning go. The highly structured nature of the object and its rays, its size, its horizontal trajectory, its presence in a foggy area with low stratiform clouds free of thunderstorm activity scarcely suggest anything like ball lightning. Nor does this account suggest any meteoric phenomenon at sub-cloud altitudes.

I would regard this as just one more of a baffling array of inexplicable aerial phenomena which span so wide a range of characteristics that it is taxing to try to invent any single hypothesis to rationalize them all. *The full spectrum of UFO phenomena will, I predict, come as a shock to every scientist who takes the necessary time to look into the wealth of reports accumulated in various archives over the past two decades and more* [my emphasis, L.S.]. Official assertions to the effect that UFO reports in no way defy explanation in terms of present scientific and technological knowledge are, in my opinion, entirely unjustified. The Villela sighting seems a case in point. And meteorologists do see UFOs, as the foregoing cases should suggest.

DON'T WEATHER BALLOONS & RESEARCH BALLOONS ACCOUNT FOR MANY UFOS?

Probably the most categorical statement ever made attributing UFO observations to balloons appeared in a *Look* magazine article by Richard Wilson in February 1951, entitled, "A Nuclear Physicist Exposes Flying Saucers." Dr. Urner Liddel, then affiliated with the Navy cosmic ray research program using the large Skyhook balloons, was quoted as saying, "There is not a single reliable report of an observation (of a UFO) which is not attributable to the cosmic balloons." When one considers the large number of UFO reports already on record by 1951 in which reliable airlines pilots, military personnel, and other credible witnesses have observed unidentified objects wholly unlike a high-altitude, slowly drifting pear-shaped Skyhook, balloon, that assertion appears very curious. Nevertheless, that many persons have misidentified Skyhook balloons and even the smaller weather balloons used in routine meteorological practice is unquestioned. A Skyhook seen against the twilight sky with back-illumination yields a strangely luminous, hovering object which many observers, especially if equipped with binoculars, were unprepared to identify correctly in the 1946-51 period when Skyhook operations were tied up with still-classified programs. To this extent, Liddel's point is reasonable; but his sweeping assertion fails to fit the facts, then or now.

Actually, in official case-evaluations, one finds Skyhook balloons invoked relatively infrequently compared with "weather balloons." But in many of the latter cases, the balloon hypothesis is strained beyond the breaking point. The official criterion used (Ref. 7, p. 135) is extremely loose: "If an object is reported near a balloon launch site within an hour after the scheduled launch times, it is classed as a balloon" with no specification of heights, shapes, distances, etc. Using such a criterion, it is easy to see why so many "balloon" explanations figure in the official summaries. There are even "balloon" UFOs whose speed, when inferred from the report, comes out supersonic! The tiny candles or flashlight bulbs hung on pilot balloons for nighttracking have been repeatedly made the basis for explanations of what witnesses described as huge luminous objects at close range.

Within only days of this writing, I have checked out such a case near Tucson where four adult witnesses saw, on July 2, 1968, a half-moon-shaped orange-red object hovering for several minutes at what they estimated to be a few hundred feet above terrain and perhaps a few miles away over open desert. They watched it tip once, right itself, then accelerate and rise over a mountain range and pass off into the distance in some ten seconds. Because a weather balloon had been released earlier (actually about an hour and forty-five minutes earlier) from the Tucson airport Weather Bureau station, the official explanation, published in the local press, was that the witnesses had seen a "weather balloon." A pilot balloon of the small type (30-gram) used in this instance rises at about 600 ft./min., the tiny light on it becomes invisible to the naked eye beyond about 10,000 ft. slant-range, and the upper-level winds weren't even blowing toward the site in question. Also the angular size estimated for the observed reddish half-moon was about twice the lunar diameter, and some said about four times larger. A pilot balloon light would have to be within about 20-30 feet to appear this large. Yet such a case will enter the files (if even transmitted to higher echelons) as a "balloon," swelling the population of curious balloon-evaluations in official files.

1. Case 31. Ft. Monmouth, N.J., September 10, 1951

It is clear from Ruppelt's discussions (Ref. 5) that a series of radar and visual sightings near Ft. Monmouth on 9/10/51 and the next day were of critical importance in affecting official handling of the UFO problem in the ensuing two-year period. Many details from the official file on these sightings are now available for scientific scrutiny (Ref. 7). Here, a sighting by two military airmen flying in a [Lockheed] T-33 [training aircraft] near Ft. Monmouth will be selected from that series of events because the sighting was eventually tagged as a weather balloon. As with any really significant UFO case, it would require far more space than can be used here to spell out adequately all relevant details, so a very truncated account must be employed.

While flying at 20,000 ft. from a Delaware to a Long Island airbase, the two men in the T-33 spotted an object "round and silver in color" which at one stage of the attempted intercept appeared flat. The T-33 was put into a descending turn to try to close on the object but the latter turned tightly (the airmen stated) and passed rapidly eastward towards the coast of New Jersey and out to sea. A pair of weather balloons (probably radiosonde balloons but no information thereon given in the files) had been released from the Evans Signal Laboratory near Ft. Monmouth, and the official evaluation indicates that this is what the airmen saw.

However, it is stated that the balloons were released at 1112 EDST, and the sighting began at about 1135 EDST with the T-33 over Point Pleasant, N.J. In that elapsed time, a radiosonde balloon, inflated to rise at the 800-900 ft./min. rate used for such devices, would have attained an altitude of about 17-18,000 ft., the analysis notes. From this point on, the official analysis seems to be built on erroneous inferences. The airmen said that, as they tried to turn on the object, it appeared to execute a 120-degree turn over Freehold, N.J., before speeding out over the Atlantic. But from the upper winds for that day, it is clear that the Ft. Monmouth balloon trajectory would have taken it to the northeast, and by

1135, it would have been about over the coast in the vicinity of Sea Bright. Hence, at no time in the interval involved could the line of sight from T-33 to balloon have intersected Freehold, which lies about 15 miles WSW of the balloon release-point. Instead, had the airmen somehow seen the radiosonde balloon from Pt. Pleasant, it would have lain to about their N or NNE and would have stayed in about that sector until they passed it.

Furthermore, the size of the balloon poses a serious difficulty for the official analysis. Assuming that it had expanded to a diameter of about 15 feet as it ascended to about the 18,000-ft. level, it would have subtended an arc of only 0.6 min, as seen from the T-33 when the latter passed over Pt. Pleasant. This angular size is, for an unaided eye, much too small to fit the airmen's descriptions of what they tried to intercept. In a press interview (Ref. 40), the pilot, Wilbert S. Rogers of Columbia, Pa., said the object was "perfectly round and flat" and that the center of the disc was raised "about six feet" and that it appeared to be moving at an airspeed of the order of 900 mph. The entire reasoning on which the balloon evaluation is elaborated fails to fit readily established points in the official case-summary.

Discussion.—The possibility that a pilot can be misled by depth-perception errors and coordinate-reference errors to misconstrue a weather balloon as a fast-maneuvering object must always be kept in mind. But in the Ft. Monmouth instance, as in many others that could be discussed in detail, there is a very large gap between the balloon hypothesis and the facts. The basic sighting- report here is quite similar to many other daytime sightings by airborne observers who have seen unconventional disc-like objects pass near their aircraft.

2. Case 32. Odessa, Wash., December 10, 1952

According to an official case-summary (Ref. 7, Rept. 10), two airmen in an [Lockheed] F-94 [Starfire] "made visual and radar contact with a large, round white object larger than any known type of aircraft" near 1915 PST on 12/10/52 near Odessa. The radar operator in the F-94 had airborne radar contact with the object for 15 minutes, and during that same interval, ground radar was also tracking it. The summary states that "the object appeared to be level with the intercepting F-94 at 26,000 to 27,000 ft.," and it is pointed out that "a dim reddish-white light came from the object as it hovered, reversed direction almost instantaneously and then disappeared." It is stated that the skies were clear above 3000 ft. The official evaluation of this incident is "Possible Balloon," although the report notes that no upper-air research balloon was known to be in the area on this date. The principal basis for calling it a balloon was the observers' description of "large, round and white and extremely large," and it was remarked that the instrument package on some balloon flights is capable of yielding a radar return.

Discussion.—To conclude that this was a "Possible Balloon" just on the basis of the description, "large, round and white and extremely large," and thereby to ignore the instantaneous course reversal and the inability of a 600-mph jet to close with it over a period of 15 minutes seems unreasonable. We may ignore questions of wind speeds at the altitude of the object and the F-94 because both would enjoy the same "tail-wind effect." In 15 minutes, the F-94 would be capable of moving 150 miles relative to any balloon at its altitude. On the other hand, airborne radar sets of that period would scarcely detect a target of cross-section represented by the kinds of instrument packages hung on balloons of the Skyhook type, unless the aircraft were within something like 10 or 15 miles of it. Yet it is stated that the F-94 was pursuing it under radar contact for a time interval corresponding to an airpath ten times that distance. Clearly, categorizing this unknown as a "balloon" was incompatible with the reported details of the case.

On the other hand, there seems no reason to take seriously Menzel's evaluation of this Odessa F-94 sighting (Ref. 25, p. 62). Menzel evidently had the full file on this case, for he adds a few details beyond those in Ref. 7, details similar to those in Ruppelt's account of the case (Ref. 5):

> "Dim reddish-white lights seemed to be coming from 'windows,' and no trail or exhaust was visible. The pilot attempted to intercept but the object performed amazing feats—did a chandelle in front of the plane, rushed away, stopped, and then made straight for the aircraft on a collision course at incredible speed."

He indicates that after the pilot banked to avoid collision he could not again locate it visually, although another brief radar contact was obtained.

Having recounted those and other sighting details, Menzel then offers his interpretations:

> "In the east, Sirius was just rising over the horizon at the exact bearing of the unknown object. Atmospheric refraction would have produced exactly the phenomenon described. The same atmospheric conditions that caused the mirage of the star would have caused anomalous radar returns."

Now stars just above the viewer's horizon do scintillate and do undergo turbulent image-displacement, but one must consider quantitative matters. A refractive excursion of a stellar image through even a few *minutes* of arc would be an extremely large excursion. To suggest that a pilot would report that Sirius did a chandelle is both to forget realities of astronomy and to do injustice to the pilot. In fact, however, Menzel seems to have done his computations incorrectly, for it is easily ascertained that Sirius was not even in the Washington skies at 7:15 p.m. PST on 12/10/52. It lay at about 10 degrees below the eastern horizon.

A further quite unreasonable element of Menzel's explanation of the Odessa case is his easy assertion that the radar returns were anomalous results of the "atmospheric conditions." Aircraft flying at altitudes of 26,000 ft. do not get ground returns on level flight as a result of propagation anomalies. These extreme forcings of explanations recur throughout Menzel's writings; one of their common denominators is lack of attention to relevant quantitative factors.

3. Case 33. Rosalia, Wash., February 6, 1953

Another official case-summary of interest here is cited by Menzel (Ref. 25, p. 46). Keyhoe (Ref. 4), who studied the case-file on it much earlier, gives similar information, though in less detail. A [Convair] B-36 [Peacemaker], bound for Spokane was over Rosalia, Wash., at 1:13 a.m. when, as Menzel describes it, "the pilot... sighted a round white light below him, circling and rising at a speed estimated at 150 to 200 knots as it proceeded on a southeast course." Menzel states that the B-36 "made a sharp descending turn toward the light, which was in view for a period of three to five minutes." The light was blinking, and Keyhoe mentions that the blink-interval was estimated at about 2 seconds. Menzel concurs in the official evaluation of this as a "weather balloon," noting that a pilot balloon had been released at Fairchild AFB at 1:00 a.m., and remarking that the "winds aloft at altitudes of 7,000 to 10,000 ft. were from the northwest with a speed of about fifty knots." He says that "computations showed that the existing winds would have carried the balloon to the southeast, and it would have been over Rosalia, which is 12.5 nautical miles southeast of Fairchild, in about fifteen minutes." In fact, Rosalia lies 33 statute miles SSE of Fairchild, or about twice as far as Menzel indicates. The net drift of the balloon cannot be deduced simply from the winds in the 7-10,000-ft. layer; and, in fact, an examination of the upper-wind data for that area on February 6 indicates that the winds at lower levels were blowing out of the southwest. The trajectory of the balloon would have taken it initially east-northeast, then east, and finally curving back to the southeast as it got up to near the 10,000-ft. levels. By that time, it would have been already east of Spokane, nowhere near Rosalia. The small light (candle or flashlight bulb) used on night pibal runs is almost invisible to the naked eye beyond a few miles' distance. (A 1-candle source at 3,000 ft. is equivalent to a star of about the first magnitude. At 6 miles, then, one finds that the same source equals the luminosity of a sixth-magnitude star, the limit of human vision under the most favorable conditions. For a pilot, looking out of a cockpit with slight inside glare to spot a 1-candle source against a dark background would require that the source be only a few miles away.) At some 30 miles, the B-36 pilot could not have seen the small light on a balloon east of Spokane.

Menzel states that "the balloon carried white running lights which accounted for the blinking described, and the circling climb of the UFO is typical of a balloon's course." Neither inference is supportable. The light used on pilot balloons is a steady source: only if one were right above it, with its random swing causing intermittent occultation, would one ever perceive blinking. But then, flying at B-36 speeds, the pilot would have swept over the sector of perceptible occultation in only a matter of seconds. Yet here the pilot watched it for a reported 3-5 minutes. Furthermore, "circling climb" cannot be called "typical of a balloon's course." The balloon trajectory is controlled by the ambient wind shears and only with unusually strong directional shears would a pilot flying a straight course perceive a pilot balloon to be "circling."

In all, there appear to be so many serious difficulties with the balloon explanation for the Rosalia sighting that it is not possible to accept Menzel's statement: "Thus all the evidence supports ATIC's conclusion that the UFO was a weather balloon."

4. Case 34. Boston, Mass., June 1, 1954

At 0930 EDST, a Paris-New York TWA Constellation was passing near Boston when the cockpit crew spotted "a large, white-colored disc-like object" overhead (Ref. 41). Capt. Charles J. Kratovil, copilot W. R. Davis, and flight engineer Harold Raney all watched it for a total time of 10 minutes as they flew on their own southwestward course to New York. They would occasionally lose it behind overlying clouds. Knowing that they were flying into headwinds, they concluded that it could not be any kind of balloon, so they radioed the Boston airport control tower, which informed him that jets were scrambled and saw the object, but could not close with it.

After landing in New York, Capt. Kratovil was informed that official spokesmen had attributed the sighting to a "weather balloon" released from Grenier AFB, in New Hampshire.

Discussion.—I am still in the process of trying to locate Kratovil to confirm sighting details; but the fact that four newspaper accounts for that day give the same information about the major points probably justifies acceptance of those points. From upper-wind data for that area and time, I have confirmed the presence of fairly strong flow from the WSW aloft, whence Kratovil's press comment, "If this was a weather balloon, it's the first time I ever saw one traveling against the wind," seems reasonable. The cruising speed of a Constellation is around 300 mph, so during the reported 10 minutes' duration of the crew's sighting, they moved about 50 miles relative to the air, so it would have been impossible for them to have kept a weather balloon in sight for this long. Furthermore, it was about 1.5 hours after scheduled balloon-release time, so that even a small balloon would have either burst or passed to altitudes too high to be visible. Finally, with flow out of the southwest sector from surface to above 20,000 ft., any balloon from Grenier AFB would have been carried along a trajectory nowhere near where the TWA crew spotted the "large, white-colored, disc-like object" overhead.

5. In my files are many other "balloon" cases from the past twenty years, cases that ought never have been so labelled, had the evaluators kept relevant quantitative points in mind. To ignore most of the salient features of a sighting in order to advance an easy "balloon" explanation is only one more of many different ways in which some very puzzling UFO observations have been shoved out of sight.

WHY AREN'T UFO'S EVER TRACKED BY RADAR?

The skeptic who asks this question, and many do, is asking a very reasonable question. With so much radar equipment deployed all over the world, and especially within the United States, it seems sensible to expect that, if there are any airborne devices maneuvering in our airspace, they ought to show up on radars once in a while. They do indeed, and have been doing so for all of the two decades that radar has been in widespread use. Here, as with so many other general misconceptions about the true state of the UFO problem, we encounter disturbingly large amounts of misinformation. As with other categories of UFO misinformation, the only adequate corrective is detailed discussion of large numbers of individual cases. Only space limitations preclude discussion of dozens of striking radar-tracking incidents involving UFOs, both here and abroad; they do exist.

1. *Case 35. Fukuoka, Japan, October 15, 1948*

A very early radar-UFO case, still held as an official Unidentified, involved an attempted interception of the unknown object by an [Northrop] F-61 [Black Widow] flying near Fukuoka, Japan, at about 11:00 p.m. local time on 10/15/48. The official file on this incident is lengthy (Ref. 42); only the highlights can be recounted here. The F-61 (with pilot and radar operator) made six attempts to close with the unknown, from which a radar return was repeatedly obtained with the airborne radar. Each time the radarman would get a contact and the F-61 pilot tried to close, the unknown would accelerate and pass out of range. Although the radar return seemed comparable to that of a conventional aircraft, "the radar observer estimated that on three of the sightings, the object traveled seven miles in approximately twenty seconds, giving a speed of approximately 1200 mph."

In another passage, the official case-file remarks that "when the F-61 approached within 12,000 feet, the target executed a 180° turn and dived under the F-61," adding that "the F-61 attempted to dive with the target but was unable to keep pace." The report mentions that the unknown "could go almost straight up or down out of radar elevation limits," and asserts further that "this aircraft seemed to be cognizant of the whereabouts of the F-61 at all times..."

The F-61 airmen, 1st Lt. Oliver Hemphill (pilot) and 2nd Lt. Barton Halter (radarman) are described in the report as being "of excellent character and intelligence and are trained observers." Hemphill, drawing on his combat experience in the European theater, said that "the only aircraft I can compare our targets to is the German [Messerschmitt] ME-163 [Komet]." The airmen felt obliged to consider the possibility that their six attempted intercepts involved more than one unknown. Hemphill mentions that, in the first attempted intercept, "the target put on a tremendous burst of speed and dived so fast that we were unable to stay with it." After this head-on intercept, Hemphill did a chandelle back to his original 6000-ft. altitude and tried a stern interception, "but the aircraft immediately outdistanced us. The third target was spotted visually by myself," Hemphill's signed statement in the case-file continues.

> "I had an excellent silhouette of the target thrown against a very reflective under-cast by a full moon. I realized at this time that it did not look like any type aircraft I was familiar with, so I immediately contacted my Ground Control Station..."

which informed him there were no other known aircraft in the area. Hemphill's statement adds further that,

> "The fourth target passed directly over my ship from stern to bow at a speed of roughly twice that of my aircraft, 200 mph. I caught just a fleeting glance of the aircraft; just enough to know he had passed on. The fifth and sixth targets were attempted radar interceptions, but their high rate of speed put them immediately out of our range."

(Note the non-committal terminology that treats each intercept target as if it might have been a separate object.)

A sketch of what the object looked like when seen in silhouette against the moonlit cloud deck is contained in the file. It was estimated to be about the size of a fighter aircraft, but had neither discernible wings nor tail structures. It was somewhat bullet-shaped, tapered towards the rear, but with a square-cut aft end. It seemed to have "a dark or dull finish."

Discussion.—Ground radar stations never detected the unknown that was seen visually and contacted by airborne radar. The report indicates that this may have been due to effects of "ground clutter," though the F-61 was seen intermittently on the ground units. The airmen stated that no exhaust flames or trail were seen from this object with its "stubby, clean lines." The total duration of the six attempted intercepts is given as 10 minutes. We deal here with one of many cases wherein radar detection of an unconventional object was supported by visual observation. That this is carried as Unidentified cannot surprise one; what is surprising is that so many other comparable instances are on record, yet have been ignored as indicators of some scientifically intriguing problem demanding intensive study.

2. Case 36. Nowra, Australia, September, 1954

The first UFO case to command general press attention in the Australian area seems to have been a combined radar-visual sighting wherein the pilot of a Hawker Seafury from Nowra Naval Air Station visually observed two unknown objects near him as he flew from Canberra to Nowra (Ref. 43). Press descriptions revealed only that the pilot said "the two strange aircraft resembling flying saucers" were capable of speeds much beyond his Seafury fighter. He saw them flying nearby and contacted Nowra radar to ask if they had him on their scope; they informed him that they had three separate returns, at which juncture he described the unidentified objects.

Under instructions from the Nowra radar operator, he executed certain maneuvers to identify himself on the scope. This confirmed the scope-identity of his aircraft vs. the unknowns. As he executed the test maneuvers, the two unknowns moved away and disappeared. No explanation of this incident was offered by Naval authorities after it was widely reported in Australian and New Zealand papers about three months after it occurred.

Discussion.—It is mildly amusing that the press accounts indicated that "the pilot, fearing that he might be ragged in the wardroom on his return if he abruptly reported flying saucers, called Nowra by radio and asked whether the radar screen showed his aircraft." Only after getting word of three, not one, radar blips in his locality did he radio the information on the unknowns, whose configuration was not publicly released. This is in good accord with my own direct experience in interviewing Australian UFO witnesses in 1967; they are no more willing than Americans to be ridiculed for seeing something that is not supposed to exist.

3. Case 37. Capetown, South Africa, May 23, 1953

In November 1953, the South African Air Force released a brief announcement concerning radar-tracking of six successive passes of one or more unknown high-speed objects over the Cape. On January 1, 1967, in a transoceanic shortwave broadcast from South Africa, the authenticity of this report was confirmed, though no additional data beyond what had been cited earlier were presented. In the six passes, the target's altitude varied between 5,000 and 15,000 ft., and its closest approach varied between 7 and 10 miles. Speeds were estimated at over 1200 mph, well beyond those of any aircraft operating in that area at that time.

Discussion.—This report, on which the available information is slim, is cited to indicate that not only visual sightings but also radar sightings of seemingly unconventional objects appear to comprise a global phenomenon. By and large, foreign radar sightings are not readily accessible, and not easily cross-checked. Zigel (Ref. 38) briefly mentions a Russian incident in which both airborne and ground-based radar tracked an unidentified in the vicinity of Odessa, on April 4, 1966, the ground-based height-finding radar indicating altitudes of well over 100,000 ft. Such reports, without accessory information, are not readily evaluated, of course.

4. *Case 38. Washington, D.C., July 19, 1952*

By far the most famous single radar-visual sighting on record is the one which occurred late in the evening of July 19, and early on July 20, 1952, in the vicinity of Washington, D.C. (Refs. 2, 4, 5, 10, 24, 25). A curiously similar incident occurred just one week later. The official explanation centered around atmospheric effects on radar and light-propagation.

Just before midnight on July 19/20, CAA radar showed a number of unidentified targets which varied in speed (up to about 800 mph) in a manner inconsistent with conventional aircraft. A number of experienced CAA radarmen observed these returns, and, at one juncture, compatible returns were being received not only at the ARTC radar but also on the ARS radar in a separate location at Washington National Airport, and on still a third radar at Andrews AFB. Concurrently, both ground and airborne observers saw unidentifiable lights in locations matching those of the blips on the ground radar.

Discussion.—I have interviewed five of the CAA personnel involved in this case and four of the commercial airline pilots involved, I have checked the radiosonde data against well-known radar propagation relations, and I have studied the CAA report subsequently published on this event. Only an extremely lengthy discussion would suffice to present the serious objections to the official explanation that this complex sighting was a result of anomalous radar propagation and refractive anomalies of the mirage type. The refractive index gradient, even after making allowance for instrument lag, was far too low for "ducting" or "trapping" to occur; and, still more significant, the angular elevations of the visually observed unknowns lay far too high for radar-ducting under even the most extreme conditions that have ever been observed in the atmosphere.

Some of the pilots, directed by ground radar to look for any airborne objects, saw them at altitudes well above their own flight altitudes, and these objects were maneuvering in wholly unconventional manner. One crew saw one of the unknown luminous objects shoot straight up, and simultaneously the object's return disappeared from the ARTC scope being watched by the CAA radar operators. The official suggestion that the same weak (1.7°C) low-level "inversion" that was blamed for the radar ducting could produce miraging effects was quantitatively absurd, even if one overlooks the airline-pilot sightings and deals only with the reported ground-visual sightings.

From the CAA radar operators I interviewed, as well as from the pilots I talked to about this case, I got the impression that the propagation-anomaly hypothesis struck them as quite out of the question, then and now. In fact, CAA senior controller Harry G. Barnes, who told me that the scope returns from the unknowns

> "were not diffuse, shapeless blobs such as one gets from ground returns under anomalous propagation," but were strong, bright pips, [which is why]… "anomalous propagation never entered our heads as an explanation."

Howard S. Conklin, who, like Barnes, is still with FAA, was in the control tower that night, operating an entirely independent radar (short-range ARS radar). He told me that what impressed him about the sighting that night was that they were in radio communication with airlines crewmen who saw unidentified lights in the air in the same area as unknowns were showing up on his tower radar, while simultaneously he and Joseph Zacko were viewing the lights themselves from the tower at the D.C. Airport. James M. Ritchey, who was at the ARTC radar with Barnes and others, confirmed the important point that simultaneous radar fixes and pilot-sightings occurred several times that night. He shared Barnes' views that the experienced radar controllers on duty that night were not being fooled by ground returns in that July 19 incident.

Among the airlines crewmen with whom I spoke about this event was S. C. Pierman, then flying for Capitol Airlines. He was one of the pilots directed by ground radar to search in a specific area for air-borne objects. He observed high speed lights moving above his aircraft in directions and locations matching what the CAA radar personnel were describing to him by radio, as seen on their radars. Other airline personnel have given me similar corroborating statements. I am afraid it is difficult to accept the official explanations for the famous Washington National Airport sightings.

5. *Case 39, Port Huron, Mich., July 29, 1952*

Many of the radar cases for which sighting details are accessible date back to 1953 and preceding years. After 1953, official policies were changed, and it is not easy to secure good information on subsequent cases in most instances. A radar case in which both ground-radar and airborne-radar contact were involved occurred at about 9:40 p.m. CST on 7/29/52 (Refs. 4, 5, 7, 10, 25).

From the official case summary (Ref. 7) one finds that the unknown was first detected by GCI radar at an Aircraft Control and Warning station in Michigan, and one of three F-94s [Lockheed Starfires] doing intercept exercises nearby was vectored over towards it. It was initially coming in out of the north (Ref. 5, 25), at a speed put at over 600 mph. As the F-94 was observed on the GCI scope to approach the unknown, the latter suddenly executed a 180° turn, and headed back north. The F-94 was by then up to 21,000 ft., and the pilot spotted a brilliant multicolored light just as his radarman got a contact. The F-94 followed on a pursuit course for 20 minutes (Ref. 7) but could never close with the unknown as its continued on its northbound course.

At the time of first radar lockon, the F-94 was 20 miles west of Port Huron, Mich. The GCI scope revealed the unknown to be changing speed erratically, and at one stage it was moving at a speed of over 14,000 mph, according to Menzel (Ref. 25), who evidently drew his information from the official files. Ruppelt (Ref. 5) states that when the jet began to run low on fuel and turned back to its base, GCI observed the unknown blip slow down, and shortly after it was lost from the GCI scope.

Discussion.—This case is still carried as an official unknown. The case summary (Ref. 7) speculates briefly on whether it could have been "a series of coincident weather phenomena affecting the radar equipment and sightings of [the bright multi-star system] Capella, but this is stretching probabilities too far." Menzel, however, asserts that the pilot did see Capella, and that the airborne and ground radar returns "were merely phantom returns caused by weather conditions." No suggestion is offered as to how any given meteorological condition could jointly throw off radar at the ground and radar at 21,000 feet, no suggestion is offered to account for 180° course-reversal exhibited by the blip on the GCI scope just as the F-94 came near the unknown, no suggestion of how propagation anomalies could yield the impression of a blip moving systematically northward for 20 minutes (a distance of almost 100 miles, judging from reported F-94 speeds), with the F-94 return following along behind it. With such *ad hoc* explanations, one could explain away almost any kind of sighting, regardless of its content.

I have examined the radiosonde sounding for stations near the site and time of this incident, and see nothing in them that would support Menzel's interpretations. I have queried experienced military pilots and radar personnel, and none have heard of anything like "ground returns" from atmospheric conditions with aircraft radar operated in the middle troposphere. If Menzel is not considering ground- returns, in the several cases of this type which he explains away with a few remarks about "phantom radar returns," then it is not clear what else he might be thinking of.

One does have to have some solid target to get a radar return resembling that of an aircraft. Refractive anomalies of the "angel" type have very low radar cross-section and would not mislead experienced operators into confusing them with aircraft echoes.

6. Many other cases might be cited where UFOs have appeared on radar under conditions where no acceptable conventional explanation exists. Ref. 7 has a number of them. Hall (Ref. 10) has about 60 instances in which both radar and visual sightings were involved. A December 19, 1964 case at Patuxent River NAS is one that I have checked on. It involved three successive passes of an unknown moving at speeds estimated at about 7000 mph. It is an interesting case, one that came to light for somewhat curious reasons. A low overcast precluded any visual sightings from control tower personnel, so this is not a radar-visual case. I found no conventional explanation to account for it.

It has to be stressed that there are many ways in which false returns can be seen on radarscopes, resulting not only from ducting of ground returns but also from interference from other nearby radars, from internal electronic signals within the radar set, from angels and insects (weak returns), etc. Hence each case has to be examined independently.

After studying a number of official evaluations of radar UFO cases, I get the impression that there would probably be more radar Unknowns if there were less tendency to quickly explain them away by qualitative arguments that overlook pertinent quantitative matters. Even at that, there are too many conceded unknowns in official files to be ignored.

A famous case in UFO annals involved a [Boeing] B-29 [Superfortress] over the Gulf of Mexico, where several unknowns were tracked on the airborne scopes and were seen simultaneously by crewmen, moving under the aircraft as they passed by (Refs. 4, 10,25). This one is still carried as Unidentified in official files. Still another famous combined radar-visual case, which Hynek has termed "one of the most puzzling cases I have studied," occurred between Rapid City [South Dakota] and Bismarck [North Dakota] on August 5,1953. It involved both ground and airborne radar and ground and airborne visual sightings, but is far too long and complex to recapitulate here. Perhaps the above suffices to indicate that UFOs are at times seen on radar and have been so seen for many years. The question of why we don't hear a great deal about such sightings, especially with newer and more elaborate surveillance radars, is a reasonable question. Some of the answers to that one are posed by the statement of Dr. Robert M. L. Baker, Jr., in these proceedings. Other parts of the answer must be omitted here.

WHY AREN'T THERE NUMEROUS PHOTOS OF UFO'S IF THEY REALLY EXIST?

Here is a question for which I regard available answers as still unsatisfactory. I concede that it does seem reasonable to expect that there should, over the past 20 years, be substantially more good photos than are known to exist. Although I do not regard that puzzle as satisfactorily answered, neither do I think that it can be safely concluded that the paucity of good photos disproves the reality of the UFOs. Many imponderables enter into consideration of this question.

1. Some general considerations

If one had reliable statistics on the fraction of the population that carried loaded cameras with them at any randomly selected moment (I would guess it would be only of the order of one per cent) and had figures bearing on the probability that a UFO witness would think of taking a photo before his observation terminated, then these might be combined with available information on numbers of sightings to attempt crude estimates of the expected number of UFO photos that should have accumulated in 20 years. Then one would need to weight the data for likelihood that any given photo would find its way to someone who would make it known in scientific circles, and then this figure might be compared with the very small number of photos that appear to stand the test of the exceedingly close scrutiny photos demand.

A general rule among serious UFO investigators with whom I have been in touch is that the UFO photo is no better than the photographer (Hall). Many hoax photos have been brought forth. A UFO photo can be sold; this attracts hoax and fraud to an extent not matched in anecdotal accounts. Many photos have been clearly established as fraudulent in nature; far larger numbers seem so suspicious on circumstantial grounds that no serious investigator gives them more than casual attention.

An interesting, even if very crude check on the likelihood of securing photos of UFOs from the general populace is afforded by fireball events. On April 25, 1966, a fireball rated at about magnitude -10, arced northward across the northeastern U.S. From the total geographic area over which this fireball was visually detected, the population count is about 40 million persons. According to one account (Ref. 43), 200 visual accounts were turned in, and I infer that only 6 photos were submitted. The fireball was visible for a relatively long time as meteors go, about 30 seconds, and was, of course, at a great altitude (25 to 110 km).

That 6 photos were submitted (at time of publication of the cited article) from a potential population of sighters of 40 million might seem to argue that perhaps we really cannot expect to get many photos of UFOs. However, one of the principal reasons for citing the foregoing is to bring out the difficulties in drawing any firm conclusions. A phenomenon lasting 30 seconds scarcely permits the observer time to collect his wits and to swing into photographic action if he does have a loaded camera. UFO sightings have often extended over much longer than 30 seconds, by contrast, affording far better opportunity to think of snapping a photo. But, on the other hand, sighting a UFO in daytime at close range, judging from my own witness-interviewing experience, is a far more disconcerting and astonishing matter than viewing a brilliant meteor. Thus one can go back and forth, with so little assurance of meaningfulness of any of the relevant weight factors that the end result is not satisfactory. I simply do not know what to think about the paucity of good UFO photos, though I do feel uncomfortable about it.

2. Case 40. Corning, Calif., July 4, 1967

A case that may shed at least a bit of light on the paucity of photos involves a multiple-witness sighting near dawn at Corning, Calif., on 7/4/67. I have interviewed four witnesses who sighted the object from two separate locations involving lines of sight at roughly right angles, serving to confirm the location of the object as almost directly over Highway 5 just west of Corning. Jay Munger, proprietor of an all-night bowling alley, was having coffee with two police officers, Frank Rakes and James Overton, when he spotted the object through the front window of his place. All three rushed out to the parking lot to observe what they described as a large flattened sphere or possibly football-shaped object, with a brilliant light shining upward from the top and a dimmer light shining down from the underside. The dawn light was such that the object was visible by reflected light even though the object's beams were discernible. It appeared at first to be hovering almost motionless at a few hundred feet above ground, and all three felt it lay about over Hwy. 5 (which estimate proved correct from sightings made on the highway by the independent witnesses). Their estimates of size varied from a diameter of maybe 50 feet to about 100 ft. It was silent, and the three men all emphasized to me that the quiet morning would have permitted hearing any kind of conventional aircraft engines. All three said they had never before seen anything like it. Munger decided to phone his wife to have her see the thing, and by the time he came back out from phoning, the object had moved southward along the highway by about a quarter of a mile or so. At about that juncture, it began to accelerate, and moved off almost horizontally, passing out of sight to the south in an additional time estimated at about 10-20 seconds.

This case is relevant to the photo question since Officer Overton was on duty and had in his patrol car both binoculars and a loaded camera. When I asked him why he didn't try to get a picture of the object, he admitted that he was so astonished by the object that he never even thought of dashing for the camera. I asked Munger to go through the motions that would yield a time estimate of the period he was inside phoning, to get a rough notion of how long Overton, along with Rakes, looked at it without thinking of the camera. The time was thus estimated by Munger as about a minute and a half, possibly two minutes.

Discussion.—It may be hazardous to try to draw any conclusions from such a case, but I do think it suggests the uncertainty we face in trying to assess the likelihood of any given witness getting a photo of a UFO he happens to see. A colleague of mine at the University of Arizona was out photographing desert flowers on a day when a most unusual meteorological event occurred nearby—a tornado funnel came down from a cloud. Despite having the loaded camera at hand, despite having just been taking other pictures, and despite the great rarity of Arizona tornadoes, that colleague conceded that it wasn't till much later that the thought of getting a photo rose to consciousness, by which time the funnel was long since dissipated.

In the Trinidad, Colo., case of March 23, 1966 (Case 14 above), Mrs. Frank R. Hoch pointed out to me that she had loaded still and movie cameras inside the house, yet never thought about getting a photo. Again, the reason cited was the fascination with the objects being viewed. I think this "factor of astonishment" would have to be allowed for in any attempt to estimate expected numbers of photos, but I would be quite unsure of just how to evaluate the factor quantitatively.

3. Case 41. Edwards AFB, May 3, 1957

Occasionally, one could argue, UFOs ought to come into areas where there were persons engaged in photographic work, who were trained to react a bit faster, and who would secure some photos. One such instance evidently occurred at Edwards AFB on the morning of 5/3/57. I have managed to locate and interview three persons who saw the resultant photos. The two who observed the UFO and obtained a number of photos of it were James D. Bittick and John R. Gettys, Jr., both of whom I have interviewed. They were at the time Askania [a type of motion camera or cinetheodolite] cameramen on the test range, and spotted the domed-disc UFO just as they reached Askania #4 site at Edwards, a bit before 8:00 a.m. that day. They immediately got into communication with the range director, Frank E. Baker, whom I have also interviewed, and they asked if anyone else was manning an Askania that could be used to get triangulation shots. Since no other camera operators were on duty at other sites, Baker told them to fire manually, and they got a number of shots before the object moved off into the distance.

Bittick estimated that the object lay about a mile away when they got the first shot, though when first seen he put it at no more than 500 yards off. He and Gettys both said it had a golden color, looked somewhat like an inverted plate with a dome on top, and had square holes or panels around the dome. Gettys thought that the holes were circular not square. It was moving away from them, seemed to glow with its own luminosity, and had a hazy, indistinct halo around its rim, both mentioned. The number of shots taken is uncertain; Gettys thought perhaps 30.

The object was lost from sight by the time it moved out to about five miles or so, and they did not see it again. They drove into the base and processed the film immediately. All three of the men I interviewed emphasized that the shots taken at the closer range were very sharp, except for the hazy rim. They said the dome and the markings or openings showed in the photos. The photos were shortly taken by Base military authorities and were never seen again by the men. In a session later that day, Bittick and Carson were informed that they had seen a weather balloon distorted by the desert atmospheric effects, an interpretation that neither of them accepted since, as they stated to me, they saw weather balloons being released frequently there and knew what balloons looked like. Accounts got into local newspapers, as well as on wire services (Ref.44). An Edwards spokesman was quoted in the *Los Angeles Times* as saying, "This desert air does crazy things." An INS wire-story said, "intelligence officers at Edwards... would say almost nothing of the incident."

Discussion.—I have not seen the photos alleged to have been taken in this incident, I have only interviewed the two who say they took them and a third person who states that he inspected the prints in company with the two Askania operators and darkroom personnel. I sent all of the relevant information on this case to the University of Colorado UFO project, but no checks were made as a result of that, unless done very recently. It would be rather interesting to see the prints.

4. Photographic sky-survey cameras might be expected to get photos of UFOs from time to time. However, one finds that, in many sky-photography programs in astronomy, tracks that do not obviously conform to what is being sought, say meteor-tracks, are typically ignored as probable aircraft. Indeed, a very general pattern in all kinds of monitoring programs operates to bias the system against seeing anything but what it was built to see. Nunn-Baker satellite cameras are only operated when specific satellites are computed to be due overhead, and then the long axis of the field is aligned with the computed trajectory. Anything that crosses the field and leaves a record on the film with an orientation markedly different from the predicted trajectory is typically disregarded. Photographic, radar, and visual observing programs have a large degree of selectivity intentionally built into them in order not to be deluged with unwanted "signals." Hence one must be rather careful in suggesting that our many tracking systems surely ought to detect UFOs. There's much evidence to suggest that, if they did, the signal would be ignored as part of a systematic rejection of unwanted data. Even in the practices of the GOC (Ground Observer Corps), some units received instructions to report nothing but unidentified aircraft. (But, for examples of some UFOs that did get into the GOC net, see Hall, Ref. 10.)

Although I am aware of a few photos allegedly showing UFOs, for which I have no reason at present to doubt the authenticity (for example a series of snapshots taken by a brother and sister near Melbourne, Australia, showing a somewhat indistinct disc in various positions), I must emphasize that the total sample is tiny. Compared with that, I have seen dozens of alleged UFO photos which I regard as of dubious origin. Other UFO photos of which I am aware are still in process of being checked in one way or another. To summarize, I do have the impression that we ought to have more valid UFO photos than the small number of which I am aware.

IF UFO'S ARE REAL, SHOULDN'T THEY PRODUCE SOME REAL PHYSICAL EFFECTS?

Again, the answer is that they do. There are rather well-authenticated cases spanning a wide variety of "physical effects." Car-stopping cases are one important class. UFOs have repeatedly been associated with ignition failures and light-failures of cars and trucks which came near UFOS or near which the UFOs moved. I would estimate that one could assemble a list of four or five dozen such instances from various parts of the world. Interference with radios and TV receptions have been reported many times in connection with UFO sightings. There are instances where UFOs have been reported as landing, and after departure, holes in the ground, or depressions in sod, or disturbed vegetation patterns have been described. In many such instances, the evident reliability of the witnesses is high, the likelihood of hoax or artifice small. A limited number of instances of residues left behind are on record, but these are not backed up by meaningful laboratory analyses, unfortunately.

A physical effect that does not typically occur under conditions where the description of events might seem to call for it, relates to sonic booms. Although there are on record a few cases where fast-moving UFOs were accompanied by explosive sounds that might be associated with sonic booms, there are far more instances in which the reported velocity corresponded to supersonic speeds, yet no booms were reported. A small fraction of these can be rationalized by noting that the reporting witnesses were located back within the "Mach cone" of the departing UFO; but this will not suffice to explain away the difficulty. One feels that if UFOs are solid objects, capable of leaving depressions in soil or railroad ties when they land, and if they can dash out of sight in a few seconds (as has been repeatedly asserted by credible witnesses), they should produce sonic booms. This remains inexplicable; one can only lamely speculate that perhaps there are ways of eliminating sonic booms that we have not yet discovered; perhaps the answer involves some entirely different consideration.

If we include among "physical effects" those that border on the physiological, then there appear to be many odd types. Repeatedly, tingling and numbness have been described by witnesses who were close to UFOs; in many instances outright paralysis of a UFO witness has occurred. These effects might, of course, be purely psychological, engendered by fear; but in some instances the witnesses seem to have noted these effects as the first indication that anything unusual was occurring. A number of instances of skin-reddening, skin-warming, and a few instances of burns of very unusual nature are on record. These physiological effects are sufficiently diverse that caution is required in attempting generalization. Curiously, a peculiar tingling and paralysis seem to be reported more widely than any other physiological effects.

A person who is almost unaware of the ramifications of the UFO evidence may think it absurd to assert that people have been paralyzed in proximity to UFOs; the skeptic might find it inconceivable that such cases would go unnoticed in press and medical literature. Far from it, I regret to have to say, on the basis of my own investigations. I have encountered cases where severe bodily damage was done, or where evident hazard of damage was involved, yet the witness and his family found ridicule mounting so much faster than sympathy that it was regarded wiser to quietly forget the whole thing. At an early stage of my investigations I would have regarded that as quite unbelievable; UFO investigators with longer experience than mine will smile at that statement, but probably they will smile with a degree of understanding. I could cite specific illustrations to make all this much clearer, but will omit them for space-limitations, except for a few remarks in the next section.

IS THERE ANY EVIDENCE OF HAZARD OR HOSTILITY IN THE UFO PHENOMENA?

Official statements have emphasized, for the past two decades, that there is no evidence of hostility in the UFO phenomena. To a large degree, this same conclusion seems indicated in the body of evidence gathered by independent investigators. The related question as to potential hazard is perhaps less clear. There are on record a number of cases (I would say something like a few dozen cases) wherein persons whose reliability does not seem to come into serious question have reported mild, or in a very few instances, substantial injury as the result of some action of an unidentified object. However, I know of only two cases for which I have done adequate personal investigation, in which I would feel obliged to describe the actions as "hostile." That number is so tiny compared with the total number of good UFO reports of which I have knowledge that I would not cite "hostility" as a general characteristic of UFO phenomena. One may accidentally kick an anthill, killing many ants and destroying the ants' entrance, without any prior "hostility" towards the ants. To walk accidentally into a whirling airplane propeller is fatal, yet the aircraft held no "hostility" to the unfortunate victim.

In the UFO phenomena, we seem to confront a very large range of unexplained, unconventional phenomena and if among them we discern occasional instances of hazard, it would be premature to adjudge hostility. Yet, as long as we remain so abysmally ignorant of the over-all nature of the UFO problem, it seems prudent to make all such judgments tentative. If UFOs are of extraterrestrial origin, we shall need to know far more than we now know before sound conclusions can be reached as to hazard-and-hostility matters. For this reason alone, I believe it to be urgently important to accelerate serious studies of UFOS.

In the remainder of this section, I shall briefly cite a number of types of cases that bear on questions of hazard:

1. Car-stopping cases

In a two-hour period near midnight, November 2-3, 1957, nine different vehicles all exhibited ignition failures, and many suffered headlight failures as objects described as about 100-200 ft. long, glowing with a general reddish or bluish glow, were encountered on roads in the vicinity of the small community of Levelland, Tex. (Ref. 10, 13, 14). This series of incidents became national headline news until officially explained in terms of ball lightning and wet ignitions. However, on checking weather data, I found that there were no thunderstorms anywhere close to Levelland that night, and there was no rain capable of wetting ignitions. Although I have not located any of the drivers involved, I have interviewed Sheriff Weir Clem of Levelland and a Levelland newspaperman, both of whom investigated the incidents that night. They confirmed the complete absence of rain or lightning activity. The incidents cannot be regarded as explained.

This class of UFO effect is by no means rare. In France in the 1954 wave of UFO sightings, Michel (14) has described many such cases involving ignition-failure in motorbikes, cars, etc. Similar instances were encountered in my checks on Australian UFO cases. There are probably of the order of a hundred cases on record (see Ref. 10 for a list of some dozens). In only a very few cases has there been any permanent damage to the vehicle's electrical system. In the Levelland case, for example, as soon as the luminous object receded from a given disturbed vehicle, its lights came back on automatically (in instances where the switches had been left on), and the engines were immediately restartable. The latter point in itself makes the "wet ignition" explanation unreasonable, of course.

It is unclear how such effects might be produced. One suggestion that has been made as to ignition-failure is that very strong magnetic fields might so saturate the iron core of the coil that it would drive the operating point up onto the knee of the magnetization curve, so that the input magnetic oscillations would produce only very small output effects. Only a few oersteds would have to be produced right at the coil to accomplish this kind of effect, but when one back-calculates, allowing for shielding effects and typical distances, and assumes an inverse third-power diple field, the requisite H-values within a few feet of the "UFO diple" end, to speak here somewhat loosely, come out in the megagauss range.

Curiously, a number of other back-calculations of magnetic fields end up in this same range; but obviously terrestrial technologies would not easily yield such intensities. Clear evidence for residual magnetization that might be expected in the foregoing hypothesis does not exist, so far as I know. The actual mechanism may be quite unlike that mentioned.

How lights are extinguished is even less clear, although, in some vehicles, relays in the lighting circuits might be magnetically closed. The lights pose more mystery than the ignition. Such cases do not constitute very disturbing questions of hazard or hostility. One might argue that highway accidents could be caused by lighting and ignition failures; however, more serious highway-accident dangers are implicit in other UFO cases where no electrical disturbance was caused.

Many motorists have reported nearly losing control of vehicles when UFOs have swooped down over them; this hazard is distinctly more evident than hazard from the car-stopping phenomenon. Indeed, the number of instances of what we might term "car-buzzing" instances that have involved road-accident hazards is large enough to be mildly disturbing, yet I know of no official recognition of this facet of the UFO problem either. An incident I learned of in Australia involved such fright on the part of the passengers of the "buzzed" vehicle that they jumped out of the car before it had come to a stop, and it went into a ditch. A similar instance occurred not long ago in the U.S. For reasons of space-limitations, I shall not cite other such cases, though it would not be difficult to assemble a list that would run to perhaps a few dozen.

2. Mild radiation exposure

By "radiation" here, I do not mean exposure to radioactivity or to other nuclear radiations, but skin irritations comparable to sunburn, etc. I have interviewed a number of persons who have experienced skin-reddening from exposure to (visible) radiations near UFOs. Rene Gilham, of Merom, Indiana, watched a UFO hovering over his home-area on the evening of Nov. 6, 1957, and received mild skin-burns, for example. I found in speaking with him that the symptoms were gone in a matter of days, with no after-effects. The witnesses in a car-stopping incident at Loch Raven Dam, Md., on the night of Oct. 26, 1958, who were close to a brightly luminous, blimp-sized object after getting out of their stopped car, experienced skin-reddening for which they obtained medical attention. Without citing other such instances, I would say that these cases are not suggestive of any serious hazard, but they warrant scientific attention.

3. More serious physical injuries
   James Flynn, of Ft. Myers, Fla., in a case that has been rather well checked by both APRO and NICAP investigators, reportedly suffered unusual injuries and physical effects when he sought to check what he had taken to be a malfunctioning test vehicle from Cape Canaveral that had come down in the Everglades, March 15, 1965. I have spoken with Flynn and others who know him and believe that his case deserved much more than the superficial official attention it received when he reported it to proper authorities. He was hospitalized for about a week, treated for a deep hemorrhage of one eye (without medical evidence of any blow), and suffered loss of all of the principal deep-tendon reflexes for a number of days, according to his physician's statement, published by APRO (Ref. 45).
   An instance of more than mere skin-reddening, associated with direct contact with a landed unidentified object, reportedly occurred in Hamilton, Ontario, March 29, 1966. Charles Cozens, then age 13, stated to police and to reporters (and recounted to me in a telephone interview with him and his father) that he had seen two rather small whitish, luminous objects come down in an open field in Hamilton that evening. He moved towards them out of curiosity, and states that he finally moved right up beside them, and touched the surface of one of them to see what it felt like. It was not hot, and seemed unusually smooth. One of the two small (8 ft. by 4 ft. plan form, 3-4 feet high) bun-shaped objects had a projection on one end that the boy thought might have been some kind of antenna, so he touched it, only to have his hand flung back as a spark shot out from the end of the projection into the air. He ran, thinking first to go to a nearby police substation. But, on

looking over his shoulder after getting to the edge of the field and seeing no objects there, he decided the police might not believe him and ran to his home. His parents, after discussing the incident at some length with the frightened boy, notified police, which is how the incident became public knowledge. Two others in Hamilton saw that night seemingly similar objects, but airborne rather than on the ground.

Cozens was treated for a burn or sear on the hand that had been in contact with the projection at the moment the spark was emitted. On questioning both the boy and his father, I was left with the impression that, despite the unusual nature of the report, it was described with both straightforwardness and concern and that it must be given serious consideration. Clearly one would prefer a number of adult witnesses to an individual boy: yet I believe the case will stand close scrutiny.

There are a few other such reports of moderate injury reportedly sustained in direct physical contact with landed aerial objects for which I do not yet feel satisfied with the available degree of authentication. It would be very desirable to conduct far more thorough investigations of some foreign cases of this type, to check the weight of the evidence involved. That only a very small number of such cases is on record should be emphasized.

4. Rare instances suggesting overt hostility

In my own investigative experience, I know of only two cases of injuries suffered under what might be describable as overt hostility, and for which present evidence argues authenticity. There are other reports on record that might be construed as overt hostility, but I cannot vouch for them in terms of my own personal investigations.

In Beallsville, Ohio, on the evening of March 19, 1968, a boy suffered moderate skin burns in an incident of puzzling nature. Gregory Wells had just stepped out of his grandmother's house to walk a few tens of yards to his parents' trailer when his grandmother and mother heard his screams, ran out and found him rolling on the ground, his jacket burning. After being treated at a nearby hospital, he described to parents, sheriff's deputies, and others what he had seen. Hovering over some trees across the highway from his location, he had seen an oval-shaped object with some lights on it. From a central area of the bottom, a tube-like appendage emerged, rotated around, and emitted a flash that coincided with ignition of his jacket. He had just turned away from it and so the burn was on the back of his upper arm. In the course of checking this case, I interviewed a number of persons in the Beallsville area, some of whom had seen a long cylindrical object moving at very low altitude in the vicinity of the Wells' property that night.

There is much more detail than can be recapitulated here. My conversations with persons who know the boy, including his teacher, suggest no reason to discount the story, despite its unusual content. After checking the Beallsville incident, I checked another report in which burn-injuries of a more serious nature were sustained in a context even more strongly indicative of overt hostility. I prefer not to give names and explicit citation of details here, but I remark that there appears to me, on the basis of my present information and five interviews with persons involved, to be basis for accepting the incident as real. Partly because of its unparalleled nature, and partly because some of the evidence is still conflicting, I shall omit details and state only that the case, taken together with other scattered reports of injuries in UFO encounters, warrants no panic response but does warrant far more thorough investigation than any that has been conducted to date.

5. UFOs and other electromagnetic disturbances

There are so many instances in which close-passage of an unidentified flying object led to radio and television disturbance that this particular mode of electro-magnetic effect of UFOs seems incontrovertible. One would require nothing more than broad-spectrum electromagnetic noise to account for these instances, of course. There is a much smaller number of instances, some of which I have checked, in which power has failed only within an individual home coincident with nearby passage of a UFO. Magnetic saturation of the core of a transformer might conceivably account for this phenomenon. Then there are scattered instances in which substantial power distribution systems have failed at or very near the time of observation of aerial phenomena similar, broadly speaking, to one or another UFO phenomenon. I have personally checked on several such instances and am satisfied that the coincidence of UFO observation and power outage did at least occur. Whether there is a casual connection here, and in which direction it may run, remains quite uncertain.

Even during the large Northeast blackout, November 9, 1965, there were many UFO observations, several of which I have personally checked. I have inquired at the Federal Power Commission to secure data that might illuminate the basic question of whether these are merely fortuitous, but the data available are inadequate to permit any definite conclusions. In other parts of the world, there have also been reports of system outages coincident with UFO sightings. Again, the evidence is quite unclear as to casual relations.

There is perhaps enough evidence pointing towards strong magnetic fields around at least some UFOs that one might hypothesize a mechanism whereby a UFO might inadvertently trigger a power outage. Perhaps a UFO, with an accompanying strong magnetic field, might pass at high speed across the conductors of a transmission line, induce asymmetric current surges of high transient intensity, and thereby trip circuit breakers and similar surge-protectors in such a way as to initiate the outage. There are some difficulties with that hypothesis, of course; but it could conceivably bear some relation to what has reportedly occurred in some instances. I believe that the evidence is uncertain enough that one can only urge that competent scientists and engineers armed both with substantial information on UFO phenomena and with relevant information on power-system electrical engineering, ought to be taking a very close look at this problem. I am unaware of any adequate study of this potentially important problem. Note that a problem, a hazard, could exist in this context without anything warranting the label of hostility.

MISAPPLICATIONS OF ATMOSPHERIC PHYSICS IN PAST UFO EXPLANATIONS

1. General comments

Since the bulk of UFO reports involve objects reportedly seen in the air, it is not surprising that many attempts to account for them have invoked principles of atmospheric physics. Over the past twenty years, many of the official explanations of important UFO sightings have been based on the premise that observers were misidentifying or misinterpreting natural atmospheric phenomena. Dr. D. H. Menzel, former director of Harvard Observatory, in two books on UFOS (Ref. 24, 25), has leaned very heavily on atmospheric physics and particularly meteorological optics in attempting to account for UFO reports. More recently, Mr. Philip J. Klass, Senior Avionics Editor of *Aviation Week*, has written a book (Ref. 36) purporting to show that most of the really interesting UFO reports are a result of unusual atmospheric plasmas similar to ball lightning. Over the years, many others have made similar suggestions that the final explanation of the UFOs will involve some still not fully understood phenomenon of atmospheric physics.

As a scientist primarily concerned with the field of atmospheric physics, these suggestions have received a great deal of my attention. It is true that a very small fraction of all of the raw reports involve misidentified atmospheric phenomena. It is also true that many lay observers seriously misconstrue astronomical (especially meteoric) phenomena as UFOs. But, in my opinion, as has been emphasized above and will be elaborated below, we cannot explain-away UFOS on either meteorological or astronomical grounds. To make this point somewhat clearer, I shall, in the following, remark on certain past attempts to base UFO explanations on meteorological optics, atmospheric electricity, and radar propagation anomalies.

2. Meteorological optical explanations

Mirages, sundogs, undersuns, and various reflection and refraction phenomena associated with ice crystals, inversions, haze layers, and clouds have been invoked from time to time in an attempt to account for UFO observations. From my study of the past history of the UFO problem and from an examination of recent "re-evaluations" of official UFO explanations, I have the strong impression that many alterations of explanations for classic UFO cases that have been made in the official files in the last few years reflect the response to the writings of Menzel (especially Ref. 25). I have elsewhere (Ref. 2) discussed a number of specific examples of what I regard unreasonable applications of meteorological optics in Menzel's writings. Some salient points will be summarized here.

A principal difficulty with Menzel's mirage explanations is that he typically overlooks completely stringent quantitative restrictions on the angle of elevation of the observer's line of sight in mirage effects. Mirage phenomena are quite common on the Arizona desert, but both observation and optical theory are in good accord in showing that mirage effects are confined to lines of sight that do not depart from the horizontal by much more than a few tens of minutes of arc. Under some extremely unusual temperature conditions in the atmosphere (high latitude regions, for example), one may get miraging at elevation angles larger than a degree, but these situations are extremely rare, it must be emphasized. In Menzel's explanations and in certain of the official explanations, however, mirages are invoked to account for UFOs when the observer's line of sight may depart from the horizontal by as much as five to ten degrees or even more. I emphasize that this is entirely unreasonable. If it were the case that all UFOs were reported essentially at the observer's horizon, then one would have to be extremely suspicious that we were dealing with some unusual refraction anomalies. However, as has been shown by many cases cited above and has been long known to serious investigators of UFO phenomena, no fixed correlation exists. Some of the most interesting UFOs have been seen at close range directly overhead, quite obviously ruling out mirage explanations.

The 1947 sighting by Arnold near Mt. Rainer is explained officially and by Menzel as a mirage, yet the objects which he saw (nine fluttering discs) changed angular elevation, moved across his view through an azimuthal range of about 90 degrees, and were seen by him during the period when he was climbing his own plane through an altitude interval that he estimates to be of the order of 500 to 1000 ft. Anyone familiar with mirage optics would find it utterly unreasonable to claim that such an observation was satisfactorily explained as a mirage.

Similarly, as has been noted above, the 1948 sighting by Eastern Airlines pilots Chiles and Whitted, once explained by Menzel as a "mirage," involves quantitative and observational factors that are not even approximately similar to known mirage effects. There are some extremely rare and still not well-explained refractive anomalies in the atmosphere, such as those that have been discussed by Minnaert, but good UFO observations are so much more numerous than those types of rare anomalies that it is quite out of the question to explain the former by the latter.

Sundogs, or parhelia, are a quite well-understood phenomenon of meteorological optics. Refractions of the sun's rays on horizontally falling tabular ice crystals produce fuzzy, brownish-colored luminous spots at about 22 degrees to the left and right of the sun when suitable ice-crystal clouds are present. Rarer phenomena, produced by the moon rather than the sun, are termed paraselenae. Sundogs are relatively common, but it is probably true that many laymen are not really conscious of them as a distinct optical phenomenon. For this reason, it might seem sensible to suggest that some observers have been misled by thinking that sundogs were UFOs. However, anyone with the slightest knowledge of meteorological optics talking directly to such a witness would, within only a few moments of questioning, establish what was involved. Instead of dealing with anything like a sharp-edged "object," one would quickly find that the observer was describing a very vague spot of light which he saw to the left or right of the sun, probably very near the horizon. To blandly suggest, as Menzel has done, that Waldo J. Harris in the 10/2/61 sighting near Salt Lake City was fooled by a sundog is to ignore either all of the main features of the report or to ignore all of what is known about sundogs.

Undersuns, sub-suns, can be seen rather frequently when flying in jet aircraft at high altitudes. They are a reflection phenomenon produced by horizontally floating ice crystals, which reflect an image of the sun (or at night the moon) and can give surprisingly sharp solar images in still air where turbulence does not cause appreciable tilting of the ice crystals. Here again, it is probably true that many laymen may be sufficiently unaware of this optical phenomenon that they could be confused when they see one. But, as with sundogs, the stringent quantitative requirements on the location of this optical effect relative to the sun would permit any experienced investigator to quickly ascertain whether or not an undersun was involved in this specific sighting. The effect involves specular reflection of the sun's rays, whence the undersun is always seen at a negative angle of elevation in which the observer's line of sight to the undersun is just as far below the horizon as the sun momentarily lies above that same observer's horizon. Clearly, many of the UFO cases that have been cited in examples given above do not come anywhere near satisfying the angular requirements for an undersun.

In my own experience, I have already come across two or three reports, out of thousands that I have examined, where I was led to suspect that the observer was fooled by an undersun. "Reflections off clouds" have been referred to repeatedly in Menzel's writings, never with any quantitative discussion of precisely what he means. But the impression is clearly left that many observers have been and are continuing to be fooled by some kind of cloud-reflections. Aside from the above-described undersun, I am unaware of any "cloud-reflection phenomenon" that could produce anything remotely resembling a distinct object. Clouds of droplets or ice crystals do not provide a source of specular reflection (except in the case of horizontally-floating ice crystals observed from above with a bright luminary, such as sun or moon, in the distance—undersun). What Menzel could possibly have in mind when he talks loosely about such cloud reflections (and he does so in many different places in his books), I cannot imagine.

Inversions are invoked by Menzel, and in official evaluations, to account for certain UFO sightings. Inversions produced by radiational cooling or by atmospheric subsidence are relatively common meteorological phenomena. In some cases, quite sharp inversions with marked temperature differences in rather small vertical distances are known to occur. It is such inversion layers that are responsible for some of the most striking desert mirages of the looming type. To experience a looming mirage, the observer's eye must be located in the atmospheric layer wherein the temperature anomalously increases with height (inversion layer), and the miraged target in the object-field must also lie in or near the inversion layer. Inversion layers are essentially horizontally, and the actually-encountered values of the inversion lapse rates are such that refraction anomalies are confined to very small departures from the horizontal, as noted above under remarks on mirages. All of these points are well-understood principles of meteorological optics.

However, Menzel has attempted to account for such UFOS as Dr. Clyde Tombaugh saw overhead at Las Cruces in August 1949 in terms of "inversion" refraction or reflection effects. Since I have discussed the quantitative unreasonableness of this contention elsewhere, I will not here elaborate the point, except to say that if inversions were capable of producing the optical disturbances that Menzel has assumed, astronomers would long since have given up any attempt to study the stars by looking at them through our atmosphere.

Other atmospheric-optical anomalies have been adduced by Menzel in his UFO discussions. He has repeatedly suggested that layers of haze or mist cause remarkable enlargement of the apparent images of stars and planets. By enlargement, he makes very clear that he means radial enlargements in all directions such that the eye sees not a vertical streak of the sort well-known to astronomers as resulting from near-horizon refraction effects, but rather a circular image of very large angular size. Menzel even describes a sighting that he himself made, over Arctic regions in an Air Force aircraft, in which the image of Sirius was enlarged to an angular size of over ten minutes of arc (one-third of lunar diameter). I have discussed that sighting with a number of astronomers, and not one is aware of anything that has ever been seen by any astronomer that approximates such an instance. In fact, it would require such a peculiar axially-symmetric distribution of refractive index, which miraculously followed the speeding aircraft along as it moved through the atmosphere, that it seems quite hopeless to explain what Menzel has reported seeing in terms of refraction effects.

Since Dr. Menzel's writings on UFO's have evidently had, in some quarters, a marked effect on attitudes towards UFOs, I regard that effect as deleterious. If I felt that we were dealing here with just a slight difference of opinion about rather controversial scientific matters on the edge of present knowledge, I would withhold strong comment. However, I wish to say for the record, that I regard the majority of Dr. Menzel's purported meteorological-optical UFO explanations as simply scientifically incorrect. I could, but shall not here, enlarge upon similar critique of official explanations that have invoked such arguments.

3. Atmospheric electricity

One phenomenon in the area of atmospheric electricity to which appeal has been made from the earliest years of investigations of the UFO phenomena is that of ball lightning. For example, a fairly extensive discussion of ball lightning was prepared by the U.S. Weather Bureau for inclusion in the 1949 Project Grudge report (Ref. 6). It was concluded in that report that ball lightning was most unlikely as an explanation for any of the cases which were considered in that report (about 250). Periodically, in succeeding years, one or another writer has come up with that same idea that maybe people who report UFOs are really seeing ball lightning.

No one ever tried to pursue this idea very far, until P. J. Klass began writing on it. Although his ideas have received some attention in magazines, there is little enough scientific backup to his contentions that they are quite unlikely to have the same measure of effect that Menzel's previous writings have had. For that reason, I shall not here elaborate on my strong objections to Klass' arguments. I spelled them out in considerable detail in a talk presented last March at a UFO Symposium in Montreal held by the Canadian Aeronautics and Space Institute. Klass has ignored most of what is known about ball lightning and most of what is known about plasmas and also most of what is known about interesting UFOs in developing his curious thesis. It cannot be regarded as a scientifically significant contribution to illumination of the UFO problem.

4. Radar propagation anomalies

In the past twenty years, there have been many instances in which unidentified objects have been tracked on radar, many of them with concurrent visual observations. Some examples have been cited above. It is always necessary to approach a radar unidentified with full knowledge of the numerous ways in which false returns can be produced on radar sets. The physics of "ducting" or "trapping" is generally quite well understood. As with mirages, the allowed angle of elevation of the radar beam can only depart from zero by a few tens of minutes of arc for typically occurring inversions and humidity gradients. Ducting with beam angles in excess of a degree or so would require unheard of atmospheric temperature or humidity gradients. Care must be taken in interpreting that statement, since beam-angles have to be distinguished from angles of elevation of the beam axis. For the latter reason, a beam-axis elevation of, say, two degrees still involves emission of some radar energy at angles so low that some may be trapped, yielding "ground returns" despite the higher elevation of the axis. All such points are well described in an extensive literature of radar propagation physics.

In addition to trapping and ground return effects, spurious returns can come from insects, birds, and atmospheric refractive-index anomalies that generate radar echoes termed "angels." These are low-intensity returns that no experienced operator would be likely to confuse with the strong return from an aircraft or other large metallic object. Also, other peculiar radar effects such as interference with other nearby sets, forward scatter from weak tropospheric discontinuities (see work of Atlas and others), and odd secondary reflections from ground targets need to be kept in mind. When one analyzes some of the famous radar-tracking cases in the UFO literature, none of these propagation anomalies seem typical as accounting for the more interesting cases. (Several examples have already been discussed (cases 32, 35, 36, 37, 38, and 39).)

SUMMARY & RECOMMENDATIONS

In summary, I wish to emphasize that my own study of the UFO problem has convinced me that *we must rapidly escalate serious scientific attention to this extraordinarily intriguing puzzle. I believe that the scientific community has been seriously misinformed for twenty years about the potential importance of UFOs* [my emphasis, L.S.]. I do not wish here to elaborate on my own interpretation of the history behind that long period of misinformation: I only wish to urge the Committee on Science and Astronautics to take whatever steps are within their power to alter this situation without further delay. The present Symposium is an excellent step in the latter direction. I strongly urge your Committee that further efforts in the same direction be made in the near future. I believe that extensive hearings before your Committee, as well as before other Congressional committees having concern with this problem, are needed. The possibility that the Earth might be under surveillance by some high civilization in command of a technology far beyond ours must not be overlooked in weighing the UFO problem.

I am one of those who lean strongly towards the extraterrestrial hypothesis. I arrived at that point by a process of elimination of other alternative hypotheses, not by arguments based on what I could call "irrefutable proof." I am convinced that the recurrent observations by reliable citizens here and abroad over the past twenty years cannot be brushed aside as nonsense, but rather need to be taken extremely seriously as evidence that *some phenomenon is going on which we simply do not understand* [my emphasis, L.S.]. Although there is no current basis for concluding that hostility and grave hazard lie behind the UFO phenomenology, we cannot be entirely sure of that. For all of these reasons, greatly expanded scientific and public attention to the UFO problem is urgently needed.

The proposal that serious attention be given to the hypothesis of an extraterrestrial origin of UFOs raises many intriguing questions, only a few of which can be discussed meaningfully. A very standard question of skepticism is "Why no contact?" Here, the best answer is merely a cautionary remark that one would certainly be unjustified in extrapolating all human motives and reasons to any other intelligent civilization. It is conceivable that an avoidance of premature contact would be one of the characteristic features of surveillance of a less advanced civilization; other conceivable rationales can be suggested.

All are speculative, however; what is urgently needed is a far more vigorous scientific investigation of the full spectrum of UFO phenomena, and the House Committee on Science and Astronautics could perform a very significant service by taking steps aimed in that direction.[6] — STATEMENT OF DR. JAMES E. MCDONALD, SENIOR PHYSICIST, INSTITUTE OF ATMOSPHERIC PHYSICS, AND PROFESSOR, DEPARTMENT OF METEOROLOGY, THE UNIVERSITY OF ARIZONA, TUCSON, ARIZ.

MCDONALD'S REFERENCES

1. NICAP Special Bulletin, May, 1960: Admiral Hillenkoetter was a NICAP Advisory Board member at the time of making the quoted statement.
2. McDonald, J. E., 1967: *Unidentified Flying Objects: Greatest Scientific Problem of our Times*, published by UFO Research Institute, Suite 311, 508 Grant Street, Pittsburgh, Pennsylvania, 15219.
3. Keyhoe, D. E., 1950: *Flying Saucers Are Real*, Fawcett Publications, New York, 175 pp.
4. Keyhoe, D. E., 1953: *Flying Saucers From Outer Space*, New York,

Henry Holt & Co., 276 pp. Keyhoe, D. E., 1955: *Flying Saucer Conspiracy*, New York, Henry Holt & Co.,315 pp. Keyhoe, D. E. 1960: *Flying Saucers: Top Secret*, New York, G. P. Putnam's Sons, 283 pp.
5. Ruppelt, E. J., 1956: *The Report on Unidentified Flying Objects*, Garden City, New York, Doubleday & Co., 243 pp. (Paperback edition, Ace Books, 319 pp.)
6. *Project Grudge, 1949: Unidentified Flying Objects*, Report No. 102 AC49/15-100, Project XS-304, released August, 1949. I am indebted to Dr. Leon Davidson for making available to me his copy of this declassified report.
7. NICAP, 1968: *USAF Projects Grudge and Bluebook Reports* 1-12 (1951-1953), declassification date 9 September, 1960. Published by NICAP as a special report, 235 pp.
8. Bloecher, T., 1967: *Report on the UFO Wave of 1947*, available through NICAP.
9. Cruttwell, N. E. G., 1960: *Flying Saucers Over Papua, A Report on Papuan Unidentified Flying Objects*, 45 pp., reproduced for limited distribution; parts of this report have been reproduced in a number of issues of the APRO Bulletin.
10. Hall, R. H., 1964: *The UFO Evidence*, Washington, D.C., NICAP, 184 pp.
11. Olsen, P. M., 1966: The reference for Outstanding UFO Sighting Reports, Riderwood, Maryland, UFO Information Retrieval Center, Inc., P. O. Box 57.
12. Fuller, J. G., 1966: *Incident at Exeter*, New York, G. P. Putnam's Sons, 251 pp. (Berkeley Medallion paperback, 221 pp.)
13. Lorenzen, C. E., 1966: *Flying Saucers*, New York, Signet Books, 278 pp. Lorenzen, C. E. and L. J., 1967: *Flying Saucer Occupants*, New York, Signet Books, 215 pp. Lorenzen, C. E. and L. J., 1968: *UFOs Over the Americas*, New York, Signet Books, 254 pp.
14. Michel, A., 1958: *Flying Saucers and the Straight-Line Mystery*, New York, Criterion Books, 285 pp. Michel, A., 1967: *The Truth About Flying Saucers*, New York Pyramid Books, 270 pp. (Paperback edition of an original 1966 book.)
15. Stanway, R. H., and A. R. Pace, 1968: *Flying Saucers*, Stoke-on-Trent, England, Newchapel Observatory, 85 pp.
16. Vallee, J., 1965: *Anatomy of a Phenomenon*, Chicago, Henry Regnery Co., 210 pp. (Paperback edition, Ace Books, 255 pp.)
17. Vallee, J., and J. Vallee, 1966: *Challenge to Science*, Chicago, Henry Regnery Co., 268 pp. (Also in paperback.)

18. Lore, G. I. R., Jr., and H. H. Denault, Jr., 1968: *Mysteries of the Skies*, Englewood Cliffs, New Jersey, Prentice-Hall Inc., 237 pp.
19. Fort, C., 1941: *The Books of Charles Fort*, New York, Henry Holt & Co., 1125 pp.
20. Stanton, L. J., 1966: *Flying Saucers: Hoax or Reality?*, New York, Belmont Books, 157 pp.
21. Young, M., 1967: *UFO: Top Secret*, New York, Simon & Schuster, 156 pp.
22. *Time Magazine*, July 14, 1947, p. 18.
23. Fuller, C., 1950: *The Flying Saucers—Fact or Fiction?*, Flying Magazine, July 1950, p. 17.
24. Menzel, D. H., 1953: *Flying Saucers*, Cambridge, Harvard University Press, 319 pp.
25. Menzel, D. H., and L. G. Boyd, 1963: *The World of Flying Saucers*, Garden City, New York, Doubleday & Co., 302 pp.
26. Shalett, S., 1949: *What You Can Believe About Flying Saucers*, Saturday Evening Post, April 30, 1949, and May 7, 1949.
27. CSI Newsletter, No. 11, February 29, 1956 (Civilian Saucer Intelligence of New York).
28. *Flying*, June 1951, p. 23.
29. Davidson, L., 1966: *Flying Saucers: An Analysis of the Air Force Project Bluebook Special Report No. 14*, Ramsey, New Jersey, Ramsey-Wallace Corp.
30. American Society of Newspaper Editors, 1967: *Problems of Journalism*, Proceedings of the 1967 Convention of the ASNE, April 20-22, 1967, Washington, D.C., 296 pp.
31. Keyhoe, D. E., 1950: *Flight 117 and the Flying Saucer*, True Magazine, August 1950, p. 24.
32. *Salt Lake Tribune*, Tuesday, October 3, 1961, p. 1.
33. *UFO Investigator*, Vol. 3, No. 11, Jan.-Feb. 1967.
34. LANS, 1960: *Report on an Unidentified Flying Object Over Hollywood, California, Feb. 5, 1960 and Feb. 6, 1960*, Los Angeles NICAP Subcommittee, 21 pp., mimeo.
35. *UFO Investigator*, Vol. 1, No. 12, April 1961.
36. McDonald, J. E., 1968: *UFOs-An International Scientific Problem*, paper presented at a Symposium on Unidentified Flying Objects, Canadian Aeronautics and Space Institute, Montreal, Canada, March 12, 1968.
37. Darrach, H. B., Jr., and Robert Ginna, 1952: "Have We Visitors from Space?", *Life Magazine*, April 7, p. 80 ff.
38. Zigel, F., 1968: "Unidentified Flying Objects," *Soviet Life*, February, 1968, No. 2 (137), pp. 27–29.

39. Klass, Philip J., 1968: *UFOs-Identified*, New York, Random House, 290.
40. *International News Service*, datelined Sept. 12, 1951, Dover, Del.
41. *New York Times*, June 2, 1954; *New York World Telegram*, June 1, 1954; *New York Post*, June 1, 1954; *New York Daily News*, June 2, 1954.
42. Official file on October 15, 1948 Fukuoka case, *Project Bluebook*.
43. *Melbourne* (Australia) *Sun*, December 16, 1954; *Melbourne Herald*, December 16, 1954; *Auckland Star*, December 16, 1954.
44. *Los Angeles Times*, May 9, 1957; *New York Journal-American*, May 10, 1957.
45. APRO *Bulletin*, May-June, 1965, p. 1-4.

CHAPTER THREE

STATEMENT OF CARL SAGAN

MR. ROUSH. Our next participant is Dr. Carl Sagan.

Dr. Sagan is associate professor of astronomy in the Department of Astronomy and Center for Radiophysics and Space Research in Cornell University, having just recently left Harvard University. He has written over 100 scientific papers, and several articles for Encyclopedia Britannica, Americana. He is coauthor of several books. Dr. Sagan, we are delighted you are participating with us in this symposium this morning and you may proceed.

☛ DR. SAGAN. Thank you very much, Congressman Roush.

As I understand what the committee would like from me, is a discussion of the likelihood of intelligent extraterrestrial life, and since this estimate is to be made in this symposium, clearly it is the hypothesis that unidentified objects are of extraterrestrial origin which the committee must have in mind.

I'm delighted to tell about contemporary scientific thinking along these lines, but let me begin by saying that I do not think the evidence is at all persuasive, that UFO's are of intelligent extraterrestrial origin, nor do I think the evidence is convincing that no UFO's are of intelligent extraterrestrial origin.

I think as each of the preceding speakers has mentioned, but perhaps not sufficiently emphasized, that the question is very much an open one, and it is certainly too soon to harden attitudes and make any permanent contentions on the subject.

I find that the discussion, like elsewhere, is best evaluated if we consider the question of life on earth. I suppose that if you had all your prejudices removed and were concerned with the question of whether the earth was populated by life of any sort, how would you go about finding out?

If, for example, we were on some other planet, let's say Mars, and looking at the Earth, what would we see? Fortunately we now have meteorological satellite photographs of the earth at various resolutions, so we can answer the question. The first large slide.

This is a photograph of the earth. That is the full earth, which you are looking at which is primarily cloud cover. This is the Pacific

Ocean. You can see southern California in the upper right, and, as advertised by the local chamber of commerce, you can see it is cloud free. [Laughter.]

Now, it is clear that very little information about the earth, much less possibility of life on it, is obtained by a picture at this resolution.

The next large slide is a TIROS photograph of the earth at about 1-mile resolution, that is, things smaller than a mile cannot be seen, and very prolonged scrutiny of the entire eastern seaboard of the United States shows no sign of life, intelligent, or otherwise.

We have looked at several thousand photographs of the earth, and you may be interested to see that there is no sign of life, not only in New York or Washington, but also in Peking, Moscow, London, Paris, and so on.

The reason is that human beings have transformed the earth at this kind of scale very little, and therefore the artifacts of human intelligence are just not detectable photographically in the daytime with this sort of resolution.

The next slide shows one of the few successful finds of intelligent life on earth that we made; down toward the lower left you can see a kind of grid, a kind of crisscross pattern, a rectangular area. This is a photograph taken near Cochran, Ontario, in Canada. What we are looking at are swaths cut by loggers through the forest. They cut many swaths in parallel, then another parallel sequence of swaths at right angles. Then the snow fell, heightening contrast, so that is the reason for the tic-tac-toe pattern. The sequence of straight lines there is anomalous. You would not expect it by geological processes. If you found that on another planet you would begin to expect there is life there. This is a photograph at about a tenth of a mile resolution, and is far better than the best photographs we have of Mars. The photographs we have of Mars are, of course, better than of any other planet. Therefore, to exclude intelligent life on another planet photographically is certainly premature. We could not exclude life on earth with this same sort of resolution.

However, there are other reasons why intelligent life on the other planets of this solar system are moderately unlikely.

To continue this sequence of photographs, I should say there are only about one in a thousand photographs where this resolution of the earth gives any sign of life.

The next photograph, however, shows a resolution about three times better. That is a Gemini capsule in the lower left-hand corner and we are looking at the vicinity of the Imperial Valley in

California. You are just on the verge of resolving the contour patterns of fields, for agricultural purposes.

The next slide shows us an area between Sacramento and San Francisco, which has a very clear geometric pattern. It is quite obvious that this is the result of some intelligent activity on the earth.

You can see an airport, a railway, the monotonous pattern of housing developments in the upper right. You can see the patterns of contour fields. And this is such a highly geometrized picture, that it is clearly the result of some intelligence.

However, a photograph taken of this same area, only let's say 100,000 years ago, when there certainly was lots of life on earth, would show none of these features, because these are all the signs of our present technical civilization.

So even though the earth was full of life, and human beings were very much in evidence 100,000 years ago, none of this would be detectable by such photography. To detect individual organisms on earth, we have to have a photographic resolution about 10 times better than this, then we occasionally see things like these in the next slide. All those little dots casting shadows are cows in a field in California.

There are other ways of detecting intelligent life on the earth. From the vantage point of Mars, detecting, say, the lights of cities at night, is extremely marginal, and in fact the only way of doing it would be to point a small radio telescope at the earth, and then as the North American Continent turned toward Mars, there would be this blast of radio emission from domestic television transmission that prolonged scrutiny would indicate some sign of intelligent life on the earth.

In fact, it is radio communications which is the only reasonable method of communications over very large distances. It is a remarkable fact that the largest radio telescope on the earth at the present time, the Arecibo dish in Puerto Rico, is capable of communicating with another dish, similarly outfitted if one existed at the incredible distance of 1,000 light years away, a light year being about 6.6 trillion miles, and the distance to the nearest star being a little over 4 light years.

Now, let me then go to the question of the cosmic perspective of where we are.

We are, of course, sitting on a planet, the third from the Sun, which is going around the Sun, which is a star-like, and the other stars visible on a clear night to the naked eye. The first small slide will give an impression of what happens when you point a moderate telescope in the direction of the center of the Milky Way Galaxy.

This is a photograph of a star cloud. You are looking at tens of thousands of suns here. In fact, the number of suns in our galaxy is about 150,000 million.

They are collected into a disk-shaped pattern, shown in the next slide; the next slide will show a photograph of the nearest galaxy like our own. That fuzzy spiral thing in the middle is M-31, that is also known as the Great Galaxy Andromeda, and if that were a photograph of our galaxy, we would be situated extremely far out, in fact, a little far off the slide, very much in the galactic boon docks. The Sun is nowhere near the center of the galaxy. It is a very out-of-the-way rural location we happen to be in.

Now, in collection of 150,000 million stars in the Milky Way Galaxy, our sun is just one, and there are at least billions of other galaxies, and the last slide, will show you what happens if you point a telescope away from the obscuring dust and stars in the galaxy. You then start seeing dozens of other galaxies, everyone of those funny-shaped spiral and irregular-shaped things there, and some of the spherical shaped ones, are other galaxies, each of which are containing about 100 billion stars as well.

So it is clear that there are in the accessible universe, some hundreds of billions of billions of stars, all more or less like our own.

Now, if we want to assess the likelihood that there are intelligent civilizations somewhere in advance of our own, on planets of other stars in our own galaxy, we have to ask questions which cover a variety of scientific subjects, some of which are fairly well known, some of which are extremely poorly known. For a numerical assessment of whether there is likely intelligence in other parts of the galaxy in a form we do not have at present, let me indicate the kinds of things we know. It depends on the rate of star formation.

It depends on the likelihood that the given star has planets. It depends on the likelihood at least one of those planets is at a position from the essential star which is suitable for the origin of life. It depends on the likelihood that the origin of life actually occurs on that planet. It depends on the probability life once arisen on that planet will evolve to some intelligence. It depends on the likelihood that intelligence, once emerged, will develop a technical civilization. And it depends on the lifetime of the technical civilization, because technical civilization of a very short lifetime, will result in very few technical civilizations being around at any given time. We know something about some of these. There is some reason to believe that planets are a reasonable likely accompaniment of star formation, that the solar system in other words is a fairly common event in the galaxy and is not unique. There are laboratory experiments on the origin of life, in which the early conditions on earth have been duplicated in the laboratory. It turns out that at least the molecules fundamental to living systems, are produced relatively easy, physics and chemistry apparently made in such a way that the origin of life may be a likely event.

Beyond that it is difficult to do laboratory experiments, because evolution takes billions of years, and scientists aren't that patient. Therefore, it is just a question of intelligent and knowledgeable estimates.

Here, some scientists believe that the evolution of intelligence and technical civilization is very likely. Others believe it is a very remarkable and unusual event and by the merest fluke did it happen here.

I don't think that this is the place to go into this very difficult question in any great detail. Let me merely say that much more important than these uncertainties is the question of the life of a technical civilization, judging from the events on the earth, one might say the likelihood of our civilization lasting only a few decades more, might be a fairly high probability, and if that is typical of other civilizations, then it is clear there aren't any other humans around.

On the other hand, if civilizations tend to have very long lifetimes, it may be there are large numbers of technical civilizations in the galaxy.

Now, one thing is clear, which is this: If there are other technical civilizations, any random one of them is likely to be vastly in advance of our own technical civilization. For example, we are only 10 or 15 years into having the technology of interstellar communication by radio astronomy. It is unlikely there is any other civilization in the galaxy that is that backward in their technical expertise.

MR. MILLER. Doctor, didn't Sir Bernard Lovell receive electrical pulses he can't explain?

DR. SAGAN. Yes, sir. There are now five objects in the heavens called pulsars, which are objects which are sending out radiation which is modulated with a frequency of about one per second; also there are submodulations. There are a variety of hypotheses to explain these things, some of which involve the oscillations of very old stars. There are certain difficulties with each hypothesis. The first suggestion made by the British at Cambridge, when they encountered this phenomenon was perhaps it was a beacon of some extraterrestrial civilization. That is not now their favored hypothesis. It is not clear that that is totally absurd, but in fact the scientific method to be used in that case is rather similar to the one to be used in this case. That is, it is a puzzling phenomenon. One therefore excludes all physical explanations that one possibly can before going to the much more hypothetical possibility of intelligence being involved.

So, that is the present state of work in that field. For data gathering to get better information, and the refinement of the purely physical hypothesis.

Well, I was saying that if there are other civilizations, many of them are likely to be far in advance of our own, and this, therefore, raises the question of how likely it is that they can traverse interstellar space and come from planets or some other star to here.

I should first emphasize that the distances between the stars are absolutely huge. Light, faster than which nothing can travel, takes 4½ years to get from here to the nearest star.

MR. ROUSH. Excuse me, isn't that a rather arbitrary statement?

DR. SAGAN. I don't think so. Perhaps you can tell me why you think it might be, then I can tell you why I think it isn't.

MR. ROUSH. In my opening statement I referred to the new audacity of imagination John Dewey had spoken of. I'm thinking of imaginative terms, not factual terms.

DR. SAGAN. Let me say in a sentence, why most physicists believe no material object can travel faster than light. That takes us into questions of the theory of relativity, which has had previous encounters with congressional committees, and perhaps we don't want to go into that in very great detail.

But the essential point is, that in making a few, very few assumptions, one of which was, the one we are talking about, nothing goes faster, Einstein was able to then derive a whole body of predictions which are confirmed in vast detail. Therefore, if someone says that is not a good idea, that things can travel faster than light, then they have to come up with a physical theory which explains everything we know in a way that is consistent with the idea that you can travel faster than light. No one has succeeded in doing that. Many physicists, have tried. Therefore, the present belief is that you can't. But that, of course, is a time-dependent statement. It may be that this isn't the ultimate truth.

In physics, as in much of all science, there are no permanent truths. There is a set of approximations, getting closer and closer, and people must always be ready to revise what has been in the past thought to be the absolute gospel truth [my emphasis, L.S.]. If I might say, to revise opinions, is one which is frequent in science, and less frequent in politics. [Laughter.]

So, in the context of contemporary science, I'm obviously speaking in that context, one cannot travel faster than light.

So the distances between the stars are extremely large. Of course, any contemporary space vehicle would take a ridiculous amount of time to get from here to anywhere else, but we are not talking about contemporary space vehicles. The question, "Is there any conceivable method of traveling from one place to another very close to the speed of light, and therefore get reasonable transit times?" involves extrapolations of technology of a very difficult sort. However, let me merely say at least some people who have looked into the subject have concluded that it is not out of the question, even with contemporary principles of science, to imagine vehicles capable of traveling close to the speed of light, between the stars.

This doesn't mean that it happens. There may in fact be insuperable engineering difficulties we don't know about, but there is nothing in the physics that prohibits interstellar space flight.

So any estimate of how likely it is that we would be visited by an extraterrestrial intelligent civilization, depends not only on how many of them are there, but on what kind of transport they have, and how often they launch their space vehicles, even very optimistic estimates for all these numbers, gives a conclusion that an advance civilization comes here very rarely. But I again emphasize the great uncertainty in any of these numerical estimates, as they involve parts of science we don't know very much about.

So, to conclude what I understand is the main reason why this committee has asked me to testify, it is not beyond any question of doubt that we can be visited. There are great difficulties from our present point of view. They are not insuperable. And if Dr. McDonald, for example, were to present me with extremely convincing evidence of an advanced technology in a UFO, I could not say to him that is impossible, because I know you can't get from there to here, or I can't say to him that is impossible because I know there aren't any other guys up there.

On the other hand, I would of course demand very firm evidence before I would say, well, that seems to be a very likely hypothesis.

So I would like to spend just a few minutes to come more closely to the subject of this symposium.

First of all, I think it is clear to the committee, but this point should be emphasized very strongly, that there are very intense, predisposing, emotional factors in this subject.

There are individuals who very strongly want to believe that UFO's are of intelligent extraterrestrial origin. Essentially to my view, for religious motives; that is, things are so bad down here, maybe somebody from up there will come and save us from ourselves. This takes all sorts of subtle and not so subtle forms. There are also predisposing emotional factors in the other direction; people who very much want to believe UFO's are not of intelligent extraterrestrial origins, because that would be threatening to our conception of us as being the pinnacle of creation. We would find it very upsetting to discover that we are not, that we are just a sort of two-bit civilization.

It is clear that the scientific method says you don't take either of those views, and you simply keep an open mind and pursue whatever facts are at hand with as many diverse hypotheses as possible, and try to eliminate each suggested hypothesis, and see if you are lucky with any one.

I might mention that, on this symposium, there are no individuals who strongly disbelieve in the extraterrestrial origin of UFO's and therefore there is a certain view, not necessarily one I strongly agree with—but there is a certain view this committee is not hearing today, along those lines.

Finally, let me say something about the question of priorities, which Congressman Rumsfeld asked us for, and the question of significance.

Now, the possibility of discovering something about extraterrestrial life, life originated on some other planet, is of the very highest interest for biology and in fact for all science. A bona fide example of extraterrestrial life even in a very simple form, would revolutionize biology. It would have both practical and fundamental scientific benefits, which are very hard to assess, it would truly be immense.

Now, if the answer to this sort of profound scientific question lies right at hand, it would be folly to ignore it. If we are being visited by representatives of extraterrestrial life, just stick our heads in the sand, would be a very bad policy, I think.

On the other hand, to mount a major effort to investigate these things, I think requires some harder evidence than is now at hand.

It is clear that if such an effort were mounted, some information on atmospheric physics would be forthcoming. I think some information on psychology would certainly be forthcoming.

I have the impression that the capability of human populations to self-delusion, has not been accorded appropriate weight in these considerations. There is an interesting book published about a century ago by [Charles] McKay called *Extraordinary Popular Delusions and the Madness of Crowds*, which I commend to the committee. It goes into such things as alchemy, and witchcraft. After all, there have been centuries in which these things were considered to be as obviously true as anything, and yet we now know that this is really nonsense.

So the possibility of these sort of delusions having a kind of contemporary guise as UFO's should not be thrown out altogether. I do not think that explains most or all of the unidentified settings.

Since the funds are so painfully tragically short for science today, the priority question boils down to this: In the search for extraterrestrial life there is a high risk, high possibility, that is the one we are talking about today; namely, UFO's—there is a high risk that they are not of extraterrestrial origin, but if they are, we are sure going to learn a lot.

Compared to that, there is a moderate risk, significant return possibility, and that is, looking for life even simple forms on nearby planets, and searching for intelligent radio communications by the techniques of radio astronomy. Here it is clear there will be significant paydirt of one sort or another for what I gather is a comparable sort of investment.

So if Congress is interested, and I'm not sure it is, I think it might very well ought to be, but if Congress is interested in a pursuit of the question of extraterrestrial life, I believe it would be much better advised to support the biology, the Mariner, and Voyager programs of NASA, and the radio astronomy programs of the National Science Foundation, than to pour very much money into this study of UFO's.

On the other hand, I think a moderate support of investigations of UFO's might very well have some scientific paydirt in it, but perhaps not the one that we are talking about today.

Mr. Chairman that concludes my statement except that I request that you include for the record a statement entitled "Unidentified Flying Objects" that I prepared for the Encyclopedia Americana [which follows].[7] — STATEMENT OF DR. CARL SAGAN, DEPARTMENT OF ASTRONOMY, CORNELL UNIVERSITY, ITHACA, N.Y.

UNIDENTIFIED FLYING OBJECTS
by Carl Sagan

Unidentified flying objects (UFO's) is the generic term for moving aerial or celestial phenomena, detected visually or by radar, whose nature is not immediately understood. Interest in these objects stems from speculation that some of them are the products of civilizations beyond the earth, and from the psychological insights into contemporary human problems that this interpretation provides.

Observations. Unidentified flying objects have been described variously as rapidly moving or hovering; disc-shaped, cigar-shaped, or ball-shaped; moving silently or noisily; with a fiery exhaust, or with no exhaust whatever; accompanied by flashing lights, or uniformly glowing with a silvery cast. The diversity of the observations suggests that UFO's have no common origin and that the use of such terms as UFO's or "flying saucers" serves only to confuse the issue by grouping generically a variety of unrelated phenomena.

In the United States, popular interest in unidentified flying objects began on June 24, 1947, when a group of rapidly moving, glistening objects was observed from the air in daytime, near Mt. Rainier, Washington. The observer, a Seattle resident, dubbed them "flying saucers." The sighting received extensive publicity. Somewhat similar sightings have been reported ever since. The differences among these observations, however, are as striking as the observations themselves.

Investigations. Because of its national defense responsibility, the U.S. Air Force investigates reports of unidentified flying object over the United States. The number of sightings investigated in 1947-65 is shown in the following table.

Reported sightings of UFO's, 1947–65

Year: 1947; number: 79
Year: 1948; number: 143
Year: 1949; number: 186
Year: 1950; number: 169
Year: 1951; number: 121
Year: 1952; number: 1,501
Year: 1953; number: 425
Year: 1954; number: 429
Year: 1955; number: 404
Year: 1956; number: 778
Year: 1957; number: 1,178
Year: 1958; number: 473
Year: 1959; number: 364
Year: 1960; number: 557
Year: 1961; number: 591
Year: 1962; number: 474
Year: 1963; number: 399
Year: 1964; number: 572
Year: 1965; number: 886[8]

Evaluation of these reports is difficult. Observations frequently are sketchy, and different reports of the same phenomenon are often dissimilar, or even irreconcilable. Observers tend to exaggerate. Deliberate hoaxes, some involving double-exposure photography, have been perpetrated. After allowances are made for these factors, the accepted scientific procedure is to attempt an explanation of the observations in terms of phenomena independently observed and understood. Only if an observation is rigorously inexplicable in terms of known phenomena does the scientist introduces alternative hypotheses for which there is no other evidence. Such hypotheses must still be consistent with all other available scientific information.

The identity of most UFO's has been established as belonging to one of the following categories: unconventional aircraft; aircraft under uncommon weather conditions; aircraft with unusual external light patterns; meteorological and other high-altitude balloons; artificial earth satellites; flocks of birds; reflections of searchlights or headlights off clouds; reflection of sunlight from shiny surfaces; luminescent organisms (including one case of a firefly lodged between two adjacent panes of glass in an airplane cockpit window); optical mirages and looming; lenticular cloud formations; ball lighting; sun dogs; meteors, including green fireballs; planets, especially Venus; bright stars; and the aurora borealis.

Radar detection of unidentified flying objects has also occurred occasionally. Many of these sightings have been explained as radar reflections from temperature inversion layers in the atmosphere and other sources of radar "angels."

Considering the difficulties involved in tracking down visual and radar sightings, it is remarkable that all but a few percent of the reported UFO's have been identified as naturally occurring—if sometimes unusual—phenomena. It is of some interest that the UFO's which are unidentified do not fall into one uniform category of motion, color, lighting, etc., but rather run through the same range of these variables as the identified UFO's. In October 1957, Sputnik 1, the first earth-orbiting artificial satellite, was launched. Of 1,178 UFO sightings in that year, 701 occurred between October and December. The clear implication is that Sputnik and its attendant publicity were responsible for many UFO sightings.

Earlier, in July 1952, a set of visual and radar observations of unidentified flying objects over Washington, D.C., caused substantial public concern. Government concern was reflected in the creation in November of that year of a special panel to evaluate these reports. The panel was established by the Office of Scientific Intelligence of the Central Intelligence Agency, and was headed by the late Professor H. P. Robertson of the California Institute of Technology. The Robertson panel, after a thorough investigation of the UFO reports to that date, concluded that all were probably natural phenomena, wrongly interpreted.

The most reliable testimony is that of the professional astronomer. Professor Jesse L. Greenstein of Mount Wilson and Palomar Observatories has pointed out that a vehicle 100 feet (30.5 meters) in diameter, at an altitude of 50 miles (80.5 km), would leave a broad track on photographic plates of the sky taken with large telescopes. This track could be distinguished easily from those of ordinary astronomical objects such as stars, meteors, and comets. Nevertheless, it appears that such tracks or unambiguous visual observations of classical UFO's have never been made by professional astronomers.

For example, in the Harvard Meteor Project performed in New Mexico during the period 1954-58, extensive photographic observations were made by Super-Schmidt cameras, with a 60° field of view. In all, a surface area of about 3,000 square miles (7,700 sq km) was observed to a height of about 50 miles (80 km) for a total period of about 3,000 hours. Visual and photographic observations were made which could detect objects almost as faint as the faintest objects visible to the naked eye. These observations by professional astronomers were made in a locale and period characterized by extensive reports of unidentified flying objects. No unexplained objects were detected, despite the fact that rapidly moving objects were being sought in a study of meteors. Similar negative results have been obtained by large numbers of astronomers and help to explain the general skepticism of the astronomical community towards flying saucer reports.

A series of puzzling and well-published flying saucer sightings in the mid-1960's again led to the appointment of a government investigating panel, this time under the aegis of the Air Force Scientific Advisory Board. It is significant that this panel was convened not at the request of the operational or intelligence arms of the Air Force, but in response to a request by the Air Force public relations office. The panel, under the chairmanship of Brian O'Brien, a member of the board, met in February 1966, and restated the general conclusions of the Robertson panel. It was recommended that the Air Force make a more thoroughgoing effort to investigate selected UFO reports of particular interest, although the probability of acquiring significant scientific information (other than psychological) seemed small. The O'Brien panel suggested that the Air Force establish a group of teams at various points within the United States in order to respond rapidly to UFO reports. The panel recommended that each team should consist of a physical scientist familiar with upper atmospheric and astronomical phenomena, a clinical psychologist, and a trained investigator.

In October 1966 the University of Colorado was selected by the Air Force Office of Scientific Research to manage the program, and to prepare a thoroughgoing analysis of the UFO problem. The National Academy of Sciences agreed to appoint a panel to review the report when it is completed in early 1968.

Hypotheses of Extraterrestrial Origin. Repeated sightings of UFO's, and the persistence of the Air Force and the responsible scientific community in explaining away the sightings, have suggested to some that a conspiracy exists to conceal from the public the true nature of the UFO's. Might not at least a small fraction of the unexplained few percent of the sightings be space vehicles of intelligent extraterrestrial beings observing the earth and its inhabitants?

It now seems probable that the earth is not the only inhabited planet in the universe. There is evidence that many of the stars in the sky have planetary systems. Furthermore, research concerning the origin of life on earth suggests that the physical and chemical processes leading to the origin of life occur rapidly in the early history of the majority of planets. From the point of view of natural selection, the advantages of intelligence and technical civilization are obvious, and some scientists believe that a large number of planets within our Milky Way galaxy—perhaps as many as a million—are inhabited by technical civilizations in advance of our own.

Interstellar space flight is far beyond our present technical capabilities, but there seem to be no fundamental physical objections to it. It would be rash to preclude, from our present vantage point, the possibility of its development by other civilizations. But if each of, say, a million advanced technical civilizations in our galaxy launched an interstellar spacecraft each year (and even for an advanced civilization, the launching of an interstellar space vehicle would not be a trivial undertaking), and even if all of them could reach our solar system with equal facility, our system would, on the average, be visited only once every 100,000 years.

UFO enthusiasts have sometimes castigated the skeptic for his anthropocentrism. Actually, the assumption that earth is visited daily by interstellar space-craft is far more anthropocentric—attaching as it does some overriding significance to our small planet. If our views on the frequency of intelligence in the galaxy are correct, there is no reason why the earth should be singled out for interstellar visits. A greater frequency of visits could be expected if there were another planet populated by a technical civilization within our solar system, but at the present time there is no evidence for the existence of one.

Related to the interstellar observer idea are the "contact" tales, contemporary reports of the landing of extraterrestrial space vehicles on earth. Unlike the UFO reports, these tales display a striking uniformity. The extraterrestrials are described as humanoid, differing from man only in some minor characteristic such as teeth, speech, or dress. The aliens—so the "contactees" report—have been observing earth and its inhabitants for many years, and express concern at "the present grave political situation." The visitors are fearful that, left to our own devices, we will destroy our civilization. The contactee is then selected as their "chosen intermediary" with the governments and inhabitants of earth, but somehow the promised political or social intervention never materializes.

Psychological Factors. The psychologist Carl Jung pointed out that the frequency and persistence of these contact tales—not one of which has been confirmed by the slightest sort of objective evidence—must be of substantial psychological significance. What need is fulfilled by a belief that unidentified flying objects are of extraterrestrial origin? It is noteworthy that in the contact tales, the spacecraft and their crews are almost never pictured as hostile. It would be very satisfying if a race of advanced and benign creatures were devoted to our welfare.

The interest in unidentified flying objects derives, perhaps, not so much from scientific curiosity as from unfulfilled religious needs. Flying saucers serve, for some, to replace the gods that science has deposed. With their distant and exotic worlds and their pseudoscientific overlay, the contact accounts are acceptable to many people who reject the older religious frameworks. But precisely because people desire so intensely that unidentified flying objects be of benign, intelligent, and extraterrestrial origin, honesty requires that, in evaluating the observations, we accept only the most rigorous logic and the most convincing evidence. At the present time, there is no evidence that unambiguously connects the various flying saucer sightings and contact tales with extraterrestrial intelligence.

DR. SAGAN. Thank you.

MR. [KEN] HECHLER. Dr. Sagan, there have been some recent experiments at Green Bank, W.Va., with its 300-foot telescope, in an attempt to synchronize this with the Arecibo dish, in such a way as you might in effect produce almost a 2,000-mile diameter collecting surface for trying to receive signals from the pulsars. I wonder if this isn't the type of specific activity in radio astronomy that could utilize some additional support in order to ascertain the truth about terrestrial life and signals therefrom?

DR. SAGAN. Congressman Hechler, as a member of the faculty at Cornell that runs the observatory, I would find some problem answering that.

MR. HECHLER. But not of West Virginia, however?

DR. SAGAN. That is right. The study of pulsars, as I indicated to Chairman Miller, is relevant. The development of a long base-line parameter of the sort you talked about is of great interest to many areas of radio astronomy, and conceivably to the area we are talking about.

However, there has not been since Project OXMA, which occurred in Green Bank some 7 years or so ago, any systematic effort in this country to look for signals of intelligent extraterrestrial origin.

There is at the present time a fairly major effort under way in the Soviet Union, but at least in this country there are no such efforts directed specifically to this question.

It may be if we ever do detect intelligent signals from elsewhere, it will be an accidental byproduct of some other program. There is at the present time no effort to search for extraterrestrial signals.

CHAIRMAN MILLER. Are they trying to do things in Australia?

DR. SAGAN. To the best of my knowledge there is no such work being done.

CHAIRMAN MILLER. The Mills-Cross program is also connected with Cornell, isn't it?

DR. SAGAN. The Cornell-Sydney Astronomy Center, yes, sir.

CHAIRMAN MILLER. Is that all, Mr. Hechler?

MR. HECHLER. I was hoping you would suggest something more specific, for our future consideration.

DR. SAGAN. Let me say, and again let me emphasize that it is by no means demonstrated that radio astronomical searches for extraterrestrial intelligence have anything whatever to do with UFO's, but if we were interested, as some of us are, in examining the possibility of extraterrestrial intelligence, sending signals to Earth, then relatively modest programs, of say less than a million dollars, could be organized, using largely existing instruments with only small modifications in the things you hook up to the radio telescope, which would be ideal for this purpose.

There are in fact many radio astronomers who are privately interested in this sort of thing, but it carries something of the same sort of stigma that both the previous speakers mentioned about UFO's. It is unconventional. It is in many senses radical. Many astronomers prefer to have nothing to do with it.

MR. PETTIS. Mr. Chairman.

MR. ROUSH. Mr. Pettis.

MR. PETTIS. I would like to ask the doctor, or any other member of the panel. Is there any indication that any other Government, particularly the Russians, are interested in this subject?

DR. SAGAN. I cannot speak about the UFO program. Perhaps Dr. Hynek can say something about that.

As far as the question that I just mentioned, the radio search for extraterrestrial intelligence, there is a state commission in the Soviet Union, for the investigation of cosmic-radio intelligence. There is a fairly major effort that has been mustered for the last few years along these lines.

And there is only some information about that; that we have gotten out of the Soviet Union.

I don't know anything about their activities on UFO's. Perhaps Dr. Hynek would like to comment on that.

DR. HYNEK. May I, Mr. Chairman, preface my remark, in answer to that, by pointing out a danger here that we may be putting the cart before the horse in the consideration of extraterrestrial intelligence.

Speaking of horses, suppose someone comes here and tells us, or announces to us there is a report of a horse in the bath tub.

I think that it would be rather pointless to then ask, what is the color of the horse, what does he eat, how could be have gotten there, who installed the bath tub? The question is is there a horse in the bath tub? This is a question I think we should direct ourselves to first. Is there anything to these reports?

Now, coming to the question of the Russian situation, I do know from my visits behind the Iron Curtain, or as they like to speak of it, the Socialist countries, there have been sightings behind the Iron Curtain. In fact, if you were to have good translations, it would be difficult to distinguish between a UFO report from Russia, from Brazil, from Argentina, from Japan, or from the United States. There is a rather rough pattern.

Now, the Russians, to the best of my knowledge, have given no official recognition to the problem, but I do know, from personal information, that there is sort of a ground-swell interest, or a latent interest, that pops up here and there, but apparently they have as much difficulty in getting official recognition as we do.

MR. ROUSH. I would first point out that I realize that a visit to Russia doesn't necessarily make a person an expert or give him all the information. A year ago June I did visit Russia. I had conversations with a few of their people, including, my pronunciation may not be correct, Dr. Millionshchikov and the head of their weather bureau, I believe it is Petrov, and several others, and I repeatedly asked the question, "Do you believe in unidentified flying objects?" In each instance they merely laughed. That was the response that I got. Since then, however, I have observed there have been papers published in Russia discussing the phenomena, and discussing it in scientific terms.

It seems to me that any discussion such as ours today raises the question of the existence of extraterrestrial life. That is one reason we asked Dr. Sagan to come here. I'm not real sure, Dr. Sagan, whether you stated whether there is or whether there is not extraterrestrial life. I was watching for that, and I don't believe I heard you say it.

DR. SAGAN. Congressman Roush, I have enough difficulty trying to determine if there is intelligent life on Earth, to be sure if there is intelligent life anywhere else. [Laughter.]

If we knew there was life on other planets, then we would be able to save ourselves a lot of agony finding out. It is just because the problem is so significant, and we don't have the answers at hand we need to pursue the subject. I don't know. It beats me.

MR. ROUSH. I believe you coauthored a book with a Russian, is that correct?

DR. SAGAN. That is correct.

MR. ROUSH. Does Dr. Shklovskii share your views?

DR. SAGAN. I think he shares my restraint.

I think both of us would say we think this is an extremely important subject, that we are on the frontier of being able to find out, but that neither of us knows whether there is or isn't life out there. Let me say if it turns out there isn't life on Mars, that is almost as interesting as if we find there is life on Mars, because then we have to ask, what happened different on Mars than on the Earth, so that life arose here and not there. That will surely give us a very profound entry into the question of follow-up of evolution and the cosmic context.

MR. ROUSH. Suppose we discover there is life on Mars, in some form, wouldn't this almost cinch your case, and you could say there is extraterrestrial life?

DR. SAGAN. Yes, sir; it certainly would, but not cinch our case about extraterrestrial intelligence. Conceivably, there might be a low form on Mars. If there is Martian life, it is of interest how low it is. If there is intelligence on Mars—but we don't know there is intelligence on Mars—then we don't have to grasp that evolution process.

MR. ROUSH. I would like to finish this morning's session just by telling of a cartoon I saw which I think Dr. Hynek perhaps saw and enjoyed as much as I did. It showed a flying saucer hovering over the Earth, with little green men looking down, and one turned to the other and said, "Do you suppose it is swamp gas?" [Laughter.]

DR. HYNEK. That is a good statement to close the session on.

MR. ROUSH. We shall reconvene at 2 o'clock this afternoon. (Whereupon, at 12:15 p.m., the hearing was recessed to reconvene at 2 p.m.)[10]

SAGAN'S BIBLIOGRAPHY

Jung, Carl G., *Flying Saucers: A Modern Myth of Things Seen in the Skies* (New York 1959).

Menzel, D. H. and Boyd, L. G., *The World of Flying Saucers: A Scientific Examination of a Major Myth of the Space Age* (New York 1963).

Ruppelt, Edward J., *The Report on Unidentified Flying Objects* (New York 1956).

Shklovsky, Iosif S., and Sagan Carl, *Intelligent Life in the Universe* (San Francisco 1966).

Tacker, Lawrence J., *Flying Saucers and the U.S. Air Force* (Princeton 1960).

Vallee, Jacques, Anatomy of a Phenomenon (Chicago 1965).

CHAPTER FOUR

STATEMENT OF ROBERT L. HALL

(Afternoon Session.)
MR. ROUSH. The committee will be in order.
This afternoon we are going to hear first from Dr. Robert L. Hall. Dr. Hall is professor and head of the Department of Sociology at the University of Illinois, and has been since 1965. He too has a distinguished career. Dr. Hall, we are glad to welcome you as a participant in this symposium, and you may proceed.

☛ DR. HALL. Thank you, Mr. Roush.
First I should like to state a few of the rather well-established facts as they would be seen by a social psychologist. I find that when I do so, there is a great deal of redundancy. You have heard most of these facts before, so I will make my presentation brief.
Fundamentally what we know that everyone can agree upon is that a great many people all over the world keep reporting some quite puzzling flying objects. In these reports there are certain recurring features, and the people so reporting often have all the characteristics of reliable witnesses.
Second, the next main thing we know is that there are several strongly, often bitterly competing systems of belief about how to explain these observations, and some rational men seem to fall into line supporting each of these positions.
This in itself is of course of great interest to a social psychologist. Inevitably he is interested in how systems of belief grow and are maintained.
The third major factual thing that can be quite well agreed upon is that to a very large extent these alternative explanations, these systems of belief, have become rooted in organizations of people who have become committed to defending their respective positions. This greatly complicates the problem of arriving at a generally accepted explanation. In that sense, in addition to any other problems that have been defined here, clearly we have a social psychological problem also.
These are very briefly the main outlines of the facts as I see them. Now, how are these explained?

There are certain things that everyone seems to agree upon, or nearly everyone, I believe. First that a great many of these observations can be quite clearly identified as mistakes on the part of the observer, misidentifications of familiar objects, hoaxes, and a miscellaneous collection of similar things.

Beyond that point, there comes to be a good deal of divergence in explanations, to say the least. Perhaps the major views now can be classified simply as follows: First, that these are technological devices or vehicles of some sort entering our atmosphere from the outside.

Second, that this is some new, as yet ill-understood natural phenomena, something like a form of plasma, that we do not understand, and so on.

The third major hypothesis to explain the hard-core cases that are not otherwise agreed upon, is that they too are simply a result of mass hysteria, and its resulting misidentifications.

This hypothesis I will address myself to particularly very soon, because obviously a social psychologist has a special interest in this possibility.

The three major topics that I believe I should address myself to are, first, what has brought about this complicated situation of strongly opposed beliefs that seem to resist the factual evidence, and are not responsive to each other?

Second, what are the probable consequences from the point of view of a sociologist or social psychologist of each of the major explanations?

And third, I would like to comment quite explicitly on the hypothesis that mass hysteria and hysterical contagion is common in many of the cases.

I believe I should start with the mass hysteria hypothesis. To begin with, I think there is very strong evidence that some of the cases do result from hysterical contagion in the sense that this has often been used by social psychologists. Once people are sensitized to the existence of some kind of a phenomenon (whether indeed it really exists or not), when there is an ambiguous situation requiring explanation, when there is emotion or anxiety associated with this, resulting from the uncertainty, there are precisely the conditions that have been observed repeatedly as resulting in what I shall call "improvised news." Lacking well-verified facts and explanations, people always seem to generate the news and the explanations that will reduce the ambiguity, thereby reduce the anxiety they have about uncertain situations.

There are many well-documented cases of this kind of mass

hysteria and hysterical contagion. I believe it will be out of place for me to go into lengthy discussions of these episodes, but I shall comment on a few ways in which we can examine the observations of unidentified flying objects to assess whether this is a reasonable hypothesis for the hard-core cases.

One of the first of these is one thoroughly familiar to attorneys, social psychologists having no monopoly on an interest in the credibility of testimony, but this is one of the principal means obviously of establishing whether we should reasonably believe certain explanations.

The criteria, as most of you know, involve such things as the established reputation of the witnesses, the quality and details of the report, whether there are apparent motives for distortion or prevarication, whether there was preexisting knowledge of the thing being reported, whether there were multiple witnesses and whether there was contact among these multiple witnesses, whether observation was through more than one medium (for example, direct visual observation confirmed by radar), whether there were verifiable effects that could be observed after the reporting by witnesses, recently of the events being reported, the duration of the period in which the witness was able to observe the phenomenon; how the witnesses reacted, whether they had intense anxiety and emotion themselves, which might interfere with their observation, and so forth.

These are some of the major factors, and a closely related factor in assessing the credibility of the testimony is of course an assessment of the care in gathering the testimony by interviewers themselves.

How does the testimony on hard-core UFO cases look with reference to these criteria? I should say that there is a substantial subset of cases which look very good on these criteria, which make it very difficult to say that the witnesses involved were victims of hysterical contagion, grossly misinterpreting familiar things.

For example, there is the Red Bluff, Calif., case in 1960, where two policemen observed for 2 hours and 15 minutes constantly, apparently without tremendous anxiety or concern, an object hovering, moving about, going through gyrations. Twice it approached their police car. When they tried to approach it, it would retreat.

They radioed in and requested that this object be confirmed on radar, and it was confirmed by local radar stations at approximately the same location.

Ultimately, after a couple of hours of observation, they watched this object move away, join a second similar object, and then disappear. They then went to the sheriff's office, where two deputies were present who had also seen this phenomenon, and gave similar descriptions.

Now applying the criteria to a case such as this, in most respects it is very convincing. These are police officers of good reputation. Their report was prompt, thorough, careful, and in writing—and I have read the report in full. There is much detail in it of a sort that could be cross-checked with the other witnesses from the sheriff's office. There are no apparent motives for prevarication or distortion. It was a long period of observation.

I cannot establish very clearly what prior interest or information these witnesses had, but I find no indication that they had any. There was confirmation of the observation from more than one medium of observation—both visual and radar.

This is the kind of case that leads me to regard the hypothesis of hysterical contagion as being quite inadequate to account for these observations. It is not a lone case; there are many others.

There were trained ground observers near White Plains, N.Y., in 1954, who observed an object which they described as having the apparent size of the moon, while simultaneously they saw the moon, which was not full that night. They watched this for 20 or 30 minutes, then it moved away to the southeast.

Two radar stations established fixes confirming the visually reported location. Jets were scrambled from two bases to intercept. The ground observers were able to see the jet trails approaching. Both the pilots of the jets and the ground observers report that as the jets approached, this object changed color and moved up very rapidly and disappeared, and at that point radar contact was also lost.

Once again this is the kind of report that seems to me to fit the customary criteria of credibility to a very considerable degree. It is very difficult to claim that these multiple observers, trained for the type of observation they were making, confirmed independently through more than one channel, were victims of hysterical contagion.

Dr. McDonald, I believe, referred briefly to the Levelland, Tex. cases in 1958, of interference with automobile ignition, in which there were 10 separate sightings in that one evening, apparently with no opportunity for the citizens involved either to read the news, hear the news of this, nor to talk with one another. They uniformly reported the same general shape. They uniformly reported—a great many of them reported also interference with automobiles ignition and headlights. This was an effect which at that time had not been observed and publicized a great deal. It subsequently has become publicized.

Now, how do these cases differ from the well-known, documented cases of mass hysteria and hysterical contagion? In general those episodes have not persisted as long as the active interest in unidentified flying objects. It lasted a week or a few weeks, and it had not been too difficult to find reasonably acceptable explanations.

In the second place, they have not generally involved a prolonged observation of a phenomenon by people who were calm, not emotionally upset. A characteristic example of hysterical contagion would be the recent study by Back and Kerckhoff, supported by the National Science Foundation. The book reporting on this study is called "The June Bug." It was a case of hysterical contagion among the employees of a factory in North Carolina.

It is one of the most thoroughly reported and studied incidents of this sort. It resembles the kind of thing we are talking about in almost no respect. I find it very difficult to find elements in common, other than the fact that some people believed something that was difficult to verify.

The employees were convinced that they were being bitten by poisonous insects, resulting in fainting and other symptoms such as rashes. All medical officers, all careful research on this, was unable to turn up any hard evidence that such an insect was present, or that there was any standard medical accounting for these symptoms. But these were people in close constant contact, sharing a particular set of problems and frustrations that raised their level of anxiety.

The epidemic can be interpreted as a convenient way of escaping the problem of coping with very difficult circumstances. I have said that I think in isolated cases you can find a similar thing in observations of unidentified flying objects, but if we look at the hard-core, well-documented cases, I see practically no resemblance.

Another important thing to note about the witnesses in the best sightings of UFO's is that very commonly—as has been mentioned, I believe, by Dr. McDonald—they first try to explain their observation in some very familiar terms. This is the well-known and labeled psychological process of "assimilation." People first try to assimilate their observation into something understood and known and familiar.

This is quite contrary to the kind of argument frequently built into the hypothesis of hysterical contagion, namely, that characteristically witnesses are eager, are motivated, to see strange objects.

Another important thing to notice about the witnesses in these cases is of course their reluctance to report. We have had some mention of that. This, for one thing, counters the argument of publicity seeking as a motive in some of the best cases. It incidently runs contrary to most experience of social psychologists engaged in public opinion research, in polling, and contrary to the experience of experienced precinct workers in politics. Those people who have not tried this kind of thing expect people not to want to talk to them, but when you start ringing doorbells, the striking thing about the American people is it is often difficult to stop them from telling you what they believe. Yet in instances of unidentified flying objects, there has often been a marked reluctance to talk about them.

I can illustrate this anecdotally simply to make my point. When I was on the faculty at the University of Minnesota, a student came to me, having heard that I had some interest in this question. He informed me that his father, a colonel, an artillery colonel in Korea this was at the time of the Korean conflict—had flown over a hill in Korea in his observer plane, and found (right next to him virtually) a characteristic unidentified flying object with the usual kind of configuration. It had promptly retreated upwards. It had frightened him, but he was an experienced and trained observer, so he took notes on it; he recorded it. When he returned he was so ridiculed and laughed at for a long period of time that he completely gave up trying to have this taken seriously. He refused to talk about it.

I urged this student to get his father to report this to some of the private organizations that might take it seriously, and he apparently was unable to do so. The ridicule suppressed the opportunity for this information.

I have encountered similar things in academic colleagues from a variety of fields, finding they are very interested and wanting to hear about this, but are afraid to talk about it.

In order to support the hysterical contagion hypothesis, it seems to me we need to present some plausible evidence:

First, that there is a very ambiguous situation. This we can all agree upon.

Second, that there is a great deal of anxiety and concern about it. This appears clearly to be the case.

Third, some plausible evidence of contact among the witnesses, either directly by conversing with one another, or indirectly by being exposed to the same information, the same stimuli. In cases that I have studied. I find that this third element is the one that is often lacking, that there are often witnesses who appear not to have had prior knowledge, not to have had contact with one another, not to have been exposed, as far as we can determine, to the same news information.

I might throw in here, in reference to a remark Dr. Hynek made, that the public is indeed very unwilling to accept the kinds of casual and bland explanations that have been offered. This has been my experience also, and is indeed an index of the amount of concern and anxiety about this, it appears to me.

Now I will turn to another subject. I might summarize in one sentence that in my eyes the hypothesis that the hard-core cases of observed UFO's is hysterical contagion is highly improbable. The weight of evidence is strongly against it.

Now I would like to address the question of what has brought about this situation of strongly opposing beliefs that seem not to become reconciled with one another. On this I will have to digress first to explain briefly what I mean by a system of beliefs in social psychology.

Perhaps the best way to explain that is to say that just as nature abhors a vacuum, nature abhors an isolated belief. Neither a belief nor the person who holds it can normally persist very long in isolation. The beliefs become organized in such a way that, for one person, his various beliefs support one another, and people gather together in organizations to lend each other support in their beliefs. This is the sense in which we have highly developed systems of belief which come to resist change, to resist evidence.

The circumstances under which systems of belief such as this characteristically arise are, as I mentioned in passing before, a situation of ambiguity about a matter of importance on which there is not reliable, verified information in which people have confidence. Clearly the antidote is simple. It is to get good, reliable information which people have confidence in.

This is probably the only way to weaken the irrational elements that are strongly resistant.

Finally, I want to comment to some extent on the probable consequences of each of the most important explanations that has been offered, and what might be done in the public interest in each instance to counter the negative aspect of these consequences.

Let's suppose to start with that these are extraterrestrial devices of some sort visiting our atmosphere. If this is the case, we for one thing have to concern ourselves with the possible consequences of contact with civilizations which are technologically very advanced and whose values we know nothing about. It is very tempting to the anthropomorphic, to attribute human characteristics to any such life form hypothesized, and to imagine, like humans, they might be hostile and might cause us some danger.

I know of no hard evidence of danger, of threat, from the cases reported. But we do not have any inkling, if indeed these are extraterrestrial devices, as to their purpose. We have no hard evidence as to their purpose, their intent, their motives, so to speak.

Consequently, I find it extremely difficult to even speculate in an intelligent way about what might result from contact with them. I can say a very great risk of contact, if this is the case, is the risk of panic, and panic is often very harmful to us mere humans, as in theater fires and so on.

Once again from all knowledge in sociology and social psychology, the best way to counter this risk of panic is not to issue reassuring statements, but to find sound information in which people have confidence which can reduce their anxiety about the situation, and explain it adequately. This to me has been one of the most unfortunate and possibly dangerous aspects of this problem, that the ridicule, the tendency not to take the problem seriously, to issue reassurances rather than good information, has in my opinion only maximized the risk of panic, at least under this hypothesis, and I believe under the others as well.

Another risk, if these are extraterrestrial devices is clearly the risk of misinterpreting the devices as hostile devices from another country on earth, which might trigger indeed a devastating nuclear war. Once again, the same conclusions follow about the need for good information.

MR. ROUSH. Might not another conclusion be that if there should be something to this, again, if there should be perhaps it would bring all the people of the world together for a better understanding, a common purpose, and a common stand, which probably would relieve us of some of our own anxieties?

DR. HALL. This is indeed within the range of possibility, though I hesitate to speculate on the probability.

MR. ROUSH. You don't have to speculate. Go ahead.

DR. HALL. The final comment about probable effect, if these are indeed extraterrestrial devices, is of course the possibility of learning something of great technological value from them. The possible value of contact for purposes of advancing our knowledge of our technology.

Let's turn then to another hypothesis, which is this is a natural phenomenon which we do not understand, something like plasma. In this case, I think we have precisely the same risk of panic through misinterpretation resulting in precisely the same recommendation for the need for understanding to reduce the risk of panic.

I think we have precisely the same risk of misinterpretation as hostile aircraft, with again the same resulting recommendation.

I think we have again the same possible great value from understanding the phenomenon in order to advance our knowledge.

The third major hypothesis, explanation, which I cited above, is that even the most solid and plausible cases reported are results of mass hysteria and hysterical contagion. I simply note that if this is the case, I regard it as *prima facie* evidence that we badly need to improve our understanding of mass hysteria, of the process of belief formation, of the means by which we might control the kinds of anxiety that produce this problem.

In this situation there is still the dangerous risk of panic, even if there is no physical phenomenon underlying these reports. There is still the risk of misinterpretation of hostile aircraft, and I would submit that there is still the great potential benefit from studying it thoroughly and scientifically, in this case the gain being a gain in sociological and psychological knowledge, which would be of obvious importance if all of this is caused simply by mass hysteria.

I have a few conclusions and recommendations which I have written out. I will try to tie these to what others have said as I go along.

My first conclusion would be that no matter what explanation you accept, we have here a rare opportunity for gaining some useful knowledge by a thorough detached study of UFO reports, and a systematic gathering of new information, hopefully with good instrumentation, and good, well-trained interviewing teams.

My second conclusion would be that hysteria and contagion of belief can account for some of the reports, but *there is strong evidence that there is some physical phenomena underlying a portion of the reports* [my emphasis, L.S.].

Third, I would conclude that because of the lack of trustworthy information the systems of conflicting beliefs has been built up to account for a very ambiguous set of circumstances. Each of these positions is sometimes defended beyond the point of rationality.

Fourth, I would repeat my earlier statement as a conclusion, that whether or not there is a physical phenomenon underlying a portion of the reports, we clearly have a social psychological problem of subduing these irrational systems of belief, defense of beliefs, of lowering the anxiety about these reports, and of reducing the ambiguity about their nature.

The recommendations that I had written out were two—excuse me, were three, and overlap considerably with the comments of my colleagues. I would say that the most important matter is to promote the fullest possible free circulation of all the available information about this phenomena. This should help reduce risks of panic and other dangerous irrational actions. It should help to weaken these systems of belief, the irrational elements in them. Here I would say indifference, or disinterest on the part of national leaders can retard our learning about this phenomenon, and open interest and encouragement can help.

I believe you are performing a fine service in having this kind of open inquiry. *This whole matter badly needs to be treated as something deserving serious study* [my emphasis, L.S.].

The second recommendation I have to make concerns some general lines of research that would seem to me called for. One of these seems to me would be to take the 100 or 200 cases per year that seem to be reliably reported and reasonably well documented, and to study them carefully for recurring patterns, with emphasis on the way they react to their environment, the way they react to light sources, the way they react to presence of humans and so on.

The second form of research would be, I think to study explicitly those portions of the problem that do result from mass hysteria, apparently. These need to be studied intensively, quite apart from the question of the physical phenomena, to improve our understanding of mass hysteria and panic, and its possibly dangerous consequences.

In doing this I think it is terribly important that particular observations be studied by the scientists of a variety of disciplines, that the study of the hysteria hypothesis not be separated from the others. If it is, there is a tendency to make this hypothesis the garbage can for otherwise unexplained sightings.

The third type of study that seems to me terribly important, but my colleagues at the table can speak with more authority than I, is the systematic gathering of new cases with good scientific instrumentation, the kind of work in quantitative evidence that would give us much more to go on.

The third recommendation I had to suggest was that possibly in addition to a careful scientific investigation and study of this phenomenon, it might be fruitful to set up formally an adversary proceeding modeled after our system of jurisprudence. There is a tendency for us academics to sit on fences as long as we possibly can, and I think that if there were several teams of investigators who were assigned the responsibility much the way a prosecuting attorney or defense attorney is, assigned the responsibility to make the strongest possible case for one of the systems of explanation, that this would challenge the others, and force them to find more solid evidence.

It would try to benefit from some of the valuable features we have in our system of jurisprudence.

That concludes my presentation, except to comment briefly on how this relates to the suggestion of my colleagues. I would certainly enthusiastically agree with Dr. Hynek's suggestion of a board of inquiry, or some competent group to study the phenomenon.

I would certainly agree with Dr. McDonald's view that a variety of approaches would be fruitful, that a single study has many disadvantages. I have taken an interest for a number of years in the problems of the support of academic institutions by Government, and I think that we are most likely to proceed to some good knowledge rapidly if we don't put all our eggs in one basket.

I certainly agree with Dr. Sagan's view that there are these very intense predisposing emotional factors for each of these beliefs. Somehow we need to weaken those.

Finally, on the idea of UN cooperation, this had not occurred to me, but I think it is an excellent idea. If it is possible to establish some detached international agency that can bring about free, open flow of information, and some cooperation internationally in investigating this, it would be helpful.

Thank you, Mr. Roush.

MR. ROUSH. Thank you, Dr. Hall.

Are there questions? [No replies.]

(The prepared statement of Dr. Hall follows)

PREPARED STATEMENT BY ROBERT L. HALL

From the point of view of a social psychologist, UFO reports present us with a most interesting and challenging situation. To a social psychologist the known facts appear to be facts about people and the things that they are saying and doing. First, many people, all over the world, including reliable and knowledgeable witnesses, keep reporting puzzling flying objects, and the reports have certain recurrent features. Second, several competing systems of belief have grownup to explain these reports, with some rational men supporting each of several different explanations. Third, as any sociologist would predict, the systems of belief have, to a large extent, become rooted in complex organizations of people: some organizations have been created to defend a particular position about UFOS; some organizations whose main purposes are remote from UFOs have been drawn into the controversy and found themselves committed to defending a position.

Nearly all rational observers appear to be agreed that the great majority of reported sightings of unidentified flying objects (UFOs) can be explained as misidentifications of familiar phenomena, with an occasional hoax contributing to the confusion. However, there are approximately 100 to 200 cases per year, based upon apparently sound testimony, with recurrent features of appearance, movement, and reaction to the environment. Strong disagreement arises over these cases.

One major area of disagreement is the question whether any novel physical phenomenon underlies these reports or whether they are simply a miscellaneous collection of familiar phenomena, misidentified because of mass hysteria and misperceived as having recurrent features because of a process of hysterical contagion. Among those who believe that there is a physical phenomenon, there are, in turn, several alternative explanations as to what it is. A substantial number argue that there are technological devices or vehicles entering our atmosphere from the outside. A substantial number argue that there is a novel natural phenomenon, as yet ill understood, such as a form of plasma or "ball lightning." There are other explanations supported by some people, such as the belief that "space animals" are swimming around in our atmosphere, or that these objects are secret devices manufactured somewhere on earth. In my judgment these last two explanations fit the available evidence so poorly that I shall not deal with them further. We might, then, label the three major hypotheses: (1) Mass hysteria and contagion; (2) Extraterrestrial devices; (3) New natural phenomenon.

My comments, as a social psychologist, will be organized around three major questions: (1) What has brought about this situation of competing systems of belief, strongly held and often unresponsive to the observed facts, and how can we modify the situation? (2) Is the mass hysteria hypothesis a plausible one, and can it account adequately for the known facts? (3) For each of the major explanations, what would be the probable consequences if the explanation were true, and what actions or precautions might be taken in the public interest?

How did the present situation come about? Much sociological research on rumor and belief systems indicates that ambiguity about an important matter begets improvised news. To the extent that trusted information is not available, systems of belief are generated to fill the gap. A recent scholarly work by [Tamotsu] Shibutani describes a rumor as a kind of improvised news which "... arises in situations of tension when ordinary communication channels are not operating adequately." (Shibutani, 1966, p. 57). Shibutani further argues that people are always being confronted with new circumstances which are not clearly and adequately treated by trusted channels of information, and therefore rumors are a normal and important part of men's efforts to adapt to their environment (p. 161, 182-183). Alternative explanations of UFO reports have arisen because of a lack of sound, authoritative information in which people have confidence. This is a normal and usual reaction to such situations of ambiguity.

In order to complete my answer about how the present situation came about, I must digress briefly to explain what I mean by a "system of belief." Just as nature abhors a vacuum, nature abhors an isolated belief. Neither a belief nor the person who holds it can normally persist long in isolation. Each person's beliefs tend to become organized into an interdependent system of beliefs which support one another. Also people who share important beliefs typically become organized into social groups in which members support one another's beliefs. Hence a particular belief, such as the belief that there is no new physical phenomenon underlying UFO reports, is intricately tied in with two systems—a system of related beliefs by the same person, and a social system of people who share similar beliefs. Many social psychologists have analyzed and documented this kind of phenomenon (e.g., Festinger, 1957; Simmons, 1964; Smith, Bruner, and White, 1956).

In circumstances such as those described, the ambiguous situation is often associated with widespread anxiety, and the belief systems which arise characteristically contain elements of hysteria which may increase the likelihood of panic or other irrational action (see Smelser, 1963). New beliefs which are improvised to reduce the ambiguity often are assimilated into preexisting belief systems, such as the beliefs of religious cults, so that, in effect, the ambiguous situation is used to manufacture support for preexisting beliefs. Once a situation of competing belief systems is established, probably the only way to modify it very much is by attacking the conditions which brought it about—that is, the lack of authoritative, trusted information.

My second major question, stated above, was: "Is the mass hysteria hypothesis a plausible one, and can it account adequately for the known facts?" First, let me reiterate the facts which we are trying to explain: numerous reliable reports, and recurrent features in these reports. When a large number of people report observations that share many details of appearance and behavior, one of three things must be the case: either they are observing the same phenomenon, or they have been exposed to the same sources of information which have influenced them to expect to see certain things, or they have been in mutual contact and influenced one another in some fashion. If the mass hysteria hypothesis is to be held plausible, we need to show that separate people reporting the same details have been in touch with each other or with some common source of information. The independence of the separate observers becomes a crucial question.

In determining the plausibility of this hypothesis, a second major concern is the credibility of testimony. Much of our legal system is based upon the assumption that we can, under appropriate conditions, accept human testimony as factual. Social psychologists are certainly not alone in having developed criteria for assessing the credibility of testimony; attorneys are thoroughly familiar with such criteria. In assessing testimony we customarily consider such questions about the witness as his reputation in his community, whether he has any apparent motive for prevarication or distortion, whether he has previous familiarity with the things reported. In addition we consider internal characteristics of his report, such as the recency and duration of the events reported, the number and type of specific observable details reported as apart from reports that are primarily interpretations, the inclusion of details that are independently verifiable (such as physical effects). Also testimony is, of course, more credible if there are multiple witnesses, especially ones who are completely independent of one another; if there have been different means of observation (e.g., both visual and auditory, or unaided observation and observation through instruments); and if the testimony is gathered by qualified, careful interviewers.

In my judgment there are many reports of UFOS that meet the above criteria quite well—better, indeed, than many court cases which a judge and jury accept. In some of these cases, no familiar explanation can be found that fits the evidence. I shall digress briefly to describe a few cases.

Consider the case of two police officers near Red Bluff,

California, on August 13, 1960. They saw a large object descending and at first thought it was an airliner about to crash. They jumped from their patrol car and noticed that the object made no discernible noise. They watched it descend to an estimated 100 or 200 feet, then reverse itself at high speed, and finally stop and hover at an estimated 500 feet. They described details of shape, color, and movement. They radioed the sheriff's office to contact a local radar base and were informed that the radar base reported an unidentified radar return at the same location as their visual observation. They reported details of the object's behavior and their own. They tried to approach the object and it retreated; when they remained stationary, the object approached their car. They reported that the object retreated when they turned on the patrol car's red light. After prolonged observation the object began to move away, and they followed slowly. They saw it join another similar object and finally disappear over the horizon. Altogether they watched the object for about two hours and fifteen minutes. Their report was prompt, thorough and written, and contained details which are contained in many other UFO reports. Immediately after losing sight of the object, the officers returned to the sheriff's office and met two deputies who reported the same observation. The officers were men of good reputation, and there is no indication of prior interest in UFOs nor prior knowledge of the kinds of details reported (e.g., the red light beam emitted by the object and radio interference each time it came near). These men have subsequently been contacted by people with scientific training and have confirmed various details of their report.

Another case of interest occurred near White Plains, New York, in late summer, 1954, and was reported by James Beatty, an experienced ground observer corps supervisor in an Air Force Filter Center. At about 9:50 p.m. an observer team about 20 miles southeast of Poughkeepsie saw an object similar in apparent size to the moon. At the same time they could see the moon, which was not full. They watched the object for about 20 or 30 minutes and then it moved slowly southeastward. According to the report of the supervisor, two radar stations had fixes corresponding to the visual sighting, and jets were scrambled from two airbases. As the jets approached the object, both the pilots and the ground observers report that the object changed color, moved upward at very high speed, and disappeared. At this point radar contact was lost, too.

In the vicinity of Levelland, Texas, on the night of November 2-3, 1957, there were ten separate sightings by several people, including police officers, over a period of approximately 2½ hours.

The descriptions were similar in several important details of visual appearance. Several observers independently reported that their cars' engines and headlights quit working when the object was close. This kind of effect has been frequently reported but had not been publicized prior to this group of reports in Levelland. In most instances it is clear that the witnesses around Levelland were going about their usual business and were surprised by the sighting; they had not been alerted to watch for a strange object.

These are only three cases out of many (see Hall, 1964, and U.S. Air Force, 1968). They are reported only sketchily here, but much more detail is available (see Hall, 1964). I introduce them only to illustrate the kinds of evidence available relevant to the hypothesis of mass hysteria to account for UFO sightings. I am forced to the conclusions that there are many sightings by multiple observers and that many observers are reliable and independently report similar details. In many instances it appears highly unlikely that they could have been exposed to similar detailed information in advance (e.g., the electrical interference effects at Levelland).

Social psychologists have studied a number of cases of mass hysteria and hysterical contagion (Cantril, 1940; Johnson, 1945; Kerckhoff & Back, 1968; Medalia & Larsen, 1958). In my judgment the "hard-core" reports of UFOs do *not* resemble those documented cases. Those cases were generally short-lived—a day, a week, or at most a few weeks; UFO reports have persisted for decades, at least, despite much ridicule and very little recent press coverage of serious cases. The documented cases of mass hysteria have not involved calm, prolonged observations such as the police officers near Red Bluff, California. The documented cases have had some plausible indication that the people involved have been in touch with one another (Kerckhoff & Back, 1968) or previously exposed in common to the information that they incorporate into a report (e.g., Johnson, 1945; Medalia & Larsen, 1958). The documented cases have not been worldwide, as are UFO reports. They have not involved phenomena that were simultaneously observed through such different media as direct visual contact and radar contact. In documented cases of mass hysteria I do not know of evidence of people reluctant to report; in UFO sightings there are numerous such cases. The hypothesis of mass hysteria does not, in my judgment, fit the "hard-core" reports very satisfactorily.

The third, and last, of the three major questions which I raised at the beginning was: "For each of the major explanations, what are the likely consequences, and what actions or precautions might we take in the public interest?"

First, let us suppose that there are extraterrestrial devices entering our atmosphere from the outside. We must then concern ourselves with the possible consequences of contact with technologically advanced civilizations whose values, or intentions, or motives are totally unknown to us. It appears to me an almost impossible task to predict the probable effects of contact between our earthly civilization and another civilization without making some clear-cut assumptions about their values and motives. I have not been able to find any rational basis for defending particular assumptions of this kind, and I shall not attempt the task. We must also be concerned with the risks of panic resulting in people hurting one another, even if the assumed extraterrestrial visitors mean no harm. This risk could be markedly reduced by preparing the public for the eventuality—by treating it as a serious possibility that must be discussed. The greatest risk of panic would come from a dramatic confrontation between the assumed "visitors" and a collection of humans who were unprepared and had been told that their leaders did not believe such visitors existed. Another risk is that we might misinterpret such devices as weapons of another country and thereby accidentally trigger nuclear war. If these are extraterrestrial devices, we have, of course, a great opportunity to learn from their technology, which would appear to be very advanced in certain respects by our terrestrial standards.

Second, let us suppose that this is a novel natural phenomenon which we do not understand. Under this assumption we still run the risk of panic if a crowd of people are confronted with a case of the phenomenon without any preparation. We still run the risk of misinterpreting an occurrence as a hostile weapon system. Also, it is a reasonable assumption that we might reap scientific and technological benefit from understanding such a puzzling thing that appears to involve some kind of concentration of energy.

Third, let us suppose that the whole persistent business of UFO reports over the years is strictly a social psychological phenomenon—a new and extreme case of mass hysteria and hysterical contagion. In this event the underlying anxiety must indeed be massive and the risk of panic accordingly very great unless we can introduce trusted information and reduce the ambiguity and anxiety. Under this assumption—if atmospheric and astronomical observations can be so badly misinterpreted and so badly reported by many people of good reputation and good education—then I would judge that we run great risk of misinterpreting those same phenomena as hostile weapons, and we must prepare for this risk. Most important, if this whole business is a social psychological phenomenon, then this is *prima facie* evidence of the urgent need to improve our understanding of the processes of mass hysteria, belief formation, and means of controlling the kinds of anxiety that generate such a problem. In this event the UFO reports present an unsurpassed natural laboratory for research on mass hysteria, human response to ambiguity, standards for assessment of human testimony, and other related matters.

CONCLUSIONS

1. No matter which explanation is correct, we have a rare opportunity for gaining useful knowledge by a thorough, detached study of UFO reports and a more systematic gathering of new evidence.

2. Hysteria and contagion of belief can account for some of the reports of UFOs, but the weight of evidence suggests strongly that there must be some kind of physical phenomenon which underlies a portion of the reports.

3. Because of the lack of trustworthy information about UFO reports, systems of conflicting belief have been built up to account for this ambiguous set of circumstances, and each position is sometimes defended beyond the point of rationality.

4. Whether or not there is a physical phenomenon underlying a portion of the reports, we now have, in addition to any other problem, a social psychological problem of subduing irrational defense of beliefs, lowering anxiety about the reports, and reducing ambiguity about the causes of the reports.

5. Our lack of understanding of UFO reports forces us to run unnecessary risks of panic and of accidental triggering of nuclear war.

RECOMMENDATIONS

1. The most important and urgent matter is to promote the fullest possible circulation of all available information about UFOs and to encourage systematic gathering of new evidence. This should help to reduce the risks of panic and other dangerous irrational actions. This should also help to weaken the irrational elements incorporated into opposing systems of belief. Indifference or disinterest on the part of national leaders can retard our learning about the phenomenon at hand; open interest and encouragement can help. The whole matter needs to be treated as something deserving serious study.

2. At least three lines of serious research should be undertaken:

a) For the 100 to 200 cases per year that are reliably reported and well documented, we need to study carefully reports of recurring patterns of behavior by the phenomenon, including its apparent reaction to other events in the environment with emphasis upon establishing independence or non-independence of separate witnesses. b)Those portions of the problem that result from mass hysteria need to be studied intensively to improve our understanding of mass hysteria toward the end of controlling its potentially dangerous consequences. c) Some systematic means of monitoring and observing should be developed so as to add well documented new cases with specific reports of details obtained independently from different observers.

3. Serious consideration might be given to the idea of setting up a formal adversary proceeding, modeled after our system of justice. Just as courts have attorneys assigned to build the best possible case for the prosecution and others to build the best case for defense, we might have a staff assigned to build the strongest possible case for each of the three major explanations of UFO reports. If each had to confront the others and answer their criticisms, we would probably force a clearer focus on the crucial points that need to be settled.

My closing comment returns to my starting point. The situation that we face in UFO reports is an exciting and challenging one which presents a rare scientific opportunity, no matter whose interpretation and explanation you may accept.[11]— STATEMENT OF DR. ROBERT L. HALL, HEAD, DEPARTMENT OF SOCIOLOGY, UNIVERSITY OF ILLINOIS, CHICAGO, ILL.

HALL'S REFERENCES

Bauer, R. A. and Gleicher, D. B., 1953. "Word-of-mouth communication in the Soviet Union." *Public Opinion Quarterly*, 17, 1953, 297–310.

Cantril, H., 1940. *The invasion from Mars*. Princeton: Princeton University Press,1940.

Festinger, L., 1957. *A theory of cognitive dissonance*. Evanston: Row-Peterson,1957.

Hall, R. H. (ed.), 1964. *The UFO evidence*. Washington, D.C.: N.I.C.A.P., 1964.

Johnson, D. M., 1945. "The 'phantom anaesthetist' of Mattoon: a field study of mass hysteria." *Journal of Abnormal and Social Psychology*, 1945. 40, 175–186.

Kerckhoff, A. C., and Back, K., 1968. *The June bug*. New York: Appleton-Century-Crofts, 1968.

Medalia, N. Z., and Larsen, O. N., 1958. "Diffusion and belief in a collective delusion: the Seattle windshield pitting epidemic." *American Sociological Review*, 1958, 23, 180-186.

Shibutani, T., 1966. *Improvised news: a sociological study of rumor.* Indianapolis and New York: Bobbs-Merrill, 1966.

Simmons, J. L., 1964. "On maintaining deviant belief systems: a case study." *Social Problems*, 11, 1964, 250-256.

Smelser, N., 1963. *Theory of collective behavior.* New York: Free Press of Glencoe, 1963.

Smith, M. B., Bruner, J. S., and White, R. W., 1956. *Opinions and Personality.* New York: John Wiley & Sons, 1956.

U.S. Air Force, 1968. Projects Grudge and Bluebook Reports 1-12 (1951-1953) Washington, D.C.: N.I.C.A.P., 1968.

CHAPTER FIVE

STATEMENT OF JAMES A. HARDER

MR. ROUSH. Our next participant is Dr. J. A. Harder.

Dr. Harder, we are delighted that you can participate. We are getting into another area here now. Again, as with the other gentlemen, Dr. Harder has a distinguished career behind him and probably an even more distinguished career ahead of him.

Dr. Harder, will you proceed.

☛ DR. HARDER. Thank you, Mr. Chairman.

Your committee has asked me to comment on the problem of propulsion as raised by some reports, and to whatever potential benefits there might be to the aerospace programs from an intense scrutiny of UFO phenomena.

I am very glad for this opportunity to present to your committee some of my views on the problems of unidentified flying objects and to indicate some of the areas in which I think a closer investigation of this problem might provide us with scientific clues that would give us important impetus to basic and applied research in the United States.

As Dr. Hall has said, there have been strong feelings aroused about UFO's, particularly about the extraterrestrial hypothesis for their origin. This is entirely understandable, in view of man's historic record of considering himself the central figure in the natural scene; the extraterrestrial hypothesis tends inevitably to undermine the collective ego of the human race. These feelings have no place in the scientific assessment of facts, but I confess that they have at times affected me.

Over the past 20 years a vast amount of evidence has been accumulating that bears on the existence of UFO's. Most of this is little known to the general public or to most scientists. But on the basis of the data and ordinary rules of evidence, as would be applied in civil or criminal courts, *the physical reality of UFO's has been proved beyond a reasonable doubt* [my emphasis, L.S.]. With some effort, we can accept this on an intellectual level but find a difficulty in accepting it on an emotional level, in such a way that the facts give a feeling of reality. In this respect, we might recall the attitude many of us have toward our own deaths: We accept the facts intellectually, but find it difficult to accept them emotionally.

Indeed, there are flying saucer cultists who are as enthusiastic as they are naive about UFO's—who see in them some messianic symbols—they have a counterpart in those individuals who exhibit a morbid preoccupation with death. Most of the rest of us don't like to think or hear about it. This, it seems to me, accurately reflects many of our attitudes toward the reality of UFO's—natural, and somewhat healthy, but not scientific.

In my remaining statements you will note that I have tacitly assumed the reality of UFO's as a hypothesis underlying my assessment of the importance of this subject for scientific study.

1. THE UFO PROPULSION PROBLEM

By way of introducing the propulsion problem of UFO's, I will review a sighting near the city of Corning, in northern California, during the night of August 13, 1960, by two California highway patrolmen. During that night, and several succeeding nights, there were many reports of UFO's over northern California, but this particular event is important not only because of the fact that it has been well authenticated but because of the relatively long time and close nature of the observations. My condensed description that follows is from the official report filed the next day by the two officers (see appendix I and II) from a half-hour taped interview conducted 3 days later by myself and Dr. Carl Johannessen, of the University of Oregon; from a letter written by Officer Charles A. Carson to Walter N. Webb, Charles Hayden Planetarium, Boston, Mass., dated November 14, 1960; and from a telephoned interview conducted by Dr. James [E.] McDonald with Mr. Carson on October 27, 1966.

APPENDIX I
ADDITIONAL WITNESSES TO THE EVENTS OF AUGUST 13, 1960

When Officer [S.E.] Scott radioed the Tehama County Sheriff's Office, two deputies, Fry and Montgomery, traveled to Los Molinos, somewhat northeast of the first sighting, from which point they observed the UFO simultaneously, though at a greater distance, with Scott and Carson. The night jailer, from a point further north, saw it, and marched his several prisoners out onto the roof of the jail, each of whom saw it. Subsequently, due to newspaper publicity, Carson received a number of letters from motorists who had seen the object while traveling along Highway 99E. After Scott and Carson returned to the Sheriff's Office, they called the radar base again, and talked with the operator while all of them listened in on various extensions. He described to them some of the movements of the object as he had observed it on the radar screen; these corresponded to the movements they had observed.

Several months later, in a personal interview conducted on January 3, 1961, Captain Blohl, Area Commander at Red Bluff, stated that both Scott and Carson had worked for him for three or four years, and that he had the highest regard for their honesty and devotion to duty. They were not publicity minded, he said. Of Scott, he remarked, "Scotty would rather take a knife or a gun away from a man than make a speech." In the unwanted publicity that was forced upon them, Blohl stated that their story stood up unaltered—that there were no baubles during question periods.

APPENDIX II
REPORT OF OFFICERS CARSON AND SCOTT TO STATE HIGHWAY PATROL AREA COMMANDER

To: Area Commander, Red Bluff.
August 13, 1960.
From: C. A. Carson, No. 2358, S. E. Scott, No. 1851.
Subject: Unidentified Flying Object.

SIR: Officer Scott and I were eastbound on Hoag Road, east of Corning looking for a speeding motorcycle when we saw what at first appeared to be a huge airliner dropping from the sky. The object was very low and directly in front of us. We stopped and leaped from the patrol vehicle in order to get a position on what we were sure was going to be an airplane crash. From our position outside the car the first thing we noticed was an absolute silence. Still assuming it to be an aircraft with power off we continued to watch until the object was probably within 100 to 200 of the ground when it suddenly reversed completely, at high speed and gained approx. 500' altitude. There the object stopped. At this time it was clearly visible to both of us, and obviously not an aircraft of any design familiar to us. It was surrounded by a glow making the round or oblong object visible. At each end, or each side of the object, there were definite red lights. At times about 5 white lights were visible between the red lights. As we watched, the object moved again and performed aerial feats that were actually unbelievable. At this time we radioed Tehama County Sheriff's Office requesting they contact the local radar base. The radar base confirmed the U.F.O.—completely unidentified.

Officer Scott and myself, after our verification, continued to watch the object. On two occasions the object came directly towards the patrol vehicle. Each time it approached the object turned, swept the area with a huge red light. Officer Scott turned the red light on the patrol vehicle towards the object and it immediately went away from us. We observed the object use the red beam approximately 6 or 7 times, sweeping the sky and ground areas. The object began moving slowly in an easterly direction and we followed. We proceeded to the Vina Plains Fire Station where we again were able to locate the object. As we watched it was approached by a similar object from the south. It moved near the first object and both stopped, remaining in that position for sometime, occasionally emitting the red beam. Finally both objects disappeared below the eastern horizon. We returned to the Sheriff's Office and met Deputy Fry and Deputy Montgomery, who had gone to Los Molinos after contacting the radar base. Both had seen the U.F.O. clearly and described to us what we saw. The night jailer also was able to see the object for a short time, each described the object and its maneuvers exactly as we saw them. We first saw the object at 2350 hours and observed it for approx. 2 hours and 15 minutes. Each time the object neared us we experienced radio interference.

Sir, we submit this report in confidence, for your information, we were calm after our initial shock and decided to observe and record all we could of the object. Charles A. Carson, No. 2358. S. E. Scott, No. 1851.

Officers Scott and Carson were searching for a speeding motorcyclist along Hoag Road, east of Corning, Calif., between U.S. Highway No. 99W and 99E when they saw what at first appeared to be a huge airliner dropping from the sky. This was at 11:50 p.m. They stopped and leaped from the patrol car in order to get a position on what they were sure was going to be an airplane crash. From their position outside the car the first thing they noticed was an absolute silence. Still assuming it to be an aircraft with power off, they continued to watch until the object was probably within 100 to 200 feet off the ground, whereupon it suddenly reversed completely, traveling at high speed back up the 45-degree glide path it had been taking, and gaining about 500-feet altitude.

This observation was from a distance of one-half to 1 mile. They said it was about the size of a [Douglas] DC-6 [airliner] without wings; Officer Carson later made a sketch which shows an elliptical object 150 feet long and 40 feet high.

It was a very clear night, with no clouds, and as the object hovered for about a minute they got a good look at it. It was obviously not an aircraft of any design familiar to them, they said. It was surrounded by a white glow, making the object visible. At each end there were definite red lights, and at times five white lights were visible between the two red lights. They called the night dispatching office at the county sheriff's office and asked that other cars be sent, and that all other cars in the area be alerted. They also asked the radar base be notified.

The object then drifted westward toward them, losing altitude, and got within some 150 yards of them, easy pistol range, before drifting eastward again. During this time it performed aerial feats that seemed unbelievable. It was capable of moving in any direction—up, down, back, and forth. At times the movement was very slow, and at times completely motionless. It could move at extremely high speeds, and several times they watched it change direction or reverse itself while moving at unbelievable speeds.

As the object moved away from them toward the east, they followed at a judicial distance, encouraged by the expectation that they were to be joined by other officers. At that time they also radioed the Tehama County Sheriff's Office requesting that they contact the local radar base. By telephone the radar operator confirmed the UFO and stated that it was unidentified.

The two officers drove the next day to the local radar base, were refused permission to talk to the radar operator that had been on duty, and were given what Carson described as the "ice water treatment" by the commanding officer.

There follow many interesting details of their hide-and-seek chase with the object over the next 2 hours along the back roads of northern California, trying to get close enough to this thing to get a better observation. It seemed always to know they were there and always kept about half a mile away.

However, when we restrict our attention to the propulsion problem, the significant facts are: (1) there was no observable noise, (2) the UFO could hover—seemed to float as if it were in water—and move in any direction without altering its orientation, (3) it could sustain very high accelerations and move very rapidly, (4) it was able to hover or to move relatively slowly for at least 2 hours under circumstances that precluded suspension by aerodynamic lift forces.

What can we learn about the propulsion of UFO's from the information provided by the observations of these two police officers? Mainly, it is negative information. From the silence it seems impossible that it could have been supported by a jet or rocket reaction. There are further considerations involving specific impulse, energy, et cetera, that we need not go into here, that provide compelling arguments against any conventional way of counteracting the earth's gravitational field. There remains a slight possibility of developing sufficient reactive force by expelling relativistic neutrinos, for they would not be intercepted by the earth under a UFO and would not be noticed.

Expelling neutrons would have this same advantage, but in the quantities required they would induce far more radioactivity than has ever been measured at sites where UFO's have come close to the ground or have been reported to have landed.

Fortunately, there has been at least one observation that tends to provide a bit of positive information. Mr. Wells Allen Webb, an applied chemist with a master of science degree from the University of California, was 1 mile north of Spain Flying Field, 7 miles east of Yuma, Ariz., just off U.S. Highway No. 80, when his attention was drawn to the sky to the north by some low-flying jet aircraft. Then he noticed a small white cloud-like object in an otherwise cloudless sky.

He watched for about 5 minutes as it traveled eastward: as it reached a spot north-northeast of his location, it abruptly altered shape from being oblong and subtending about half the angle of the full moon—about 15 minutes of arc—to be circular and subtending about 5 minutes of arc. Webb was wearing Polaroid glasses and noted that there appeared around the object a series of dark rings, the outermost of which was about six times the diameter of the central white or silvery object, or about the diameter of the full moon. The object or cloud then decreased in apparent diameter, as if it were traveling away from him, and disappeared in another few minutes. During this time Webb repeatedly took off his glasses and then put them back on, noting each time that the rings appeared only when he was wearing the glasses. He did not know what to make of the sighting, but took notes, including the fact that it was about 10 in the morning. The date was May 5, 1953.

One of the first things to note about the situation as described in the account is that the dark rings were observed with Polaroid glasses, but not without them. The second thing is that, from the orientation of the observer relative to the position of the sun at that time of day, the blue scattered light from the part of the sky that formed the background for the object was polarized.

To this fortunate circumstance we must add the fact that Mr. Webb was curious about clouds, the effect of viewing them with polarized light, and took notes of what he observed. He did not, however, realize that he was observing the rotation of the plane of polarization of the blue light in the vicinity of the object. This was the interpretation I made some 8 years later upon reading his account.

MR. [JOHN W.] WYDLER. How would you define UFO's as you are using it in this paper before us.

DR. HARDER. I don't know how I could define it without being circular.

MR. WYDLER. That is the conclusion to which I came. You state on the very first page or you more or less say you are going to tacitly assume the reality of UFO's, merely an "unidentified flying object." I think we can assume their reality without worrying much about it. It is only if they have some particular interplanetary significance that might become a real problem, the way we look at it, isn't that so? We all agree there are unidentified flying objects. I think you are defining them as interplanetary. I don't see you really come out and say that, but I think you hinted at it.

DR. HARDER. Well, if my interpretation of these rings is correct, it is certainly nothing we have been able to accomplish on earth.

MR. WYDLER. Are you saying, when you use this term, for the purposes of your statement, in your testimony, you are assuming they are of an interplanetary nature?

DR. HARDER. Yes, that is right.

MR. WYDLER. All right.

DR. HARDER. In my statement, which is available to the transcriber, I have gone through a little bit of argument suggesting why the outer of the three rings represents light that had been rotated through 90 degrees, so it would not pass through the polarizer, if it is polarized glasses. The next ring represented light that had been rotated 90 plus180 degrees. If you have Polaroid glasses and look at the right part of the blue sky, any afternoon, you can seen that the light is polarized, and as you rotate your Polaroid glasses there is an alternate darkening-lightening, as you go through 180 degrees.

We can assume, to begin with, that the plane of the polarizer in his glasses was parallel to the plane of the undisturbed polarized light from the general direction of the object. If then something affected the light so as to turn its plane of polarization through 90 degrees, the portion that had been originally polarized would not pass through the glasses. Likewise, for light that had had its plane of polarization turned through 90 plus 180 degrees, 90 plus 360 degrees, and so on, there would be a partial extinction of light.

On this basis, the outer dark ring was due to the rejection by the polarizing filter of the glasses of light which had had its plane of polarization turned through 90 degrees, the next outermost band by light that had been turned through 270 degrees, et cetera.

This interpretation is strengthened by Webb's observation that the dark rings were narrower than the brighter areas between them; this is what should be expected on the basis of the above explanation.

What hypotheses can be constructed that might account for this unusual observation? There are at least two that have interesting implications for the propulsion problem. First by the Faraday effect, a magnetic field parallel to the path of the light could so rotate the plane of polarization. A quick calculation using the properties of the atmosphere shows that a field of 200,000 gauss, operating over a distance of 130 feet—40 meters—could turn the plane 90 degrees; this is indeed a very intense and extensive magnetic field and, of course, would only account for one ring. Three rings would require a million gauss over the same distance.

We have been able to achieve these field strengths in the laboratory for only fractions of seconds over very small distances. However, the principal argument against this hypothesis is the conclusion that were such a field brought at all close to the surface of the earth its effect would be to induce very strong remnant magnetism in nearly every piece of iron within several hundred yards. This has not been found.

We have been able to achieve that kind of field strength for fractions of seconds only over short distances on earth, or at least we, on earth.

However, there has been a suggestion made earlier that a very strong magnetic field might so saturate certain iron cores of electrical machinery as to explain some of the observed phenomena of electrical malfunctioning.

Despite the above-described observation, there is little reason to believe that magnetic fields, of themselves, could be of much use in propelling a spacecraft, although there has been much uninformed speculation about this in popular UFO publications. The simple reason is that we cannot produce a north pole without at the same time producing a south pole. This is a consequence of fundamental theory. Such a dipole cannot exert a force in conjunction with a uniform magnetic field, such as the earth may be assumed to have in a given locality, though it can produce a force in a nonuniform field.

To go beyond the above discussion would be rather speculative, but it is just here that we find a stimulus and challenge to scientific theory. It is almost circular to say that when we find a phenomenon we understand but vaguely we have also found a means of advancing our understanding; this has been particularly true in astronomy.

Concerning the propulsion of UFO's, a tentative hypothesis would be that it is connected with an application of gravitational fields that we do not understand.

Gravitation remains one of the enigmas of modern science, although there have been some advances in its understanding, beyond general relativity, in the past decade. There are theoretical grounds for believing there must exist a second gravitational field, corresponding to the magnetic field in electromagnetic theory, and that the interaction between these two fields must be similar to that between the electric and magnetic fields.

This interaction and its exploitation forms the basis for our modern electrical generators and motors. Without the interaction, we would be back to the days of electrostatic attraction and of permanent magnets—two phenomena that can produce only very weak forces when operating individually. Some day perhaps we will learn enough to apply gravitational forces in the same way we have learned to apply electromagnetic forces. This will depend upon advances in many fields of science. Some of the things required will be enormously increased sources of power from atomic fusion; very intense magnetic fields and current densities, perhaps from superconducting sources; and extremely strong materials to contain mechanical forces. Some of these advances are approaching, or are on the horizon. Others we have yet to see clearly.

May I close this part of our discussion by recalling the statement that the most important secret of the atomic bomb was that it worked. This gave the crucial impetus to other nations in their own efforts to duplicate the research of the United States. In the UFO phenomena we have demonstrations of scientific secrets we do not know ourselves. *It would be a mistake, it seems to me, to ignore their existence* [my emphasis, L.S.].

I have further comments on UFO's and high-strength materials, but perhaps the committee would rather interrupt at this point before I go on to that second subject?

MR. ROUSH. Any questions? I think you better go ahead, Dr. Harder, because if we get started questioning it is impossible to stop these people.

MR. [W. H.] BOONE. Mr. Chairman, may I ask one question?

MR. ROUSH. Go ahead, Mr. Boone.

MR. BOONE. Have you concluded that what you have just told us is true, we should not ignore their existence?

DR. HARDER. I have no doubt of the veracity of the observer who saw this thing in the sky; I know him personally.

MR. BOONE. I didn't question the observer, I questioned your remark, and the magnetic, if you will, electromagnetic interactions, and so forth, when you said we undoubtedly must admit the existence of these I am sorry I can't quote you exactly. But your last sentence there is what I refer to.

It does seem like an obvious conclusion resulting from all the previous remarks you said about some supernatural, if you will—

DR. HARDER. Oh, heavens, I never suggested that, I hope.

MR. BOONE. Well, let me say, science fiction propulsion system, then.

DR. HARDER. Well, sir, what we have been discussing this morning, and this afternoon, is perhaps closer to science fiction than anything. I hope it is more science than fiction, however.

MR. ROUSH. Go ahead, Dr. Harder.

DR. HARDER. The instances in which physical fragments of UFO's have been found are disappointingly few. To my knowledge, there is only one well-authenticated finding, and that was in Brazil, in 1957. The story of its discovery is contained in chapter 9 of the *Great Flying Saucer Hoax*, written by Dr. Olavo T. Fontes.

Briefly, several small metallic fragments were recovered by some fishermen near the coastal town of Ubatuba, Sao Paulo, after they saw what they described as a brilliant explosion of a flying disc. Some of the fiery fragments were extinguished in the water near the shore, where they were recovered.

Fontes acquired three of the fragments that weighed less than a tenth of an ounce each, and had one of them analyzed at the Mineral Production Laboratory in the Brazilian Agriculture Ministry. The results of the first analysis was that the substance was magnesium of an unusually high degree of purity, and that there was an absence of any other metallic element.

On the basis of the first examination a second spectrographic test was conducted, using the utmost care and the most modern instruments.

The second report was again marked by references to the "extreme purity" of the sample. Even impurities that are sometimes detected due to contamination from the carbon rod used as an electrode were absent. A further test, using X-ray diffraction, failed to turn up any other metallic component.

One of the pieces was flown to California and was analyzed. I have the report here. They used neutron activation analysis and discovered a total of one-tenth of 1 percent of other metallic elements than magnesium, 500 parts per million zinc, that included zinc, which is interesting, and small amounts of barium and strontium.

Certainly this metal is of extraordinary purity, certainly far beyond the capacity of fishermen at Ubatuba to produce.

What could be the use of such high-purity magnesium in the context of a spacecraft? One clue lies in its crystalline structure. It is close packed hexagonal structure, and is in this regard similar to the high-strength metals beryllium and titanium. Hexagonal crystals have but one slip plane, and this tends to make them brittle but strong.

One of the reasons for slip along crystal planes is that local imperfections in the crystal, or foreign atoms, create lines of stress concentration that move rapidly through the crystal, producing deformation.

If these imperfections, or dislocations, could be eliminated, the theoretical strength of the crystal lattice itself might be approached. This strength is on the order of millions of pounds per square inch for any materials. Carefully prepared ¼-inch diameter glass rods, etched to remove microscopic surface cracks and then lacquered, have withstood stress of 250,000 psi for 1 hour. Fused silica fibers have been stressed to 2 million psi.

Thus, foreign atoms within a crystal lattice are focal points for dislocations—points of stress concentration where the crystal lattice itself tears and slips. We can imagine that a high-purity crystal, free of surface and internal imperfections, would achieve fantastic strengths. Indeed, with the advent of iron whiskers, and boron fiber reinforced composites, we are already approaching some of these strengths, but only for extremely small diameter fibers.

Should, by any good fortune, further samples of UFO material be found, there may be further clues that would spur on research into high-strength materials, and perhaps give us hints of how to achieve superstrength in materials that are larger than the tiny fibers we have produced so far.

Needless to say, if we persist in denying the reality of UFO's we will not be looking for such samples, and may indeed reject them as having no importance when they are brought to our attention.

That is the conclusion of my prepared statement. I would like to comment on some of the suggestions as asked by Congressman Rumsfeld earlier.

I conclude in some of my colleagues' recommendations that a multiple-faceted exploration be made of this subject, preferably at several institutions simultaneously.

I have some suggestions as to how we could acquire additional scientific data even at the present time.

This is a three-point program which involves first the establishment of an early warning network, which the Colorado project began last February. Then to take advantage of one of the characteristics of UFO sightings that they, in many instances, are seen on one or two successive nights.

We could have prearranged instrument packages which are arranged for instant transportation to locations where UFO's have been sighted. If the budget for such a program were low, you might be able to borrow such things and have them ready at various universities where the instruments were otherwise occupied for research.

That would be the second point of this investigation.

The third point would be the cooperation of the Air Force for logistics and high-speed transport to crucial areas on a 24-hour basis.

Now, that three-point program may well bring to us physical data that so far has appeared only in anecdotal, still from essential amateurs who happened just accidentally to be at the right place. It was truly a fortunate accident when Mr. Webb was there to make the observation I described earlier.

MR. ROUSH. Does that conclude your statement?

DR. HARDER. That concludes my statement.[12] — STATEMENT OF DR. JAMES A. HARDER, ASSOCIATE PROFESSOR OF CIVIL ENGINEERING, UNIVERSITY OF CALIFORNIA AT BERKELEY, CALIF.

MR. ROUSH: Are there any questions? [No replies.]

[As Dr. Harder's "prepared statement" is almost identical in wording to his statements above, I have excluded it here. L.S.].

CHAPTER SIX

STATEMENT OF ROBERT M. L. BAKER, JR.

MR. ROUSH. Let's go to our next participant.
Our next participant is Dr. Robert M. L. Baker, Jr.
Dr. Baker, again we recognize your own eminence in your field, and we are very happy to have you here as a participant in this symposium. You may proceed.

☛ DR. BAKER. Fine, thank you, Mr. Roush.
I should like to preface my remarks by stating my preference for the term "anomalistic observational phenomena," as opposed to the term "unidentified flying objects."
MR. ROUSH. I observed you were going to say that and I wonder about some of my Hoosiers back home using those terms.
DR. BAKER. It comes trippingly off the tongue.
MR. ROUSH. It might not only cause some Hoosiers but some laymen some problems. It might be easier to say UFO's. You may go ahead.
DR. BAKER. I call it AOP.
From the data that I have reviewed and analyzed since 1954, it is my belief that there does exist substantial evidence to support the claim that an unexplained phenomenon—or phenomena—is present in the environs of the earth, but that it may not be "flying," may not always be "unidentified," and, perhaps, may not even be substantive "objects." In the following statement I will—

> (1) Present a summary of the analyses that I have accomplished to date—those that have led me to believe that anomalistic phenomena exist;
> (2) Explain the probable inadequacy of our current terrestrial sensors in observing and/or defining the characteristics of the anomalistic phenomena;
> (3) Suggest a number of tentative hypothetical sources for the phenomena, and the justification for their scientific study;
> (4) And, finally, I will make specific recommendations concerning the necessity for new types of closely related

observational and study programs which might be implemented in a fashion that would permit the detection and quantitative analysis of the anomalistic phenomena.

Several appendixes accompany this report. The first two are in response to Congressman Roush's invitational letter of July 10, 1968, and consist of my biographical sketch and a listing of my bibliography, respectively. The third appendix relates directly to my specific recommendations, and was included with the kind permission of Dr. Sydney Walker III. The fourth appendix presents three reprints of articles (Baker (1968a) and (1968b) and Walker (1968)) that are pertinent to the subject matter of this report.

PART I
ANALYSES OF ANOMALISTIC OBSERVATIONAL PHENOMENA
- UTAH & MONTANA FILMS -

My initial contact with anomalistic observational phenomena—AOP—came in 1954 when I was a consultant to Douglas Aircraft Company in Santa Monica, Calif., serving as special assistant to Dr. W. B. Klemperer, director of Douglas' research staff. The data consisted of two short film clips: one taken in Montana—termed by us as the Montana film—and one taken in Utah—called by us the Utah film. These films were provided to us by the Air Technical Intelligence Center—ATIC, now the Foreign Technology Division—FDT—at Wright Patterson Air Force Base; 35-millimeter prints were furnished by Green-Rouse Productions of Samuel Goldwyn Studios.

Both films had been taken by apparently reliable and unbiased men using amateur movie cameras and, in each case, there was a credible, substantiating witness present. The films exhibited the motion of rather fuzzy white dots, but the Montana film was remarkable in that foreground was visible on most of the frames.

Preliminary analysis excluded most natural phenomena. More detailed study indicated that the only remaining natural phenomenon candidate for the Utah film was birds in flight, and for the Montana film it was airplane fuselage reflections of the sun. After about 18 months of rather detailed, albeit not continuous, study using various film-measuring equipments at Douglas and at UCLA, as well as analysis of a photogrammetric experiment, it appeared that neither of these hypothesized natural phenomena explanations had merit, and a report was published by me (Baker (1956) and forwarded to Brig. Gen. Harold E. Watson, commander, ATIC. Since the description of the circumstances of the filmings and the analyses of the data provided on the films is rather lengthy, and have since been published in the open literature,[13] it does not seem unreasonable to repeat the analyses here.

- FLORIDA FILM -

During the course of this study we also had the opportunity to view some gun-camera photographs taken over Florida. Unfortunately, we could not retain this film, and did not have time available to accomplish a comprehensive analysis. Like the Montana and Utah films, this film also exhibited only white-dot images; however, since a foreground was present, a competent study could have been carried out. Dr. Klemperer and I agreed on the preliminary conclusion—not supported by detailed analyses—that, again, no natural phenomenon was a likely source for the images.

- VENEZUELA FILM -

In June of 1963 I received a movie film clip from a Mr. Richard Hall that had purportedly been taken from an aircraft (Douglas DC-3 airliner) near Angel Falls, Venezuela, at about 12:15 p.m. This film clip was 8-millimeter color film, exposed at 16 frames per second and showed a very bright yellow, slightly pear-shaped object that disappeared in a cloud bank after about 60 or 70 frames. At the time I was the head of the Lockheed Aircraft Co.'s Astrodynamics Research Center. We had developed a small group of photogrammetrists consisting of Dr. P. M. Merifeld and Mr. James Rammelkamp, and were able to undertake a study of the film. Initially, Merifeld and Rammelkamp found little of interest on the film. After their preliminary examination, I expended considerable effort in further analysis. Again, I was only able to draw the conclusion that the yellow object was no known natural phenomenon; but we could make a quantitative determination of angular rates and accelerations, and the bounds of distance, linear velocity, and acceleration, the film was lost (except for a microphotograph exhibiting the object on one frame). There was, however, no question in my mind as to the anomalistic character of the images.

- CALIFORNIA FILM -

In January 1964, Mr. Zan Overall showed me three cinetheodolite films which had been taken simultaneously by three different cameras of a Thor-Able Star launching at Vandenberg AFB (project A4/01019). These films depicted a white object moving vertically (relative to the film frame) against a clear, blue-sky background. The object was about as bright as the booster's second-stage exhaust, and passed the booster at about one-third degree per second. Rough estimates of the direction of the Sun—based on shadows on early frames and the winds aloft—indicated by the motion of the rocket's exhaust plume) were made. These, together with the brightness of the object, and its rate of ascent, seemed to rule out balloons, airplanes, lens flare, mirages, et cetera. Since one of the cinetheodolites was at a site some distance from the other two, a parallax determination of the actual distance and speed of the object could be determined rather easily. Because the films were on loan from the Navy, I was unable to carry out the necessary study and a determination of the precise character of the phenomenon (natural or anomalistic) could not be made. In 1967, I discussed the matter with Prof. William K. Hartmann of the University of Arizona, and Prof. Roy Craig of the University of Colorado. At that time, they were involved in the Colorado UFO Study Group, and indicated that they would attempt to obtain the film for further analysis. Although I am confident that they made a conscientious effort to obtain the films, apparently they were unsuccessful (as of 6 months ago, at least).

- PROBABLY NONANOMALISTIC FILMS -

In addition to the foregoing film clips—which seemed to involve data that were the result of anomalistic phenomena—the Montana film in my opinion, certainly was anomalistic and all of the other films except for the California film, most probably were anomalistic—I have also had the opportunity to view approximately a half dozen other films, purportedly of "UFO's." The images on these films appeared possibly to be the result of natural phenomena, such as reflections on airplanes, atmospheric mirages, optical flares, birds, balloons, insects, satellites, et cetera. For example, a recent (February 1968) set of two films were taken, using professional motion picture equipment, by a Universal Studio crew on location. Although rather peculiar in appearance, the objects thus photographed could have conceivably been the result of airplane reflections.

To this date my analyses of anomalistic motion picture data have been rather ungratifying. Although I am convinced that many of the films indeed demonstrated the presence of anomalistic phenomena, they all have the characteristic of rather ill-defined blobs of light, and one can actually gain little insight into the real character of the phenomena. For example, linear distance, speed, and acceleration cannot be determined precisely, nor can size and mass. As I will discuss in a moment, this situation is not particularly surprising, since, without a special-purpose sensor system expressly designed to obtain information pertinent to anomalistic observational phenomena, or a general-purpose sensor system operated so as not to disregard such data, the chance for obtaining high-quality hard data is quite small.

PART 2
INADEQUACIES OF EXISTING
SENSOR EQUIPMENT & SYSTEMS

The capabilities of astronomical optical sensors have been dealt within a thorough fashion by Page in 1968. The Prairie Network for Meteor Observations (McCrosky and Posen (1968)) is a good example of a wide-coverage optical system, but as is so often the case, and as Page (1968) pointed out. "R. E. McCrosky of the Smithsonian Astrophysical Observatory informed me that no thorough search (for anomalistic data) has been carried out." Even so, some astronomical photographs are bound to exhibit anomalistic data. Again quoting from Page (1968), "W. T. Powers of Northwestern University Astronomy Department informed me that 'several' of the Smithsonian-net photographs show anomalous trails." As I have already pointed out (Baker (1968b) to be found in appendix 4), the majority of our astronomical equipment (e.g., conventional photographic telescopes, Baker-Nunn cameras, meteor cameras, Markowitz Dual-rate Moon Cameras, et cetera) are special purpose in nature, and would probably not detect the anomalous luminous phenomena reported by the casual observer if they were indeed present. Their photographic speed, field of view, et cetera, impose severe restrictions on their ability to collect data on objects other than those they have been specifically designed to detect. As already noted in the quotes from Page (1968), even if such data were collected, the recognition of their uniqueness or anomalous character by an experimenter is improbable. Examples abound, in the history of celestial mechanics, of minor planets being detected on old astronomical plates that had been measured for other purposes, and then abandoned.

Our radar and optical space surveillance and tracking systems are even more restrictive and thus, even less likely to provide information on anomalistic phenomena than are astronomical sensors. The Signal Test Processing Facility (STPF) radar at Floyd, N.Y. is a high-performance experimental radar having a one-third degree beam width. For lockon and track, an object would have to be pinpointed to one-sixth degree, and even if the radar did achieve lockon, an erratically moving object could not be followed even in the STPF radar's monopulse mode of operation. For this reason only satellites having rather well-defined paths (i.e., ephemerides), which have been precomputed, can be acquired and tracked.

Our three BEMEWS radars propagate fans of electromagnetic energy into space. If a ballistic missile or satellite penetrates two of these fans successively, then it can be identified. Since astrodynamical laws govern the time interval between detection fan penetrations for "normal" space objects, all other anomalistic "hits" by the radar are usually neglected, and even if they are not neglected, they are usually classified as spurious images or misassociated targets, and are stored away on magnetic tape, and forgotten.

One space surveillance site operates a detection radar (FPS-17) and a tracking radar (FPS-79). If a new space object is sensed by the detection radar's fans, then the tracking radar can be oriented to achieve lockon. The orientation is governed by a knowledge of the appropriate "normal" object's astrodynamic laws of motion, or by an assumption as to launch point. Thus, if an unknown is detected, and if it follows an unusual path, it is unlikely that it could, or would, be tracked. Furthermore, the director of the radar may make a decision that the unknown object detected is not of interest (because of the location of the FPS-17 fan penetration or because of the lack of prior information on a possible new launch). In the absence of detection fan penetration (the fan has a rather limited coverage), the FPS-79 tracking radar is tasked to follow other space objects on a schedule provided by the Space Defense Center, and again there is almost no likelihood that an anomalistic object could, or would, be tracked.

The NASA radars, such as those at Millstone and Goldstone, are not intended to be surveillance radars, and only track known space objects on command. Again the chances of their tracking anomalistic objects are nearly nil. The new phased-array radar at Eglin AFB (FPS-85) has considerable capability for deploying detection fans and tracking space objects in a simultaneous fashion. Such versatility raises certain energy-management problems—that is, determining how much energy to allocate to detection and how much to tracking—but this sensor might have a capability (albeit, perhaps, limited) to detect and track anomalistic objects. The problem is that the logic included in the software associated with the FPS-85's control computers is not organized in a fashion to detect and track anomalistic objects (I will indicate in a moment how the logic could be modified). Furthermore, the FPS-85, like the other surveillance radars is usually tasked to track a list of catalogued space objects in the Space Defense Center's data base and the opportunity to "look around" for anomalistic objects is quite limited.

There are a number of other radar surveillance systems such as a detection fence across the United States. In the case of this fence, we have a situation similar to BMEWS, in which the time interval between successive penetrations (in this case separated by an orbital period for satellites) must follow prescribed astrodynamical laws. If they do not, then the fence penetrations are either deleted from the data base or classified as "unknowns," or "uncorrelated targets," filed, and forgotten.

There is only one surveillance system, known to me, that exhibits sufficient and continuous coverage to have even a slight opportunity of betraying the presence of anomalistic phenomena operating above the Earth's atmosphere. The system is partially classified and, hence, I cannot go into great detail at an unclassified meeting. I can, however, state that yesterday (July 28, 1968) I traveled to Colorado Springs (location of the Air Defense Command) and confirmed that since this particular sensor system has been in operation, there have been a number of anomalistic alarms. Alarms that, as of this date, have not been explained on the basis of natural phenomena interference, equipment malfunction or inadequacy, or manmade space objects.

PART 3
HYPOTHETICAL SOURCES FOR ANOMALISTIC OBSERVATIONS & JUSTIFICATION FOR THEIR STUDY

In Baker and Makemson (1967), I discussed the usual candidates for the natural sources of anomalistic observations. For example, some scanning radars—such as airport radars—pick up anomalistic returns termed "angels." A variety of explanations have been proposed, variously involving ionized air inversion layers, etc. (see Tacker (1960)) and even insects (see Glover, et al. (1966)). With respect to human observation of anomalistic luminous phenomena, some rather strong positions have been taken by such authorities as Menzel (1953), who feels that the predominant natural phenomenon is atmospheric mirages; by Klass (1958a), who feels that the predominant natural phenomenon is related to ball lightning triggered by high-tension line coronal discharge, jet aircraft, electrical storms, etc.; by Robey (1960), who feels that the observations are of "cometoids" entering the earth's atmosphere, etc. The list of hypothetical sources for anomalistic observational phenomena is long indeed, but *from the photographic data that I have personally analyzed, I am convinced that none of these explanations is valid* [my emphasis, L. S.]

The analyses that I have carried out to date have dealt with observational evidence that I term "hard data"—that is, permanent photographic data. Although I will not discuss in detail the analyses of eyewitness reports (which I term "soft data"),[14] Powers (1967), McDonald (1967), Hynek (1966), and others have concluded that overwhelming evidence exists that a truly anomalistic phenomenon is present.

Of course, there are numerous others who have come to a completely opposite conclusion; in fact, it becomes almost a matter of personal preference: it is possible for one to identify all of the anomalistic data as very unusual manifestations of natural phenomena. No matter how unlikely it is, anything is possible—even a jet plane reflecting the sun in direct opposition to the laws of optics. I'm sometimes reminded of the flat earth debates that I organized 10 years ago in my elementary astronomy courses at UCLA. Some students became so involved in justifying their positions—either flat or spherical—that they would grasp at even the most improbable argument in order to rationalize their stand.

MR. ROUSH. Dr. Baker, I'm sorry to interrupt, but I'm going to have a brief recess here.

DR. BAKER. Certainly.

MR. ROUSH. There is a motion to recommit the military construction bill, and I would like to vote on it. None of my colleagues are here right now, so we will declare a very brief recess, and I shall return as quickly as I can.

(Whereupon a short recess was taken for a floor vote.)

MR. ROUSH. The committee will be in order.

Dr. Baker, you may proceed.

DR. BAKER. Thank you.

Personally, I feel that it is premature for me to agree that the hard and soft data forces the scientific community to give overriding attention to the hypothesis that the anomalistic observations arise from manifestations of extraterrestrial beings. On the other hand, I strongly advocate the establishment of a research program in the area of anomalistic phenomena—an interdisciplinary research effort that progresses according to the highest scientific standards; that is well funded; and that is planned to be reliably long term. The potential benefit of such a research project to science should not hinge solely on the detection of intelligent extraterrestrial life; it should be justified by the possibility of gaining new insights into poorly understood phenomena, such as ball lightning, cometoid impact, and spiraling meteorite decay.

There is practical value in such research for the Military Establishment, as well. Let us suppose that something similar to the "Tunguska event" of 1908 occurred today, and that it was Long Island in the United States, rather than the Podkamenaia Tunguska River Basin in Siberia that was devastated by a probable comet impact. Would we misinterpret this catastrophic event as the signal for world war III? What if another "fireball procession," such as occurred over Canada on February 9, 1913, repeated itself today, and the low-flying meteors were on nearly polar orbits that would overfly the continental United States. Would we interpret the resulting surveillance data as indicating that a fractional orbital bombardment system (FOBS) had been initiated in Russia? My knowledge of our Air Force sensors, both current and projected (see Baker and Ford (1968)), indicates that they are sufficiently sophisticated so that they would probably not react prematurely and signal a false alarm—although a careful study of this point should be made. On the other hand, there may exist other anomalistic sources of data that might give rise to a false alarm and perhaps provoke us either to deploy our countermeasures, or even to counterattack.

Before I enumerate the specific benefits this research might confer upon various scientific disciplines, allow me to digress briefly on the subject of soft data. The primary reason that I have avoided the introduction of soft data into my photographic studies and have not involved myself in the analysis of eyewitness reports (such as the excellent ones given by Fuller (1966)), is that I have been unable to develop a rational basis for determining the credibility level for any given human observer. Although they lie outside the field of my own scientific competence, I feel that credibility evaluations of witnesses would form an important adjunct to any serious study of anomalistic phenomena (see Walker (1968) included in app. 4 of this report). The soft data must involve some useful information content, and it would be extremely unrealistic to neglect it entirely. For this reason, I have included appendix 3 by Dr. Walker, which presents a logical procedure for establishing a credibility level for observers. Walker's report of a hypothetical case integrates the results of general medical, neuroopthalmologic, neurologic, and psychiatric evaluations, and develops a logical basis for assigning an overall credibility score.

Dr. Robert L. Hall is, of course, eminently qualified to comment on the question of eyewitness testimony at this seminar.

If serious studies can be initiated, with the objectives of detecting, analyzing, and identifying the sources of anomalistic observational phenomena, then I feel that the following scientific benefits can be expected:

(1) *Meteoritics.*—Although there are a number of excellent meteor observation nets operating today, data collected on erratically moving phenomena (including rapid determination of the location of any "landings" or impacts) would add significantly to the coverage and analyses of meteorites and, possibly, entering comets. Furthermore, the timely recovery of meteoritic debris at the subend point of fireballs would be most valuable.

(2) *Geology.*—It has been pointed out by Lamar and Baker (1965), that there exist residual effects on desert pavements that may have been produced by entering comets. Furthermore, any geological or material evidence of the impact or "landing" of extraterrestrial objects would be of great interest. As Dr. John O'Keefe (1967), Assistant Chief, Laboratory for Theoretical Studies of NASA GSFC indicated,

> "Would it not be possible to get some scraps of these ('UFO') objects for examination? For instance, a scrap of matter, however small, could be analyzed for the kind of alloys in terrestrial foundries. A piece of a screw, however small, would be either English, Metric, or Martian. I am impressed by this because I looked at some tens of thousands of pictures of the Moon and found that the very small amount of chemical data has more weight in interpreting the past history of the Moon than the very large amount of optical data. It doesn't seem possible that objects ('flying saucers') of this size can visit the Earth and then depart, leaving nothing, not even a speck, behind. We could analyze a speck no bigger than a pinhead very easily."

I concur with O'Keefe's remarks, and if there exist "landings" associated with the anomalistic phenomena, then a prompt and extremely thorough investigation of the landing site must be accomplished before geological/material evidence is dispersed or terrestrialized.

(3) *Atmospheric physics.*—One of the great mysteries today is the formation, movement, and explosion of ball lightning. As Singer (1963) noted:

> "The specific properties of ball lightning, which present particular

difficulty in experimental duplication, are formations of the sphere in air (at near-atmospheric pressure and at a distance from the source of energy) and its extensive motion. It is evident that additional clarification of both theoretical and experimental aspects is needed."

With respect to "plasma UFO's" Mr. Philip J. Klass (1968b) comments that:

"If conditions—all of the conditions—needed to create plasma-UFO's near high tension lines or in the wake of jet aircraft occurred readily we should have millions of UFO reports and the mystery would have been solved long ago. But the comparative rarity of legitimate UFO sightings clearly indicates that the ball lightning related phenomenon is a very rare one."

Even if ball lightning is not the primary source of anomalistic data (and I am not at present convinced that it is), any program investigating anomalistic observational phenomena would surely shed significant light on the ball-lightning problem.

(4) *Astronomy.*—I have already noted the possibility of cometary entry, a study of which would be valuable to the astronomer. If as some respected astronomers believe, the anomalistic observational phenomena (including perhaps, "intelligent" radio signals from interstellar space) are the results of an advanced extraterrestrial civilization, then the study of the phenomena would become a primary concern of the entire human race. The implications for astronomy are overwhelming.

(5) *Psychiatry and psychology.*—Since bizarre events have been reported, the study of eyewitness credibility, under stressful circumstances of visual input, if possible. As I will recommend later: if a competent, mobile task force of professionals could be sent into action as soon as anomalistic events are detected, then reliable evaluation of eyewitness reports (soft data) in relation to the actual hard data obtained, could be accomplished. Even if the event was only a spectacular fireball, or marsh gas, the psychiatric/medical examination of eyewitnesses would still be more informative.

(6) *Social science.*—Although not classified as a physical science, there appears to be a challenge here for the social sciences. It has been my contention throughout this report that it is not a prerequisite to the study of anomalistic observational phenomena to suppose that they result from extraterrestrial intelligence.

Nevertheless, it still is an open possibility in my mind. It seems reasonable, therefore, to undertake a few contingency planning studies. In order to extract valuable information from an advanced society, it would seem useful to forecast the approximate characteristics of such a superior intelligence—or, if not necessarily superior, an intelligence displayed by an industrial, exploratory culture of substantially greater antiquity. There exist dozens of treatises on technological forecasting; one can key estimates of technological advancement to speed of travel, production of energy, productivity, ubiquity of communications, etc. There have been many debates on the technical capabilities or limits on the capabilities of advanced extraterrestrial societies (for example, see Markowitz (1967) and Rosa, et al. (1967)). Often intermixed with these technological capabilities arguments, however, are very dubious comments concerning the psychological motivations, behavioral patterns, and unbiased projections of the social motivations of an advanced society.

Hypothetical questions are often raised such as, "if there are flying saucers around, why don't they contact us directly? I would if I were investigating another civilization." Such comments are made on extremely thin ice, for, to my knowledge, no concerted study has been carried out in the area of forecasting the social characteristics of an advanced extraterrestrial civilization.

Philosophers, social scientists, and others usually undertake studies of rather theoretical problems. (See Wooldridge (1968) and Minas and Ackoff (1964). If only a quantitative index or indices of social advancement could be developed that, say, would differentiate us from the Romans in our interpersonal and intersociety relationships (for example, tendencies toward fewer crimes of violence, fewer wars, etc.), then we might be better equipped to make rational extrapolations from our own to an advanced society. In fact, such as index, if it could be developed might even be beneficial in guiding our existing earth-based society.

(7) *Serendipity*.—In addition to the value of anomalistic phenomena studies to these specific scientific disciplines, there is always serendipity. Any scientific study of this nature is potentially capable of giving substantial dividends in terms of "spin-off." For example: in improved techniques in radar and optical sensor design and control; in giving a reliable quantitative credibility level to witness' statements in court; or in deciphering and/or analyzing anomalistic radio signals from interstellar space.

PART 4
CONCLUSIONS & RECOMMENDATIONS

For the past 16 years I have seriously (albeit sporadically) followed the analyses of "UFO" or "flying saucer" reports—both scientific and quasi-scientific. It is my conclusion that there is only so much quantitative data that we can squeeze out of vast amounts of data on anomalistic observational phenomena that has been collected to date. I believe that we will simply frustrate ourselves by endless arguments over past, incomplete data scenarios; what we need is more sophisticated analyses of fresh anomalistic observational data. We must come up with more than just a rehash of old data.

I emphasize that it is very unlikely that existing optical and radar monitoring systems would collect the type of quantitative data that is required to identify and study the phenomena. Moreover, we currently have no quantitative basis upon which to evaluate and rank (according to credibility) the myriad of eyewitness reports. Thus continuing to "massage" past anomalistic events would seem to be a waste of our scientific resources. In balance, then, I conclude that:

(1) We have not now, nor have we been in the past, able to achieve a complete—or even partially complete—surveillance of space in the vicinity of the earth, comprehensive enough to betray the presence of, or provide quantitative information on, anomalistic phenomena.

(2) Hard data on anomalistic observational phenomena do, in fact exist, but they are of poor quality, because of the inadequacies of equipment employed in obtaining them.

(3) Soft data on anomalistic phenomena also exist, but we have no quantitative procedure to evaluate their credibility and develop clear-cut conclusions on the characteristics of the anomalistic phenomena.

(4) It follows from the scientific method that an experiment or experiments should be devised, and closely related study programs be initiated expressly to define the anomalistic data better.

(5) In order to justify such an experiment and associated studies, it is not necessary to presuppose the existence of intelligent extraterrestrial life operating in the environs of the earth, or to make dubious speculations either concerning "their" advanced scientific and engineering capabilities or "their" psychological motivations and behavioral patterns.

In the light of these conclusions, I will make the following recommendations:

(1) In order to obtain information-rich hard and soft data on

anomalistic phenomena, an interdisciplinary, mobile task force or team of highly qualified scientists should be organized. This team should be established on a long-term basis, well funded, and equipped to swing into action and investigate reports on anomalistic phenomena immediately after such reports are received. Because of the relatively low frequency of substantive reports (see p. 1968), immediate results should not be anticipated, but in the interim periods between their investigations in the field, their time could be productively spent in making thorough analyses of data collected by them previously, and in "sharpening up" their analysis tools.

(2) In concert with the aforementioned task force, a sensor system should be developed expressly for detecting and recording anomalistic observational phenomena for hard-data evaluation. The system might include one or more phased-array radars (certainly not having the cost or capability of the FPS-85, but operating in a limited fashion that would be similar to the FPS-85). A phased-array radar would have the advantage over a conventional "dish" radar in that it could track at high rates and divide its energy in an optimum fashion between detection and tracking. The control system would be unique, and would necessitate the development of a sequential data processing controller that would increase the state variables describingthe object's path from a six-dimensional position and velocity estimation to a 12-dimensional acceleration and jerk estimation (Baker(1967)) in order to follow erratic motion.

In addition, the data base would have to be especially designed, to avoid manmade space objects and (if possible) airplanes, birds, common meteors, etc. It should, however, be designed to detect and track nearby cometoids, macrometeorites (fireballs), ball lightning, and any other erratic or anomalistic object within its range. Optical cameras (including spectrographic equipment) should be slaved to the radar, in order to provide more comprehensive data. Because of the aforementioned low frequency of anomalistic data, alarms from the system should not occur very frequently and could be communicated directly to the recommended task force.

(3) A proposed new-generation, space-based long-wave-length infrared surveillance sensor system should be funded and the associated software should be modified to include provisions for the addition of anomalistic objects in its data base. The specific sensor system cannot be identified for reasons of security, but details can probably be obtained from the Air Force. This sensor system, in particular, could provide some data (perhaps incomplete) on

anomalistic objects which exhibit a slight temperature contrast with the space background, on a basis of noninterference with its military mission. The system represents a promising technological development, and no other novel technique introduced in recent years offers more promise for space surveillance. In my view, the scientific principles underlying the proposed surveillance system are sound, and a developmental measurements program should be initiated.

(4) The software designed for the FPS-85 phase-array radar at Eglin Air Force Base be extended in order to provide a capability to detect and track anomalistic space objects. The relatively inexpensive modification could include the implementation of tracking techniques such as those outlined in Baker (1967). It should, however, be clearly borne in mind that only a limited amount of tracking time (about 30 percent) could be devoted to this endeavor, because of the overriding importance of the surveillance of manmade space objects which is the basic responsibility of this radar.

(5) Various "listening post" projects should be reestablished (using existing instruments) in order to seek out possible communications from other intelligent life sources in the universe. See, for example, Shklovskii and Sagan (1966), chapters 27, 28, 30, and 34.

(6) Technological and behavioral pattern forecasting studies should be encouraged in order to give at least limited insight into the gross characteristics of an advanced civilization. These studies (probably not Government funded) should include the social-psychological implications of anomalistic observational phenomena, as well as the psychological impact upon our own culture that could be expected from "contact" with an advanced civilization. (See ch. 33 of Shklovskii and Sagan (1966).)

(7) Studies should be initiated in the psychiatric/medical problems of evaluating the credibility of witness' testimony concerning bizarre or unusual events. (See app. 3 of this report.)

PART 5
AFTERWARD

All of the foregoing recommendations involve the expenditure of funds, and we are all well aware of the severe limitations on the funding of research today. On the other hand, I feel that one of the traps that we have fallen into, so far, is reliance on quick-look, undermanned and underfunded programs to investigate a tremendous quantity of often ambiguous data. I would discourage such programs as being diversionary, in regard to the overall scientific goal.

The goal of understanding anomalistic phenomena, if attained, maybe of unprecedented importance to the human race. We must get a positive scientific program off the ground; a program that progresses according to the highest scientific standards, has specific objectives, is well funded, and long term.

Thank you.[15] — STATEMENT OF DR. ROBERT M. L. BAKER, JR., SENIOR SCIENTIST, COMPUTER SCIENCES CORP., EL SEGUNDO, CALIF., AND FACULTY, DEPARTMENT OF ENGINEERING, UCLA

(The appendixes and attachments to Dr. Baker's statement follow:)

PART 1
ABSTRACTS FROM BAKER (1956) RELATED TO THE
UTAH FILM-ANALYSIS OF PHOTOGRAPHIC MATERIAL

PHOTOGRAMMETRIC ANALYSIS OF
THE "UTAH" FILM TRACKING UFO'S

Several Unidentified Flying Objects (UFO's) were sighted and photographed at about 11:10 MST on July 2, 1952, by Delbert C. Newhouse at a point on StateHighway 30, seven miles north of Tremonton, Utah (latitude 41° 50, longitude 112° 10). Mr. Newhouse, a Chief Warrant Officer in the U.S. Navy,[16] was in transit from Washington, D.C. to Portland, Oregon.

He, his wife and their two children were making the trip by car. Shortly after passing through the city of Tremonton, his wife noticed a group of strange shining objects in the air off towards the eastern horizon. She called them to herhusband's attention and prevailed upon him to stop the car. When he got out, he observed the objects (twelve to fourteen of them) to be directly overhead and milling about. He described them as "gun metal colored objects shaped like two saucers, one inverted on top of the other." He estimated that they subtended "about the same angle as B-29's at 10,000 feet" (about half a degree—i.e., about the angular diameter of the moon).

Next, he ran to the trunk of his car, took out his Bell and Howell Automaster 16mm movie camera equipped with a 3" telephoto lens, loaded it, focused it at infinity and began shooting. There was no reference point above the horizon so he was unable to estimate absolute size, speed or distance. He reports that one of the objects reversed its course and proceeded away from the rest of the group; he held the camera still and allowed this single object to pass across the field of view of the camera, picking it up later in its course. He repeated this for three passes.

During the filming, Newhouse changed the iris stop of the camera from f/8 to f/16. The density of the film can be seen to change markedly at a point about 30% through the sequence. The camera was operated at 16 fps.

The color film (Daylight Kodachrome) after processing was submitted to his superiors. The Navy forwarded the film to the USAF-ATIC where the film was studied for several months. According to Al Chop (then with ATIC and presently with DAC) Air Force personnel were convinced that the objects were not airplanes; on the other hand the hypothesis that the camera might have been out of focus and the objects soaring gulls could neither be confirmed nor denied. Mr. Chop's remarks are essentially substantiated by Capt. Edward Ruppelt, reference (1) then head of Project Blue Book for ATIC.

A 35mm reprint of the Newhouse "Utah" film was submitted to Douglas Aircraft Company for examination. Visual study of the reprints on the Recordak and astronomical plate measuring engine revealed the following: The film comprises about 1,200 frames; on most of the frames there appear many round white dots, some elliptical. The dots often seem clustered in constellations, or formations which are recognizable for as long as seventeen seconds. A relative motion plot (obtained from an overlay vellum trace on the Recordak) of two typical formations are presented. The objects seem to cluster in groups of two's and three's. On some frames they flare up and then disappear from view in 0.25 seconds or less and sometimes they appear as a randomly scattered "twinkling" of dots. The dot images themselves show no structure; they are white and have no color fringes.

Examination under a microscope shows the camera to be well focused as the edges of the images are sharp and clear on many of the properly exposed frames (of the original print). Angular diameters range from about 0.001,6 to 0.000,4 radians. Their pattern of motion is essentially a curvilinear milling about. Sometimes the objects appear to circle about each other. There are no other objects in the field of view which might give a clue as to the absolute motion of the cluster.

In the overlay trace, the frame of reference is determined by a certain object whose relative motion during a sequence of frames remains rather constant. This object is used as a reference point and the lower edge of the frame as abscissa. Assuming the camera to have been kept reasonably uncanted, the abscissa would be horizontal and the ordinate vertical. In the overlay trace, the particular frame itself is used as the reference. Assuming the camera was held steady (there is an unconscious tendency to pan with a moving object) the coordinate system is quasi-fixed. It is realized that both of these coordinate systems are in actuality moving, possibly possessing both velocity and acceleration.

No altitude or azimuth determination can be made because of lack of background. The only measurable quantities of interest are therefore the relative angular distances between the objects and their time derivatives. Graphs of two typical time variations of relative angular separation and velocity are included (in Baker and Makemson (1967)). The relative angular velocity is seen to vary from zero to 0.006,5 radians per second. The relative angular acceleration had a maximum value of 0.003,6 radians per second squared. Supposing the camera was kept stationary the average angular velocities for the object moving across the field are 0.039 and 0.031 radians per second. The angular velocities in these sequences sometimes vary erratically from 0.07 to 0.01 radians per second. This variation may be attributed in part to camera "jiggling" and in part to the object's motion. The decrease in average angular velocity could be due to the object's having regressed between filmings just as was reported by Newhouse. Also the average image diameter decreases about 30% over the entire film, indicating a possible over-all regression of the objects.

The following tabulation indicates the hypothetical transverse component of relative velocities and accelerations at various distances. It is noted that the transverse velocity may be only a fraction of the total velocity so that the numbers actually indicate minimum values.

If the object's distance was—	Its transverse velocity was—	Its transverse acceleration was—	Velocity of single object was—
100 ft.	0.65 ft./sec. or 0.44 m.p.h.	0.36 ft./sec.² or 0.11g.	3.8 ft./sec. or 2.7 m.p.h.
1,000 ft.	6.5 ft./sec. or 4.4 m.p.h.	3.6 ft./sec.² or 0.11 g.	39 ft./sec. or 27 m.p.h.
2,000 ft.	13 ft./sec. or 8.8 m.p.h.	7.2 ft./sec.² or 0.22 g.	78 ft./sec. or 54 m.p.h.
1 mile.	23 m.p.h.	0.56 g.	135 m.p.h.
5 miles.	115 m.p.h.	2.8 g.	670 m.p.h.
10 miles.	230 m.p.h.	5.6 g.	1,300 m.p.h.

The objects in the "Utah" and "Montana" films can only be correlated on the basis of two rather weak points. First, their structure, or rather lack of it, is similar. Thus as shown in the "blow-ups" there are no recognizable differences between them.[17] Second, the objects on the "Montana" film are manifestly a single pair; on the "Utah" film perhaps 30% of the frames show clusters of objects seemingly also grouped in pairs.

The weather report was obtained by the author from the Airport Station at Salt Lake City. The nearest station with available data is Corinne which reported a maximum temperature of 84°, a minimum of 47° and no precipitation. A high pressure cell from the Pacific Northwest spread over Northern Utah during July 2, the pressure at Tremonton would have a rising trend, the visibility good, and the winds relatively light. The absence of clouds and the apparently excellent visibility shown on the films would seem to be in agreement with this report. Through use of References (2) and (3), the Sun's azimuth N132° E, altitude 65°, was computed. No shadows were available to confirm the time of filming.

The image size being roughly that of the Montana film (a few of the objects being perhaps 10% larger than the largest on the Montana) the same remarks as to airplane reflections apply, i.e., they might have been caused by Sun reflections from airplanes within one to three miles to the observer, although at these distances they should have been identified as conventional aircraft by the film or the observer. No specific conclusions as to Sun reflection angles can be drawn since the line of motion of the objects cannot be confirmed. However, the reported E to W motion of the UFO's and their passing overhead coupled with the SE azimuth of the Sun would make the achievement of optimal Sun reflections rather difficult.

That the images could have been produced by aluminum foil "chaff"[18] seems possible, at least on the basis of the images shown, as very intense specular Sun reflections from ribbons of chaff might flare out to about the size of the UFO's.

Examination of film frames obtained from the photogrammetric experiment—reference Analysis of Photographic Material, Serial 01, Appendix II, show that no significant broadening is produced by flat white diffuse reflectors such as birds, bits of paper, etc. at f/16 under the conditions of the filming. Actual measurements show a slight "bleeding" or flaring of about 10% to 20%.

The rectangular flat white cardboards of the aforementioned experiments represented very roughly the configuration of birds. The light reflected by such a surface is probably greater than that from a curved feather surface of a bird. One figure shows the appearance of one and two foot birds[19] as they might appear on a 16mm frame taken with a 3" telephoto lens f/16 at a distance of 1,200', at 3,000' and at 3,300'. Many of the images on the "Utah" film have an angular diameter of 0.001, 2 radians (some as large as

0.001,16 radians), thus they might be interpreted as one foot (wing span) birds at 600' to 800', two foot (wing span) birds at 1,200' to 1,600' or three foot (wing span) birds at 2,400' to 3,200'. *At these distances, it is doubted if birds would give the appearance of round dots; also they would have been identifiable by the camera if not visually.* However, actual movies of birds in flight would have to be taken to completely confirm this conclusion. The following type of gulls have been known to fly at times over this locality: California Herring Gull (a common summer resident), Ring-Billed Gull and the Fork-Tailed Gull, see Reference (4).

The images are probably not those of balloons as their number is too great and the phenomenon of flaring up to a constant brightness for several seconds, and then dying out again cannot well be associated with any known balloon observations.

Certain soaring insects—notably "ballooning spiders" (References (5) and (6)) produce bright-moving points of light. The author has witnessed such aphenomenon. It is produced by Sun reflections off the streamers of silken threads spun by many types of spiders. Caught by the wind, these streamers serve as a means of locomotion floating the spider high into the air. They occasionally have the appearance of vast numbers of silken flakes which fill the air and in some recorded instances extend over many square miles and to a height of several hundred feet. The reflection, being off silk threads, is not as bright as diffuse reflection from a flat white board. Thus no flaring of the images could be expected. The author noted that the sections of the "web" that reflected measured from 14" to 2" for the largest specimens. Thus the images might be attributed to ballooning spiders at distances of 50 to 100 feet. However, these web reflections ordinarily show upon only against a rather dark background and it is doubted if their intensity would be great enough to produce the intense UFO images against a bright sky.

Besides the above remarks, pertinent to the actual images, several facts can be gleaned from the motion of objects. The observations are not apt to support the supposition that the objects were conventional aircraft as the maneuvers are too erratic, the relative accelerations probably ruling out aircraft at distances of over five miles. Several observers familiar with the appearance of chaff have seen the film and concluded that the persistence of the nontwinkling constellations, their small quantity, and the reported absence of aircraft overhead makes chaff unlikely. Furthermore, the single object passing across the field of view would be most difficult to explain on the basis of chaff. These same remarks would apply also to bits of paper swept up in thermal updrafts. The relative angular velocity might be compatible with soaring bird speeds at distances of less than one mile, the angular velocity of the single object could be attributed to a bird within about one thousand feet. There is a tendency to pan with a moving object—not against it—so the velocities in the table probably represent a lower bound. *The motion of the objects is not exactly what one would expect from a flock of soaring birds (not the slightest indication of a decrease in brightness due to periodic turning with the wind or flapping) and no cumulusclouds are present which might betray the presence of a strong thermal updraft.* On the other hand the single object might represent a single soaring bird which broke away in search of a new thermal—quite a common occurrence among gulls—see Reference (7).

That the air turbulence necessary to account for their movement if they were nearby insects (even the single object's motion!) is possible, can be concluded from examination of Reference (8). However, if the objects were nearby spider webs the lack of observed or photographed streamers is unusual. Furthermore, the fact that they were visible from a moving car for several minutes is hard to reconcile with localized insect activity.

The phenomenon of atmospheric mirages, Reference (9), might conceivably account for the images. Such a hypothesis is hampered by the clear weather conditions and the persistence and clarity of the images. Also no "shimmering" can be detected and the motion is steady. Again the object which breaks away would be difficult to explain.

It has been suggested that spurious optical reflections or light leaks in the camera might be responsible. Examples of such effects have been examined and found to be quite different from the UFO's (in the Utah Film).

The evidence remains rather contradictory and no single hypothesis of a natural phenomenon yet suggested seems to completely account for the UFO involved. The possibility of multiple hypotheses, i.e. that the Utah UFO's are the result of two simultaneous natural phenomena might possibly yield the answer. However, as in the case of the "Montana" analysis, no definite conclusion (as to a credible natural phenomenon) could be obtained.

MR. ROUSH. Thank you, Dr. Baker.

230 ∾ MYSTERIOUS INVADERS

PHOTOGRAPHS OF FRAMES FROM THE MOVIE FILMS THAT DR. BAKER ANALYZED

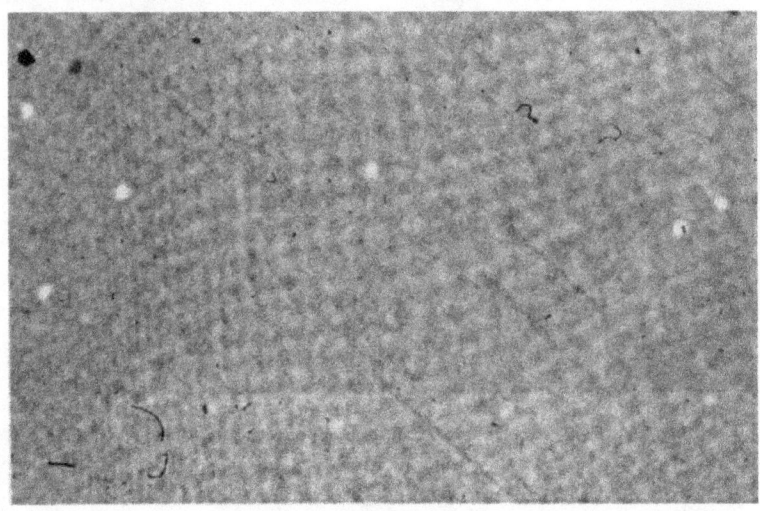

Blow up of a frame from the Utah film showing a typical formation of the objects. [Baker's caption.]

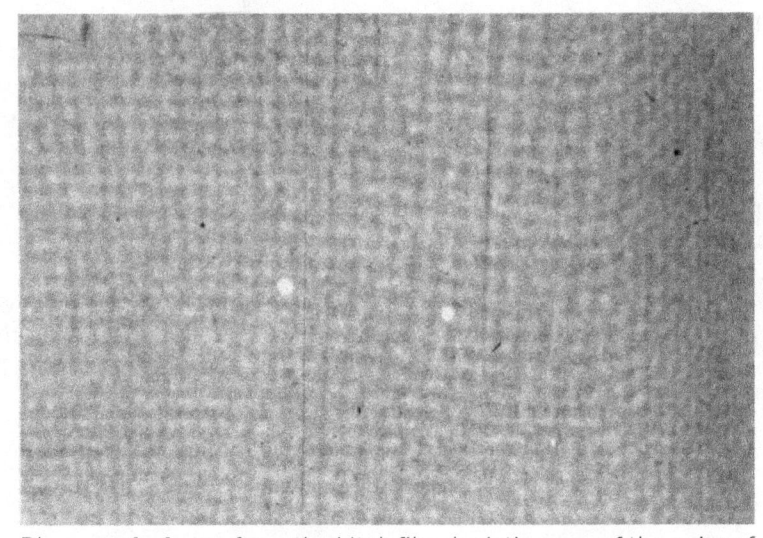

Blow up of a frame from the Utah film depicting one of the pairs of objects. [Baker's caption.]

Blow up of a frame from the Montana film depicting the two objects. [Baker's caption.]

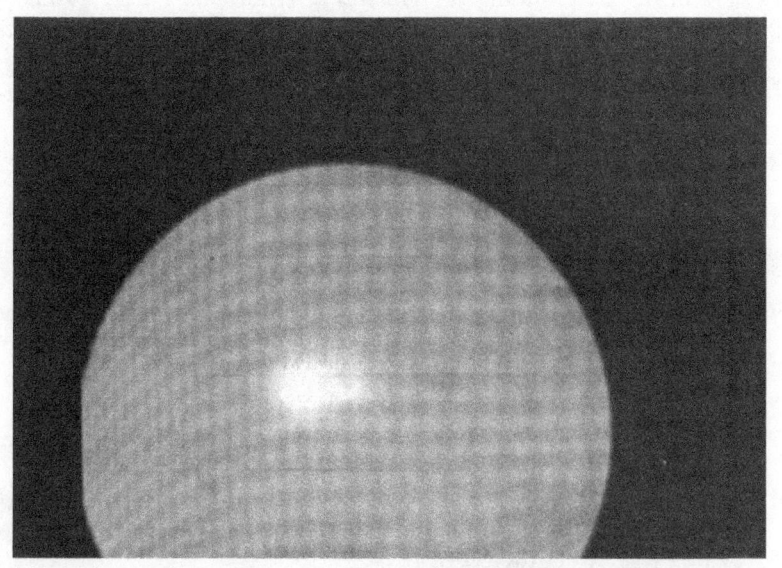

Microphotograph of one of the frames of the Argentina film that exhibits the luminosity of the yellow, pear-shaped anomalistic object. [Baker's caption.]

BAKER'S REFERENCES 1

(1) The *American Nautical Almanac* 1950.
(2) H. O. No. 214, "Tables of Computed Altitude and Azimuth for Latitudes 40° to 49°."
(3) J. Veath, J. G. *200 Miles Up*, Ronald Press Company, N.Y. Second Edition,1955, p. 111.
(4) Kaiser, T. R., *Meteors*, Pergamon Press, 1955.
(5) La Paz, L. "Meteoroids, Meteorites, and Hyperbolic Meteor Velocities," Chapt. XIX of the *Physics and Medicine of the Upper Atmosphere*.
(6) O. C. Farrington, *Meteorites*, Chicago, 1915.
(7) *Measurement of Birds*. Scientific Publications of the Cleveland Museum of Natural History, Vol. II, 1931.
(8) Kartright, F. H., *The Ducks, Geese and Swans of North America*, American Wild Life Institute, 1943.
(9) Headley, F. W. *The Flight of Birds*, Witherly and Co., 326 Holborn, London, 1912.
(10) Menzel, D. H., *Flying Saucers*, Harvard University Press, 1953.
(11) Mees, C. E. K. *The Theory of the Photographic Process*, Revised Edition, MacMillan Co., N.Y., 1954.
(12) Danjon, A., Conder, A. *Lunettes et Telescopes*, Paris, 1935.
(13) Kuiper, G. P., *The Atmospheres of the Earth and Planets*, University of Chicago Press, 1951.
(14) Ruppelt, E. J., *The Report on Unidentified Flying Objects*, Doubleday and Co., 1956.

BAKER'S REFERENCES 2

1) Baker, R. M. L., Jr. (1956) "Analysis of Photographic Material Serial 01 and 02," *Douglas Aircraft Report* dated 24 March and 26 May 1956.
2) Baker, R. M. L., Jr. (1967) *Astrodynamics: Applications and Advanced Topics*, Academic Press, New York, pp. 112-115 and pp. 376 to 392.
3) Baker, R. M. L., Jr. (1968a) "Observational Evidence of Anomalistic Phenomena," *Journal of the Astronautical Sciences*, Vol. XV, No. 1, pp. 31-36.
4) Baker, R. M. L., Jr. "Future Experiments on Anomalistic Observational Phenomena," *Journal of the Astronautical Sciences*, Vol. XV, No. 1, pp. 44-45.
5) Baker, R. M. L., Jr. and Ford, K. C. (1968) "Performance Analysis of Space-Population Cataloging Systems (U)," Secret, SAR, NOFORN Report completed under Air Force Contract

F04701-68-C-0219. 22 April 1968.
6) Baker, R. M. L., Jr. and Makemson, M. W. (1967) *An Introduction to Astrodynamics*, Second Edition, Academic Press, New York, pp. 328-330.7)
7) Fuller, J. G. (1966) *Incident at Exeter*, Putnam, New York.
8) Glover, K. M., Hardy, K. R., Konrad, T. G., Sullivan, W. N., and Michaels, A. S. (1966) "Radar Observations of Insects in Free Flight," *Science*, Vol. 154, pp. 967-972.
9) Hynek, J. A. (1966) *Science*, Vol. 154, p. 329.
10) Klass, P. J. (1968a) *UFO's Identified*, Random House, New York.
11) Klass, P. J. (1968b) Letter dated May 29, 1968.
12) Lamar, D. L. and Baker, R. M. L., Jr. (1965) "Possible Residual Effects of Tunguska-type Explosions on Desert Pavements," Presented at the 28th Annual Meeting of the Meteoritical Society in Odessa, Texas, October 21 to 24.
13) Markowitz, W. (1967) *Science*, Vol. 157, pp. 1274-1279.
14) McDonald, J. (1967) "The UFO Phenomenon: A New Frontier Awaiting Serious Scientific Exploration," (an article on an interview with Dr. McDonald by Nyla Crone), *Arizona Daily Wildcat*, April 6, pp. 6 to 8.
15) Menzel, D. H. (1953) *Flying Saucers*, Harvard University Press, Cambridge, Mass.
16) Minas, J. S. and Ackoff, R. L. (1964) "Individual and Collective Judgements," Chapt. 17 in *Human Judgements and Optimality*, edited by M. W. Shelly, II and G. L. Bryan, John Wiley and Sons, New York, pp. 351-359.
17) O'Keefe, J. A. (1967) Letter dated October 26.
18) Page, T. (1968) "Photographic Sky Coverage for the Detection of UFO's," *Science*, Vol. 160, pp. 1258-1260.
19) Powers, W. T. (1967) "Analysis of UFO Reports, 7 April, 1967, p. 11." *Science*.
20) Robey, D. H. (1960) "A Hypothesis on the Slow Moving Green Fireballs," *Journal of the British Interplanetary Society*, Vol. 17, No. 11.
21) Rosa, R. J., Powers, W. T., Valee, J. F., Gibbs, T. R. P., Steffey, P. C., Garcia, R. A. and Cohen, G. (1967) *Science*, Vol. 158, pp. 1265-1266.
22) Shklovskii, L. S. and Sagan C. (1966) *Intelligent Life in the Universe*, Holden-Day, Inc. San Francisco.
23) Singer, S. (1963) in *Problems of Atmospheric and Space Electricity*, edited by S. C. Coroniti, Elsevier Publishing Company, New York, p. 463.

24) Tacker, L. J. (1960) *Flying Saucers and the United States Air Force*, Van Nostraud, Princeton, New Jersey.
25) Walker, S., III (1968) "Establishing Observer Credibility: A Proposed Method," *Journal of the Astronautical Sciences*, Vol. XV, No. 2, pp. 92-96.
26) Wooldridge, D. E. (1968) *Mechanical Man: The Physical Basis of Intelligent Life*, McGraw-Hill, New York, Chapt. 19.

APPENDIX 1
BAKER'S BIBLIOGRAPHY, JULY 1968

00001—"Elements of Churm's Objects" (with M. W. Corn, G. L. Matlin, and Silvia Rachman), *Minor Planets Circular*, 1100, July 15, 1954.

00002—"Optimal Thrust Angle Program for Transit Between Space Points," *Douglas Aircraft Company Report* SM19180, July 1, 1955.

00003—"Keplerian Missile Trajectories Modified by Initial Thrust and Aerodynamic Drag," *Douglas Aircraft Company Report* SM-19234, August 1, 1955.

00004—"Approximation to Missile Trajectories on a Rotating Earth," *Douglas Aircraft Company Report* SM-19235, May 7, 1956.

00005—"Satellite Librations" (with W. B. Klemperer), *Astronautica* ACTA, III, Fasc. 1, 16-27, 1957.

00006—"Units and Constants for Geocentric Orbits" (with Samuel Herrick and C. G. Hilton), *American Rocket Society* Reprint No. 497-57; Proceedings of the 8th International Astronautical Congress, Barcelona, 1957, 197-235.

00007—"Orbits" (with Samuel Herrick) *Aviation Age*, March 1958, 70-77, Vol. 28, #9.

00008—"Transitional Correction to the Drag of a Sphere in Free Molecule Flow" (with A. F. Charwat), *The Physics of Fluids*, 1, No. 2, 1958, 73-81.

00009—"Drag Interactions of Meteorites with the Earth's Atmosphere," dissertation submitted in partial fulfillment of the degree of PhD at UCLA, May, 1958, xii + 183 pp.

00010—"Passive Stability of a Satellite Vehicle," *Navigation*, 6, No. 1, Spring 1958, 64-5.

00011—"Navigational Requirements for the Return from a Space Voyage," *Navigation*, 6, No. 3, Autumn 1958, 175-181.

00012—"Practical Limitations on Orbit Determination," Institute of Aeronautical Science Preprint No. 842, July 8-11, 1958, 10 pp.

00013—"Astrodynamics and Trajectories of Space Vehicles," Space Technology Lecture Series, sponsored by the Long Island IRE and the American Rocket Society, November 13, 1958.

00014—"Encke's Method and Variation of Parameters as Applied to Reentry Trajectories," *American Astronautical Society* Reprint No. 58-36, August 19, 1958, 13 pp. and *Journal of the American Astronautical Society*, 6, No. 1, 1959.

00015—"Recent Advances in Astrodynamics," (with Samuel Herrick), *Jet Propulsion*, 28, No. 10, 1958, 649-654.

00016—"Ephemeral Natural Satellites of the Earth," *Science*, 128, 1958, 1211.

00017—"Gravitational and Related Constants for Accurate Space Navigation," University of California, Los Angeles, *Astronomical Papers*, No. 24, 1, 1958, 297-338. (Same as Item 00006).

00018—"Precision Orbit Determination," (with L. Walters and E. Durand), Aeronutronic Systems, Inc. Report U-306, December 16, 1958.

00019—"Note on Interplanetary Navigation," *Jet Propulsion*, 28, No. 12, 1958, 834-835.

00020—"Accuracy Required for a Return from Interplanetary Voyages," J. British Interplanetary Soc., May-June, 1959, 93-97 (similar to Item 00011), Vol. 17, #3.

00021—"The Application of Astronomical Perturbation Techniques to the Return from Space Voyages," ARS Journal, March 1959, 29, No. 3, 207-211.

00022—"Sputtering as it is Related to Hyperbolic Meteorites," *J. Applied Physics*, 30, No. 4, April 1959, 550-555.

00023—"Transitional Aerodynamic Drag of Meteorites," *Astrophysical Journal*, 129, No. 3, May 1959, 826-841.

00024—"The Sky is No Limit for Opportunities in Astrodynamics," *IRE Student Quarterly*, May 1959.

00025—"Efficient Precision Orbit Computation Techniques," (with G. Westrom, C. G. Hilton, R. Gersten, J. Arsenault, and E. Browne) ARS Reprint, 1959.(No. 869-59).

00026—"Three-Dimensional Drag Perturbation Technique," *UCLA Astrodynamical Report* #4, July 1, 1959.

00027—"Astrodynamics," (with Samuel Herrick) *Astronautics*, 4, No. 11, pp. 30, 180-1, 1959.

00028—"Effect of Accommodation on the Transitional Aerodynamic Drag of Meteorites," *Astrophysical Journal*, 130, No. 3, 1024-1026, November 1959.

00029—"Training in Astronautics," *Space*, December 1959.

00030—"An Introduction to Astrodynamics" (with Maud

Makemson), Academic Press, New York, October 1960, 358 + xxi.
00031—"Librations on a Slightly Eccentric Orbit," *ARS Journal*, 30, No. 1, 124-26, January 1960.
00032—"Plane Librations of a Prolate Ellipsoidal Shell," *ARS Journal*, 30, No. 1, 126-128, January, 1960.
00033—"Lunar Guidance," (with Maj. J. Schmitt and C. C. Combs) in SR-183 *Lunar Observatory Study* Vol. II (S), ARDC Project No. 7987, Task No. 19769, AFBMD TR 60-44, pages II-3 to II-43, April 1960.
00034—"Orbit Determination from Range and Range-Rate Data," ARS Preprint 1220-60, May 1960.
00035—"Astrodynamics," in *Space Trajectories* (Academic Press, New York), October 1960 29-68.
00036—"Three-Dimensional Drag Perturbation Technique," *ARS Journal*, 30, No. 8, 748-753, 1960. (Same as 00026)
00037—"Review of Perturbations of Orbits of Artificial Satellites Due to Air Resistance," *ARS Journal*, July 1960, 703-704, Vol. 30, No. 7.
00038—"Review of Dependence of Secular Variations of Orbit Elements on Air Resistance," *ARS Journal*, July 1960, 675, Vol. 30, No. 7.
00039—"Efficient Precision Orbit Computation Techniques" (revised), *ARS Journal*, 30, No. 8, 740-747, 1960.
00040—"State-of-the-Art-1960 Astrodynamics," *Astronautics*, 5, No. 11, 30, 1960.
00041—"Novel Orbit Determination Techniques As Applied to Air Force Systems," paper presented to the Seventh Annual ARDC Science and Engineering Symposium, Boston, Massachusetts, November 30, 1960.
00042—"1960 Advances in Astrodynamics," *ARS Journal*, December 1960 (expanded version of Item 00038).
00043—"Analysis and Standardization of Astrodynamic Constants," (with Makemson and Westrom), *Journal of the American Astronautical Society*, VII, No. 1.
00044—"Preliminary Results Concerning Range-Only Orbit Determination," *Proceedings of the First International Symposium on Analytical Astrodynamics*, p. 61, June 29, 1961.
00045—"Perturbations," pp. 4-16 - 4-18; "Orbit Determination," pp. 8-34 - 8-38; "Navigation," pp. 27-33 27-34, *Handbook of Astronautical Engineering*, McGraw-Hill Book Company, Inc., 1961.
00046—"State of the Art - Astrodynamics," *Astronautics*, Vol. 6,

No. 12, December 1961.
00047—Review of *Methods of Celestial Mechanics*, by Dirk Brouwer and G. M. Clemence, and Review of *Physical Principles of Astronautics* by Arthur I. Berman, *The Journal of the Astronautical Sciences*, Vol. VIII, No. 4, Winter 1961.
00048—"Astrodynamics," Chapter in McGraw-Hill *Encyclopedia of Science and Technology*, McGraw-Hill Book Company, Inc., 1962.
00049—"Determination of the Orbit of the Russian Venus Probe," (with B. C. Douglas, David Newell, A. K. Stazer, R. L. Held and M. Lifson). *ARS Journal*, pp. 259-260, February 1962.
00050—"A Note on the Determination of Orbit from Fragmentary Data," (with B. C. Douglas and Mary P. Francis). *Lockheed Astrodynamics Research Report* #1, LR 15379, April 1962.
00051—Review of *Introduction to Space Dynamics* by W. T. Thomson, Review of *An Introduction to Celestial Mechanics* by Theodore E. Sterne, Review of *Fundamentals of Celestial Mechanics*, by J. M. A. Danby, *The Journal of Astronautical Sciences*, Vol. IX, No. 4, Winter 1962.
00052—"Influence of Planetary Mass Uncertainty on Interplanetary Orbits," *ARS Journal*, No. 12, Vol. 32, December 1962.
00053—"Elimination of Spurious Data in the Process of Preliminary and Definitive Orbit Determination," *Dynamics of Satellites Symposium* (Paris, May 28-30, 1962), Berlin, Springer-Verlag, 1963.
00054—"Utilization of the Laplacian Method from a Lunar Observatory," *Icarus*, Vol. 1, No. 4, January 1963.
00055—"Lunar Radio Beacon Location by Doppler Measurements," (with T. P. Gabbard), *AIAA Journal*, Vol. 1, No. 4, April 1963.
00056—Review of *Space Mechanics*, by W. C. Nelson and E. E. Loft, *Journal of Astronautical Sciences*, Winter 1963.
00057—"A Bibliography of General Perturbation Solutions of Earth Satellite Motion," by Taylor Gabbard Jr. and Eugene Levin. *Astronautics and Aerospace Engineering*, November 1963.
00058—"Review of Introduction to Celestial Mechanics," by S. W. MCuskey, *Journal of Astronautical Sciences*, Winter 1963.
00059—"Review of Space Flight," Vol. II *Dynamics*, by Kraft Enricke, *Journal of Astronautical Sciences*, Winter 1963.
00060—"Influence of Martian Ephemeris and Constants on Interplanetary Trajectories," Chapter in *Exploration of Mars*, American Astronautical Society, 1963.

00061—"Orbit Determination by Linearized Drag Analysis," (with Kurt Forster). *AIAA Preprint* No. 63-428, presented to *AIAA Astrodynamics Conference* August 19-21, 1963, Yale University, New Haven, Connecticut.

00062—"Extension of f and g Series to Non-Two- Body Forces," *AIAA Preprint* No. 64-33. Presented at the *Aerospace Sciences Meeting*, New York, New York, January 20-22, 1964, also *AIAA Journal*, July, 1964.

00063—Review of *Orbital Dynamics of Space Vehicles*, by Ralph Deutsch, Prentice-Hall, Inc., *Journal of Astronautical Sciences*, Spring 1964.

00064—*An Introduction to Astrodynamics*, (with Maud Makemson) Academic Press, New York, October 1960, third printing, 1963), Fourth Printing in preparation.

00065—"1964 State of the Art in Astrodynamics," AIAA Annual Meeting, Wash., D. C. June 19, July 2, 1964, *AIAA Preprint* No. 64-535. (Also lecture given at Univ. of Wash., Seattle, May 29, 1964, and at Boeing Scientific Research Laboratory, June 1, 1964).

00066—"Space Mechanics," Chapter in *Space/Aeronautics*. Research and Development Tech. Handbook, 1964/1965, pp. 11-13, published by Conover-Mast, 1964. (New York).

00067—"Radiation on a Satellite in the Presence of Partly Diffuse and Partly Specular Reflecting Body," presented at the Joint Symposium on the Trajectories of Artificial Celestial Bodies as Determined from Observations; Paris, France, April 20-23, 1965.

00068—"Possible Residual Effects of Meteor and Comet Explosions on Desert Pavements," with Donald L. Lamar; presented at the 28th Meteoritical Society Meeting, Odessa, Texas, October 1965.

00069—*Proc. of COSPAR/IUTAM/IAU Symp.*, Springer/Verlag, 1966 (Same as 00067).

00070—*An Introduction to Astrodynamics*, 2nd Edition, Academic Press, New York, 1967. (With M. W. Makemson)

00071—*Astrodynamics - Applications and Advanced Topics*, Academic Press, New York, 1967.

00072—"Recent Advances in Astrodynamics," 1961, (with Mary P. Francis), *UCLA Astrodynamical Report* # 13, January 1962. (Similar to 00046)

00073—Review of *Theory of Orbits* by V. Szebehely, *Journal of the Franklin Institute*, Vol. 284, No. 6, December 1967.

00074—"Observational Evidence of Anomalistic Phenomena,"

1968, *Journal of the Astronautical Sciences*, Volume XV, No. 1, pp. 31-36.

00075—"Future Experiments on Anomalistic Observational Phenomena," 1968, letter to editor, *Journal of the Astronautical Sciences*, Volume XV, No. 1, pp. 44-45.

00076—"Astrodynamics," 1968, in *Encyclopaedic Dictionary of Physics*, Pergamon Press.

00077—"Performance Analysis of Space - Population Cataloging Systems (U)," 1968, Secret, SAR, NO FORN Report completed under Air Force Contract F04701-68-C-0219. (With K. C. Ford), April 22, 1968.

00078—"Hydrofoil Sailcraft Water Conveyance Optimum Lift-off Speed," 1968, *Science*, in press.

00079—"Preliminary Orbit Determination for High-Data-Rate Sensors," 1968, *Journal of the Astronautical Sciences*, Volume XV, No. 5.

00080—"Surveillance System Sensor Mis-Association of One Object with Another," 1968, to be published.[20]

[Note: Baker's lengthy Appendix 2 entitled, "The Applied Assessment of Central Nervous System Integrity: A Method for Establishing the Credibility of Eye Witnesses and Other Observers," written by Dr. Sydney Walker III, originally appeared here. Although of interest to trained health care professionals, I have omitted it due to the fact that the article proposes to medically evaluate eyewitnesses using a sociological, physiological, pharmacological, neurological, opthalmological, and psychiatric approach, with little direct correlation to the UFO phenomena—as the title itself indicates. The author-editor, L.S.]

CHAPTER SEVEN

FREE DISCUSSION AMONG THE SIX SCIENTISTS

MR. ROUSH. I anticipated we would have difficulty keeping the members of the committee here at a time when important legislation is considered on the floor. We thought we would reserve the final few minutes for those of you who have made presentations to discuss among yourselves questions which may have been aroused by one of your colleagues' presentation today.

With that in mind, we are going to permit you to have a real free for all. Dr. Sagan.

DR. SAGAN. I just wanted to underline one point that Dr. Baker made, Congressman Roush, in his detailed presentation of the various Air Force systems. I am afraid that the main point won't come across to a lay audience, and that is that with relatively little expenditure of funds, it would be possible to significantly improve the available information.

Apparently what is now happening is that the Air Force surveillance radar is throwing away the data that is of relevance for this inquiry. In other words, if it sees something that is not on a ballistic trajectory, or not in orbit, it ignores it, it throws it in the garbage.

Well, that garbage is just the area of our interest. So if some method could be devised by the Air Force to save the output that they are throwing away from these space surveillance radars, it might be the least expensive way to significantly improve our information about these phenomena.

MR. ROUSH. Thank you.

DR. BAKER. Let me just make a comment: That is quite true. At the present time our space surveillance sensors are about 200 percent overtasked. That means they could make about 50 percent of their time available to us. They task too many space objects, their capacity is much greater than the space objects that they are tasked to watch. The space population may grow to fill this void, but currently what Dr. Sagan says is true, we could as I indicated in conclusion (4) modify our current space surveillance system.

It is not an expensive thing to modify existing radars. The

FPS-85 itself costs something like $100 million. The software modification called for here I am sure would be much less.

MR. ROUSH. Dr. Hynek.

DR. HYNEK. I would just like to concur in what Dr. Sagan has said. I understand there are several hundred UCT's a month, uncorrelated targets, that because they don't—I understand—which since they do not follow ballistic trajectory, they are tossed out. It would not be expensive to introduce a subroutine into the computer to take care of these things for a short while. I strongly second Dr. Sagan's and Dr. Baker's suggestions.

MR. BOONE. Mr. Chairman.

MR. ROUSH. Mr. Boone.

MR. BOONE. I think the gentleman should advise you too, though, when you do that, you must make a trajectory determination on each target including aircraft, which may put a terrific burden on the radar you are insisting on upgrading.

DR. HYNEK. I will certainly grant that.

DR. HARDER. I would only respond to Mr. Boone by suggesting you could reject all objects that were found, for instance, under 90,000 feet.

DR. SAGAN. That is just what I was going to say. Certain velocity and altitude limitations.

MR. BOONE. With that I agree. But I don't think we make many sightings at that altitude. We do have a problem here of what you want to look at. So in fact I think the thrust of Dr. Baker's argument here was that most of the Air Force equipment do not supply the material you would like to have.

So you are going to have to go to a much lower altitude, and you are going to have to check a much larger number of targets.

DR. SAGAN. I may have misunderstood, but my understanding was, since all of these "uninteresting," trajectory objects are thrown away, we have no way of knowing at the present time whether there are or are not large numbers of interesting objects at altitudes above 90,000 feet.

MR. BOONE. What this means is you check each one and determine its trajectory, and then throw it away, so it no longer becomes a simple task of saying "Oh, I only want to look at the unidentified ones." I have to check each one, and discard it.

DR. SAGAN. Isn't that being done already?

MR. BOONE. No, it doesn't do it below certain altitudes.

DR. SAGAN. Right.

MR. BOONE. All right. Certain targets are picked up at certain ranges, are they not?

DR. SAGAN. Right. So therefore the suggestion is that within the altitude range, that is being used anyway by the surveillance radar—

MR. BOONE. You complicate the procedure.

DR. SAGAN. Slightly.

MR. BOONE. The procedure is used but it involves the software again which is much more difficult to add to the systems than I believe is being presented. It can be done, there is no question it can be done.

DR. HARDER. I would agree the amount of effort that goes into the relative softwares, although by no means a $100 million project, it is not a very simple project.

MR. ROUSH. Dr. McDonald, do you have a comment?

DR. MCDONALD. Yes. I would underscore another one of the points, the general points that Dr. Baker made. I think it addresses itself to the question raised. Both scientists and members of the public are quite aware we have many monitoring radar systems, optical and so on.

This question is raised often, why aren't UFO's tracked? The point one is struck with in studying each of these systems in turn is the large degree of selectivity that is necessarily built into them. Good examples were cited by Dr. Baker.

It has to be kept well in mind that even systems like SAGE when they were developed necessarily had to have programed into them certain speed limits both lower and upper, certain safe requirements like if the target was on an outbound path it could be ignored. In almost every monitoring system you set up, whether for defense or scientific purposes, if you don't want to be snowed with data, you intentionally built selectivity in, and then you do not see what you are not looking for.

Consequently, this point is important, that despite our many sensing and monitoring systems, the fact that they don't repeatedly turn up what appear to be similar to UFO's, whatever we define those to be, is not quite as conclusive as it might seem.

The second comment I would make concerns Dr. Baker's remark that we should move ahead to instrumental techniques and perhaps lessen attention on the older data.

I too agree that we have much need to replace what police officers and pilots saw with good hard instrumental data, the sooner the better, but there are many fields in which once you get instrumental data, say seismology, and being to learn about the phenomenon you are studying, seismology, astronomy, meteorology, once you understand these things you do go back to exploit the knowledge that is implicit in older data. Seismologists do study old earthquake records to improve the seismicity data available. Ecologists do look at old shifts in plant and animal patterns. Astronomers do look at old eclipse information, because once you begin to understand a problem, you can then sort out much better the important material.

I would not want to see excluded entirely—in fact, I think it would be folly to exclude observations that go back 20 years, and a part of the problem we have not talked about today, still earlier observations.

DR. BAKER. Yes, I concur in that.

My message there was that if we preoccupy ourselves with continually going over past history, it is going to be frustrating. I think we can always use past history in retrospect. In order to go back, as you say, to look at the data and to put it in the proper perspective, when we learn more about the phenomena. So I agree.

MR. ROUSH. Is there any other aspect of previous presentations that any of you would like to question?

DR. BAKER. I have a question of Dr. Harder about the Ubatuba magnesium.

Was this magnesium terrestrial? In other words, it is granted that Ubatubas couldn't produce it, but could the magnesium have been produced terrestrially, and if so, in what connection would we produce and employ such magnesium here on earth?

DR. HARDER. Well, such pure magnesium is indeed produced terrestrially in connection with Grignard reagents, and produced by the Dow Chemical Company, where magnesium is produced in greater purity actually than this.

At the time in 1957, the Brazilians did not have a sample of magnesium from the U.S. Bureau of Standards that was as pure as this Ubatuba magnesium with which to compare it. I might enlarge upon the data which was produced, or which was gotten at the request of Dr. [Roy] Craig, that of the impurities found by the Colorado group, the principal one was zinc strontium with barium being a runner-up. These are very curious kinds of alloys from any terrestrial point of view.

No detected aluminum, and only three parts per million copper, and those are the most likely alloying elements from the terrestrial point of view.

DR. BAKER. Would you say that the sample was partially terrestrialized, and it might be the remnants of an ultrapure nonterrestrial alloy, or did it appear these particular impurities were in the sample from the beginning?

DR. HARDER. This was done by a neutralization analysis on a very tiny slicer. It would be hard to say to what extent over the intervening 9 years there might be some territorialization, but certainly it would not have taken out aluminum or copper. It might have added zinc or barium, although that seems somewhat unlikely.

DR. SAGAN. So some comparison analysis has been made for example of the magnesium flares. A magnesium flare has an abundance of impurities?

DR. HARDER. It would hardly be 99.9 percent purity.

DR. SAGAN. That is what I meant.

DR. HARDER. Yes, that is right.

MR. ROUSH. Dr. McDonald.

DR. MCDONALD. Both Dr. Hall and Dr. Sagan remarked in different contexts on the intense emotional factors that predispose some people to certain systems of belief, and I would like to remark on that to be sure that some perspective is maintained on that part of the problem.

In the witnesses I have interviewed—I have intentionally stayed away from those who immediately show a very strong interest in a salvation theory, or something like that—so I have cut down my sample right at the start.

I would want to leave the point strongly emphasized that though there are a few people, and some of them rather visible and vocal, who are emotional about the problem and tie it to almost religious beliefs, the body of evidence that puzzles me, that bothers me, and I think demands much more scientific attention, comes from people who are really not at all emotional about it; they are puzzled by it, they are reliable, a typical cross-section of the populace. They have not built any wild theories on it.

In fact, let me mention one important sighting in New Guinea. I didn't interview the witness in New Guinea, but in Melbourne, Australia. An Anglican Missionary, Rev. William B. Gill, was teaching the school in New Guinea, and when he and some three dozen mission personnel saw an object hovering offshore with four figures visible on top of it, even this minister didn't begin to put any religious interpretation on it. He said this is what he saw, and he wrote very careful notes about it. It is that kind of evidence, and not evidence that comes from people with emotional factors predisposing them to system beliefs that impress me.

MR. ROUSH. Let's have the psychologist speak here for just a moment.

DR. HALL. Thank you.
I welcome that clarification.

The point I was making was not that the witnesses generally are emotional and precommitted to a position at all, but that the people who are interpreting the evidence after it has been gathered are usually precommitted beyond the point of rationality, and it is a very important distinction that you brought out.

The primary problem of witnesses, it seems to me, is this reluctance to report based apparently on a feeling that they will be ridiculed that their evidence is not welcome—and I guess I can't resist telling the little story from the *Wall Street Journal*, quite recently, of a man who had five pet wallabies in Westchester County. A wallaby is a miniature kangaroo. These five wallabies escaped, and rather than upset people he didn't report this, he waited for people to tell him that they had seen them. And nothing happened for days and days.

Well, when they were finally relocated and caught then lots of people started admitting, yes, they had seen these wallabies, but after all, if you see a tiny kangaroo loping across the road in New Rochelle, you are reticent to report it.

MR. ROUSH. Dr. Hynek again.

DR. HYNEK. I think that is a most interesting point that ties in.

I think sometimes we don't ask ourselves really very fundamental questions, and that is, how is it that these reports exist in the first place?

It is not just because they are strange, because we don't have reports of Christmas trees flying upside down, or elephants doing strange things in the sky; the reports are strange, but they do have a certain pattern.

Now, I have often asked myself, well, why do the reports exist in the first place? And how many are reported?

Whenever I give a presentation to some group I frequently will ask them, well, how many of you have seen something in the skies you couldn't explain; that is a UFO, or some friend whose veracity you can vouch for?

I have been surprised to find that 10 to 15 percent, albeit it is a specialized audience, they are there already because they are interested, hence there is a selection factor, but nonetheless I am quite surprised that many respond.

Then I ask the second one, Did you ever report it to the Air Force? And maybe one or two will say that they have.

Now, why, then, should people make reports anyway, since they face such great ridicule? They do it for two reasons, those that I have talked to: One, is out of a sense of civic duty. Time and again I will get a letter saying, I haven't said this to anybody, but I feel it is my duty as a citizen to report this. And many letters come to me. In fact, even saying, please do not report this to the Air Force.

The second reason is that their curiosity finally bugs them. They have been thinking about it and they want to know what it was they saw, and many letters I get will end in a rather plaintive note, can you possibly tell me, or can you tell me whether it is possible what I saw?

Those two reasons are the "springs" of why the report is made in the first place. I don't know how much store can be put in the Gallup poll, but I understand when, about 2 years ago a poll was made on this subject, there was something like the poll reported 5 million people, 5 million Americans had seen something in the skies they could not explain. Over the past 20 years the Air Force has had some 12,000 reports. Therefore, one can logically ask, who is holding out on the other 4,988,000 reports?

I think there may be quite a reservoir of reports that simply have not come out into the open because of this natural reluctance of people to speak out.

MR. ROUSH. Dr. Hynek, your experience has been similar to mine, although much more extensive. In the 10 years I have served on this committee I have had occasion to ask various witnesses their beliefs as far as UFO's are concerned. They have included Air Force generals and Army generals, and usually they display a great interest. Sometimes they will say, I don't believe, but my wife does; some will say.

The other day I was engaged in a colloquy over on the floor of the House, not a part of the record, but just as a side conversation, with two of my colleagues who sit on this committee.

(At this point, discussion was off the record.)

MR. ROUSH. Back on the record. As a result of my experience on this committee I have been privileged to visit the tracking stations which NASA has throughout the world. Each place I have visited I have asked the question, "Have you tracked any unidentified flying object?"

Well, it is obvious they apparently don't have the ability to track, but the response was "No," everywhere except in South Africa. Then they said, "Anything we track, which we do not understand, we turn over to the Department of Defense," inferring there were some things they did not understand.

The same is true with those places in the world where there is a Baker-Nunn camera. I asked the same question of them. For the most part there was a boundless curiosity, but a negative response.

DR. HYNEK. I might respond to that, of course, in talking to them, you have represented officialdom, and they may themselves be a little afraid to say anything to a Congressman that might get them into trouble.

But I get reports subrosa that are to the effect that people, trackers, and so forth, have seen things, but they would not dare think of reporting it.

Now, that is hearsay. I am sorry it is not hearsay; it has happened to me. But it is not what I would call "solid evidence."

MR. ROUSH. Just one other comment. I serve on the board of trustees of a college back in Indiana. In the course of a year they had numerous lectures by outstanding people in their lecture series, quite outstanding people on various subjects, but they scheduled one lecture given by a student at the college on unidentified flying objects. Needless to say, he had the best attendance of the entire series.

Dr. Harder.

DR. HARDER. Following on something that Dr. Hynek said about the small percentage of actual sightings that are reported, this would suggest that the two instances that I brought out, which to my knowledge are the only extant pieces of what you might call scientific information—information containing information of a scientific nature, might well be multiplied by a factor of 10, if it were not for this ridicule bit, and furthermore, if it were not the subject of ridicule, many people would perhaps take greater care in the observations that they do make, and perhaps come up with similar kinds of anecdotal nature of somewhat more importance than just flashing lights.

For instance, the plane of polarization or—well, many kinds of observations came to us. We would have even at this point far more anecdotal information of a scientific nature and of scientific importance than we now have.

MR. ROUSH. I think those of you who have sat on this panel today have made perhaps a greater contribution than you realize in adding some respectability to the interest the American people have in this phenomena. Perhaps we can, by further activity on the part of this committee, and you on your part, and by the public reading what you have said today, cause people to be more responsive and to report what they see. Perhaps we can thereby give an air of respectability to these sightings which will permit people to go ahead without being embarrassed or ashamed of reporting what they have seen.

Does anyone else have anything here?

MR. [JAMES G.] FULTON. Mr. Chairman, sightings of UFO's

in western Pennsylvania have now increased to the point where interested citizens have established a UFO Research Institute with a 24-hour answering service, to investigate reports and sightings.[21]

MR. ROUSH. Dr. Baker, and Dr. Hall, Dr. McDonald, Dr. Harder, Dr. Hynek, and Dr. Sagan, I believe that you people have made a real contribution here, and I think the time will come when certain people will look back and read what has been done here today and realize that we have pioneered in a field insofar as the Congress of the United States is concerned. They will be very mindful that something worthwhile was done here today.

As a personal note, I would like to say this has been one of the most unusual and most interesting days I have spent since I have been in the Congress of the United States.

Thank you.

I thank each of you.

The committee stands adjourned.

(Whereupon, at 4:39 p.m., the committee was adjourned.)

In addition to the six symposium participants, the chairman, Mr. Roush, invited other scientists to submit written papers expressing their views of unidentified flying objects. Their papers follow as a part of the record of the symposium.

End of six-scientist symposium

SECTION TWO

SUBMITTED PREPARED
STATEMENTS BY SIX
ADDITIONAL SCIENTISTS

CHAPTER EIGHT

DONALD H. MENZEL

PREPARED STATEMENT BY DONALD H. MENZEL

UFO: FACT OR FICTION?

☛ Flying saucers or UFO's have been with us for a long time. June 24, 1968, marked the 21st anniversary of the sighting of nine bright disks moving rapidly along the hogback of Mount Rainier. However, similar sightings go far back in history, where they have assumed various forms for different people. Old records refer to them as fiery dragons, fiery chariots, wills-o'-the-wisp, jack-o'-lanterns, *ignis fatuus*, firedrakes, fox-fire, and even the devil himself.

And now a new legend—a modern myth—has arisen to explain a new rash of mysterious sightings. Certain UFO buffs argue that the peculiar properties and maneuvers of these apparitions, as reported by reliable people of all kinds, are so remarkable that only one explanation for them is possible. They must be vehicles from outer space, manned by beings far more intelligent than we, because the operators have clearly built vehicles with capabilities far beyond anything we can conceive of.

On the face of it, this reasoning sounds much like that of Sherlock Holmes, who said on several occasions: "It is an old maxim of mine that when you have excluded the impossible, whatever remains, however improbable, must be the truth!"

I am willing to go along with this formula, but only after we have followed Holmes and excluded every possibility but that of manned UFO's. And we must also show that no further possible solutions exist.

The believers are too eager to reach a decision. Their method is simple. They try to find someone, whom they can establish as an authority, who will support their views. They then quote and often misquote various authorities or one another until they believe what they are saying. Having no real logic on their side, they resort to innuendo as a weapon and try to discredit those who fail to support their view. The UFO magazines refer to me as the arch-demon of saucerdom!

I concede that the concept of manned spaceships is not an absolute impossibility. Neither are the concepts of ghosts, spirits, witches, fairies, elves, hob-goblins, or the devil. The only trouble with this last list is the fact they are out of date. We live in the age of space. Is it not natural that beings from outer space should exhibit an interest in us? But, when we consider that these beings if indeed they are beings have been bugging us for centuries, why should one not have landed and shown himself to the President of the United States, to a member of the National Academy of Sciences, or at least to some member of Congress?

Please don't misunderstand me. I think it is very possible that intelligent life—perhaps more intelligent than we—may exist somewhere in the vast reaches of outer space. But it is the very vastness of this space that complicates the problem. The distances are almost inconceivable. The time required to reach the earth—even at speeds comparable with that of light—range in hundreds if not thousands of years for our near neighbors. And it takes light some billions of years to reach us from the most distant galaxies, times comparable with that for the entire life history of our solar system. The number of habitable planets in the universe is anybody's guess. Any figures you may have heard, including mine, are just guesses. I have guessed that our own Milky Way may contain as many as a million such planets. That sounds like a lot, but the chances are the nearest such inhabited planet would be so distant that if we send out a message to it today we should have to wait some 2000 years for a reply. Alas, the evidence is poor for intelligent life in our solar system, though I do expect some lower forms of life to exist on Mars.

With respect to UFO's my position is simply this. That natural explanations exist for the unexplained sightings. The Air Force has given me full access to their files. There is no vast conspiracy of either the Air Force or CIA to conceal the facts from the public, as some groups have charged. The basic reason for continued reporting of UFO's lies in the possibility—just the possibility mind you—that some of them may derive from experimentation or secret development by a hostile power. And I don't mean hostile beings from outer space!

The Air Force has made its mistakes. They never have had enough scientists in the project. They have failed to follow up certain sightings of special importance. Their questionnaire is amateurish, almost cleverly designed in certain cases to get the wrong answer and lose track of the facts. The Air Force is aware of my criticism and, on a voluntary basis, I have helped them improve the questionnaire. It was not an easy job. Especially when the Air Force rejected some vital questions as "an invasion of the privacy of the individual."

From 1947 until 1954 a bewildered group of Air Force personnel tried honestly and sincerely to resolve the UFO problem. Many highly reliable persons had reported seeing "objects" moving at fantastic speeds, and apparently taking evasive action in a manner impossible for known terrestrial craft. By 1952 a sizable number of those in the Air Force group had concluded that extraterrestrial vehicles were the only explanation. Some of this unrest leaked out. Popular writers exploited these ideas and soon various UFO clubs came into existence. In 1953, a committee of scientists, headed by the late H. P. Robertson of California Institute of Technology met at CIA to consider a number of the Air Force's most convincing cases. They immediately solved many of them. Others could not be solved because of poor or insufficient data. They concluded that all cases had a natural solution. There was no evidence to support the idea that UFO's are vehicles from another world.

Nevertheless, the UFO buffs believe, almost as an article of faith, that "trained observers," such as military or airline pilots, could not possibly mistake a meteor, a planet, a star, a sundog, or a mirage for a UFO. This viewpoint is absolutely nonsense and the Air Force files bear witness to its falsity! They contain thousands of solved cases—sightings by "reliable individuals" like the pilots. But such persons have made huge errors in identification.

A huge meteor flashes in the sky! The co-pilot thinks it is going to strike the plane and takes evasive action. The pilot disagrees and he is right. The UFO proves to be a fireball or meteor a hundred miles away! Such occurrences are frequent, not rare. They have even increased with the growing number of re-entries and spectacular decay of satellite debris from the space operations of the U.S.A. and the U.S.S.R.

Distances overhead are uncommonly hard to estimate—either on the ground or in the air. A bird's feather, shining brightly in the sun and floating a mere 20 feet overhead may seem to be a distant object moving at very high speed. Conversely, a pilot may think that a bright object on the horizon, in reality a star or planet, lies just beyond his wing tip. Sometimes, a layer of warm air, sandwiched between 2 layers of cold air, can act as a lens, projecting a pulsing, spinning, vividly colored, saucer-like image of a planet. Pilots, thinking they were dealing with a nearby flying object, have often tried to intercept the image, which evades all attempts to cut it off. The distance may seem to change rapidly, as the star fades or increases in brightness. Actual "dogfights" have been recorded between a confused military pilot and a planet. I myself have observed this phenomenon of star mirage. It is both realistic and frightening.

Such observations fortified the UFO legend—that these objects "maneuver as if under intelligent control." But the pilots failed to realize that the "intelligent control" came from within themselves. And I think that Air Force personnel of Project Blue Book still do not appreciate this important UFO phenomenon.

Mirages are not the only apparitions that appear to maneuver. I think I was the first person to point out that a special kind of reflection of the sun (or moon), sometimes called a sundog (or moondog), also can perform evasive action. Layers of ice crystals are necessary, like those found in cirrus clouds. An aviator flying through cirrus sometimes sees a peculiar metallic appearing reflection, a reflection of the sun or moon. He may elect to chase it. The apparition will recede if approached, or approach if the pilot reverses his course. The object seems to execute evasive action! As the pilot runs out of ice crystals, the UFO will seem to put on a burst of speed and disappear into the distance.

But such behavior does not imply, as the UFO addicts argue, the presence of an intelligence pilot to guide it. No! It's like chasing a rainbow, which recedes as you approach it or advances as you move away.

As we look over the Air Force files, we find that some 90 per cent of the solved cases result from the presence of material objects in the atmosphere. I list some of these objects. Reflections from airplanes, banking in the sun, simulate saucers. Momentarily, a bright reflection appears and then vanishes. The plane is invisible in the distant haze. An imaginative person concludes that an interplanetary vehicle has come in fast, reversed course, and rapidly receded into the distance. Often the observers say "It couldn't have been a plane," because "no noise was heard" or because "it moved too swiftly." And yet careful study proves beyond doubt that the object was indeed an aircraft. The brilliant landing lights of a plane can almost dazzle a person on the ground. Sometimes such lights may appear to be very close only a few hundred feet away.

You'd be surprised at the variety of mundane objects that people have reported as UFO's. Balloons, child's balloons, weather balloons lighted or unlighted, and especially those enormous plastic balloons as large as a ten-story building, which carry scientific instruments to altitudes of 100,000 feet! Reflecting full sunlight while the earth below lies in dim twilight, these balloons shine more brilliantly than Venus! Advertising planes or illuminated blimps frequently become UFO's.

Birds, by day or night, often reflect light from their shiny backs. Wind-blown kites, hats, paper, plastic sacks, feathers, spider webs, seed pods, dust devils have all contributed their share of UFO sightings. Insects single or in swarms. Saucer-shaped clouds, reflections of searchlights on clouds! Special space experiments, such as rocket-launched sodium vapor releases or balloons from Wallop's Island have also produced spectacular apparitions! Ball lightning and the Aurora Borealis occasionally contribute.

Reflections from power lines, insulators, television antennas, radars, radio telescopes, even apartment windows! These, too, have produced realistic UFO's.

I could add to this list almost indefinitely. But the chief point I want to make is that simple phenomena like the above have tricked intelligent people into reporting a UFO.

But there are a few other phenomena that can produce UFO's of a type that, as far as I know, the Air Force still does not recognize.

I quote from an article on "Vision" in Volume 14 of the McGraw-Hill *Encyclopedia of Science and Technology*:

> "Any observant person can detect swirling clouds or spots of 'light' in total darkness or while looking at a homogeneous field such as a bright blue sky."

If you want to see flying saucers just look up. If you don't see them, you probably are not "observant."

I see them most clearly in a dark room or on a moonless night with the sky even darker with heavy clouds. I find stars somewhat distracting. Just lie down on your back, open your eyes and see the saucers spin. The show is free. You will almost surely see bright, irregular patches of light form. Most of them seem grey green, but I occasionally see silver or gold and occasionally red. I can imagine windows in some of them. As you move your eyes they will cavort over the sky. To speed up the action just rub your eyes like a person coming out of a sleep. Occasionally the whole field becomes large and luminous. Now, I ask you. How can you be sure that the UFO reported by an airline pilot is not one of these spurious images? And even if an alerted co-pilot confirms it, he might also be responding to a similar effect in his own eyes!

The chemistry and physiology of the human eyes are certainly responsible for many UFO sightings. The eye responds in different ways to different kinds of stimuli. A sudden burst of bright light, like that from a flash bulb, for example, exerts an enduring effect on the eye. The light from the flash produces an immediate change in the so-called visual purple of the retina. In a sense the retinal spot on which the image fell becomes fatigued. For some minutes after the flash you will be able to see a bright, usually greenish, floating spot, which could be mistaken for a UFO by someone unfamiliar with the problem.

Let me take an actual case, which is typical of a large number actually in the files of Project Bluebook. A child, going to the bathroom turns on a bright light and accidentally awakens one of his parents who is blinded by the sudden illumination. The light goes off and the parent gets up to investigate and just happens to glance out of the window. He is startled to see a peculiar spot of light floating over the trees and making irregular, jerky motions. He watches the UFO for a minute or two until it finally disappears.

He cannot be blamed for failing to realize that the erratic and often rapid movements of his UFO are those of the after-image, drifting with the similar movements of his own eye. The UFO appears in the direction he happens to be looking. That is all. And yet he may describe it graphically as a luminous object "cavorting around in the sky."

Many such stimuli are possible by day or night. Some time ago I was driving directly toward the setting sun. When I came to a stop-light and looked out the side window of the car, I was startled to see a large, black object shaped something like a dirigible, surrounded by dozens of small black balloons. I suddenly realized that they were after-images of the sun. The big one was where I had been looking most fixedly. The spots were images where my eye had wandered. A UFO buff could have sworn that he was seeing a "mother ship" and a swarm of UFO's in rapid flight.

I once had another similar experience. I suddenly glanced up and was surprised to see a whole flotilla of UFO's flying in formation across the blue sky. They looked like after-images, but I hadn't been conscious of the visual stimulus responsible. I quickly retraced my steps and found it: sunlight reflected from the shiny surface of the fender of a parked car.

I am sure that many UFO's still unknowns, belong to this class. Look fixedly at the full moon for at least 30 seconds and then turn away. A greenish balloon will swim over your head and perform maneuvers startling or impossible for any real object. I'm been able to attain the same effect with the planet Venus, when near maximum brilliance. Yet most observers will swear that such

UFO's are true objects. And the Air Force questionnaire, failing to recognize even the existence of this kind of UFO, contains not a single question that would help them to identify it. In fact the words signifying UFO, unidentified flying object, show the state of mind of the Air Force personnel who invented this abbreviation. What I am saying is that the UFO's are not unidentifiable, they are often not flying, and many are not even objects. It is this point of view—to regard the apparitions as actual solid objects—that has retarded [finding a] solution [for] so long.

After-images possess still other complicated characteristics. A colored light tends to produce an after-image with complementary color. A green flash will cause a red after-image and vice versa. Color-blind persons and persons with defective vision will often experience effects different from those of people with normal eyesight.

Another optical phenomenon that can produce an illusion of flying objects lies within the eye itself. Again, look at some uniformly bright surface—sky or ceiling. Relax your eyes. By that I mean focus your eyes on infinity. The chances are that you will see an array of dark spots. These specks, which may seem to be near like a swarm of gnats or as ill-defined objects at a distance, are either on or in your eye. They may be dust floating on the lens, minute imperfections in the cornea, or possibly blood cells on the retina. These, too, can simulate evasive and erratic movement.

The eyeball jumps a little every time you blink. Walking transmits vibrations to the eye at every step. Many individuals think they see stars, planets, or satellites oscillating when the movement is actually that of the eye itself. Here is an example.

On our return across Minnesota we had an experience which I have always remembered as illustrative of the fallacy of all human testimony about ghosts, rappings, and other phenomena of that character. We spent two nights and a day at Fort Snelling. Some of the officers were greatly surprised by a celestial phenomenon of a very extraordinary character which had been observed for several nights past. A star had been seen, night after night, rising in the east as usual, and starting on its course toward the south. But instead of continuing that course across the meridian, as stars invariably had done from the remotest antiquity, it took a turn toward the north, sunk toward the horizon, and finally set near the north point of the horizon. Of course an explanation was wanted.

My assurance that there must be some mistake in the observation could not be accepted, because this erratic course of the heavenly body had been seen by all of them so plainly that no doubt could exist on the subject. The men who saw it were not of the ordinary untrained kind, but graduates of West Point, who, if any one, ought to be free from optical deceptions. I was confidently invited to look out that night and see for myself. We all watched with the greatest interest.

In due time the planet Mars was seen in the east making its way toward the south. "There it is!" was the exclamation.

"Yes, there it is," said I. "Now that planet is going to keep right on its course toward the south."

"No, it is not," said they; "you will see it turn around and go down towards the north."

Hour after hour passed, and as the planet went on its regular course, the other watchers began to get a little nervous. It showed no signs of deviating from its course. We went out from time to time to look at the sky.

"There it is," said one of the observers at length, pointing to Capella, which was now just rising a little to the east of north; "there is the star setting."

"No, it isn't," said I; "there is the star we have been looking at, now quite inconspicuous near the meridian, and that star which you think is setting is really rising and will soon be higher up."

A very little additional watching showed that no deviation of the general laws of Nature had occurred, but that the observers of previous nights had jumped at the conclusion that two objects, widely apart in the heavens, were the same.

Those words came from a book called "Reminiscences of an Astronomer," published in 1903 by Simon Newcomb, who was in charge of the American Nautical Almanac office from 1877 until 1897. The event actually occurred in 1860. The similarity to modern UFO's is overpowering. A star cavorting across the sky! Military officers as responsible witnesses!

In his delightful book, *Light and Colour in the Open Air*, the well-known Dutch astronomer, M. Minnaert, wrote.

> Moving Stars: In the year 1850 or thereabouts, much interest was aroused by a mysterious phenomenon; when one looked intently at a star, it sometimes seemed to swing to and fro and to change its position. The phenomenon was said to be observable only during twilight, and then only when the stars in question were less than $10°$ above the horizon. A brightly twinkling star was first seen to move with little jerks, parallel to the horizon, then to come to a standstill for five or six seconds and to move back again in the same way, etc. Many observers saw it so plainly that they took it to be an objective phenomenon, and tried to explain it as a consequence of the presence of hot air striae.

But any real physical phenomenon is entirely out of the question here. A real motion of ½° per second, seen by the naked eye, would easily be magnified to 100° or more, by a moderately powerful telescope; that means that the stars would swing to and fro and shoot across the field of vision like meteors. And every astronomer knows that this is sheer nonsense. Even when atmospherical unrest is at its worst the displacements due to scintillation remain below the limit of perceptibility of the naked eye. Psychologically speaking, however, the phenomenon has not lost any of its importance. It may be due to the fact of there being no object for comparison, relative to which the star's position can be easily observed. We are not aware that our eye continually performs little involuntary movements, so that we naturally ascribe displacements of the image over our retina to corresponding displacements of the source of light.

Somebody once asked me why a very distant aeroplane appears invariably to move with little jerks when followed intently with the eye. Here the same psychological cause obviously comes into play, as in the case of the "moving" stars, and "very distant" seems to point to the fact that this phenomenon, too, occurs most of all near the horizon.

And how can we account for the fact that, suddenly and simultaneously, three people saw the moon dance up and down for about thirty minutes?

This is the phenomenon of "telekinesis," the apparent erratic motion of an object caused by the erratic motion of the human eye. I have seen a number of UFO reports in which the observer stated that the object could not have been a meteor or a satellite because it moved irregularly.

For you who wear eyeglasses there is still another way of seeing a UFO. Look directly at some bright light, with your head turned slightly to the left or right. You will probably see a faint roundish out-of-focus spot. This is light reflected from the front surface of your eyeball, back to the lens, and then back into the pupil of your eye. A bright source, to one side and slightly behind you, can also reach your eye through reflection from the internal surface of the spectacle lens.

To this moment I have not mentioned still another method of detecting saucers one not subject to the vagaries of the human eye. I mean radar, of course. Radar is a machine. It can't make mistakes. Or at least that is the common argument advanced by UFO buffs.

Radar is cursed with all the potential afflictions that any complicated electrical gadget can suffer. But let me mention only one: mirage. Let me explain briefly what a radar does. It sends out a pulse of radio waves. We know the direction, Northeast for example. We know the elevation above the horizon. An echo returns. From the interval between transmission and reception of the pulse, we know how far away the object is that reflected the pulse back to us. We think we detect a plane or a UFO in flight—because the radar directs the pulse upward.

We have no way of following the pulse in its path toward the target. A layer of warm, dry air or even a layer containing a few bubbles of warm air will bend the radar beam back to earth. The reflection may be from a distant building, a train, or a ship. No wonder that planes, sent to intercept radar UFO, find nothing. In one such case, a well-known writer on flying saucers wrote:

> "The discovery of *visible* saucers had been serious enough. The discovery now of invisible flying saucers would be enough to frighten anyone."

Small changes in the atmosphere can make the UFO seem to maneuver at fantastic speeds, executing right-angle turns or suddenly vanishing completely from the radar scope. I was very familiar with such effects from having worked with them during Naval Service in World War II. The greatest radar saucer flap of all times occurred in the hot, dry month of July 1952, when a whole fleet of UFO's were detected by radar at Washington National Airport. Subsequent research by the Weather Bureau completely confirmed what the UFO buffs pointedly refer to as my "Hot Air Theory." After all why should one be surprised to find hot air over Washington?

I know of no reliable case of simultaneous visual and radar sightings. In view of the physical properties of the eye, the surprising fact is that so few cases have been reported.

Time will not permit me to elaborate on still other relevant phenomena. For example the Air Force appears to have neglected completely the psychological angle of which mass hallucination is just one phase. Back in 1919, in Spain, a not unrelated phenomenon occurred. Thousands of people—reliable people—swore that they had seen images of saints rolling their eyes, moving their hands, dripping drops of blood, even stepping out of their panels. One person would callout, others would imagine they had seen something!

There are many similar events recorded through the ages. There are hundreds of known hoaxes, such as the ingenious one perpetrated by students of the University of Colorado. Spurred by the allotment of an Air Force grant for studying UFO's to the University of Colorado, enterprising pranksters made hot-air balloons from candles and plastic bags, the kind used for packaging dry cleaning. The show was spectacular. And it gave the University investigators a good opportunity to see how poor the evidence can be, a fact well-known to the legal profession. This is still another point that the Air Force has sometimes failed to realize. Moreover their poor questionnaire only further confused an already confused picture. A recent similar sighting south of Denver, later identified as plastic-bag balloons and candles, produced fantastic reports from "reliable" witnesses.

Several times I have used the phrase "UFO's cavorting across the sky." I did so deliberately because it seems to be a favorite phrase of my good friend Dr. J. Allen Hynek of Northwestern University and consultant to the Air Force Project Blue Book. He has sometimes expressed doubts about the UFO because stars don't "cavort" across the sky. What I have tried to show is that many kinds of optical stimuli can produce weird effects.

With all these kids of phenomena masquerading as UFO's, many of them, like those related to physiology of the human eye still practically not investigated, I think I can reasonably claim, applying the criterion of Sherlock Holmes, that we have not excluded all the impossibles. I have shown that the arguments advanced in favor of the interplanetary nature of UFO's are fallacious. Their alleged high speeds and ability to maneuver have completely natural explanations.

I think the time has come for the Air Force to wrap up Project Blue Book. It has produced little of scientific value. Keeping it going only fosters the belief of persons that the Air Force must have found something to substantiate belief in UFO's. In making this recommendation I am not criticizing the present or recent administration of the project. But it is time that we put an end to chasing ghosts, hobgoblins, visions, and hallucinations.

More than twenty years of study by the Air Force and an additional year of analysis by the University of Colorado have disclosed no tangible evidence supporting the popular view that UFO's are manned interplanetary vehicles. An irresponsible press, which has over-publicized the sensational aspects of the phenomenon, has been largely responsible for keeping the subject alive. Both newspapers and leading magazines must bear the blame for mishandling the news. But such publications are not scientific journals. They present incomplete data and draw sensational conclusions without supporting evidence.

The question of UFO's has become one of faith and belief, rather than one of science. The believers do not offer additional clear-cut evidence. They repeat the old classical cases and base the reliability of the sighting on the supposed honesty of the observer. I have shown that many honest observers can make honest mistakes.

The press has recently played up a story to the effect that, even in the U.S.S.R., an official UFO investigation has been started, under government sponsorship. Nothing could be farther from the truth! But the newspapers failed to retract after an official statement from the National Academy of the U.S.S.R. appeared in [the official communist newspaper] *Pravda*, to the effect that the reported study was the work of an unofficial and irresponsible amateur-group. The Academy statement further disclaimed any support whatever for the view that UFO's are other than badly misinterpreted natural phenomena, and certainly not manned extraterrestrial vehicles.

I am aware that a small but highly vociferous minority of individuals are pressing for further studies of UFO's supported—of course—by huge congressional appropriations. The heads of a few amateur UFO organizations urge their members to write Congress, asking for investigations of both UFO's and the Air Force. The members have responded enthusiastically, and Congress reacted by financing a special study, which led to the project at the University of Colorado. And now, when it seems likely that the report from this study will be negative, the same vociferous group is again turning to Congress with the same appeal but with no more chance of success. Time and money spent on such efforts will be completely wasted. Congress should strongly disapprove any and all such proposals, large or small. In this age, despite the doubts expressed by a very small group of scientists, reopening and reopening the subject of UFO's makes just about as much sense as reopening the subject of Witchcraft.

Within the vast field of atmospheric physics, there exist many imperfectly understood phenomena which deserve further study, such as ball lightning and atmospheric optics. But any investigations of such phenomena should be carried out for their own sake, not under the cloak of UFO's.

I express my appreciation to Congressman Roush for the invitation to present my views on UFO's. I append herewith [below] my telegram to him dated July 24, 1968.[22] — DONALD H. MENZEL, SMITHSONIAN ASTROPHYSICAL OBSERVATORY

J. EDWARD ROUSH,
Committee on Science and Astronautics,
Rayburn House Office Building,
Washington, D.C.: JULY 24, 1968.

"Received your letter of July and will contribute paper as you suggest. Am amazed, however, that you could plan so unbalanced a symposium, weighted by persons known to favor Government support of a continuing, expensive, and pointless investigation of UFOs without inviting me, the leading exponent of opposing views and author of two major books on the subject."[23] — D. H. MENZEL

CHAPTER NINE

STANTON T. FRIEDMAN

PREPARED STATEMENTS BY STANTON T. FRIEDMAN

MR. [JAMES G.] FULTON. In my [Pennsylvania] congressional district, there is the Westinghouse astronuclear plant, whose fine work is well known to the members of our committee. As I have been asked by Mr. Stanton T. Friedman, a nuclear physicist at Westinghouse who makes a hobby of investigating UFO sightings and publicly speaking on the subject, it is a pleasure to insert a statement by Mr. Friedman, "Flying Saucers Are Real" into the record at this point. He is one of the few observers with the candor to conclude and so state that "the earth is being visited by intelligently controlled vehicles" from outer space.

(Mr. Friedman's statement follows:)

☛ MR. FRIEDMAN. *After considerable study, first-hand investigation, and review of a great variety of data, I have concluded that the evidence is overwhelming that the earth is being visited by intelligently controlled vehicles whose origin is extraterrestrial* [my emphasis, L.S.]. This does not mean that I know why they are here, where they come from, how they operate, why they don't seem to be talking to us. It also does not mean that I believe that everything that people see that they cannot identify is an extraterrestrial spaceship. Quite the contrary, I believe that most things that people report as UFO's can be identified as relatively conventional phenomena seen under unconventional circumstances just as most isotopes cannot fission or fusion, most chemicals don't cure any diseases, most people cannot run a four-minute mile, and most women don't look like Brigitte Bardot. The scientific approach to any problem is to sift the information to find that which is relevant to the solution of the problem at hand. The fact that most initially strange objects in the sky and on the ground can be identified is totally irrelevant to the question of the existence of extraterrestrial spaceships. Also irrelevant are the facts that we cannot yet comfortably visit other planets, that some of us might behave differently from the way our visitors act, that we have not yet publicly been exposed to pieces of such a vehicle, or to an extraterrestrial humanoid on television.

While almost everyone has heard of flying saucers and has an opinion about them, most people, including the non-believing scientists who have made such definite statements about their non-existence, are ignorant not only of the facts concerning UFO's but also of the technology that might aid one in understanding the vehicles' motion, the possibility of interplanetary and interstellar travel, or the possibility of life on Mars.

Sightings of UFO's are relatively common and have occurred all over the world. One out of every 25 adult Americans has seen a UFO. Judging from the one detailed, official, scientific investigation that has been published, one-fifth of the sightings can be labelled as Unknowns. These Unknowns are completely separate and distinct from the 20% of the 2199 sightings which were labeled "Insufficient Information" because some vital piece of data was missing. Many of the Unknowns are reported by highly trained, competent witnesses who have close-up sightings lasting for many minutes. UFO's have been observed on radar and been subsequently labelled as Unknowns. There have been simultaneous radar and visual sightings. Comparisons between Knowns and Unknowns clearly showed definite differences in color, shape, size, velocity, maneuverability, etc. This data, which most people have never seen or even heard of, is published in a document entitled *Project Blue Book Special Report, Number 14*, which was completed in 1955 and has never been made readily available. The low percentage of Unknowns since that time is the direct result of deception on the part of the U.S. Air Force whose entire approach since that time has been based upon the assumption that everything can be identified.

The usual arguments made against "visitations" are based upon false assumptions, wrong (unanswerable) questions and faulty knowledge. "Things cannot go that fast in the atmosphere—spaceflight is impossible—trips to the stars are impossible, if they were here they would talk to us. . . etc." The typical educated non-believer focuses on the irrelevant IFO's [identified flying objects] and poor sightings by incompetent observers and carefully neglects the *unknowns* seen by competent observers. The great probability that there are civilizations thousands, perhaps millions of years, ahead of us and possessing technology about which we are probably totally ignorant is neglected. The distressing thought that we, the inhabitants of this planet, might not be worth talking to is pushed aside. The most effective filter between the facts as they are and the widespread distribution of those facts has been ridicule. *Fewer than 1% of the sightings that have occurred have been investigated or reported* [my emphasis, L.S.]. Documents containing solid data about UFO's rather than IFO's have been privately published so that most people have never seen the data that they contain. An entire mythology of false information has been widely distributed instead. Now is the time to breakthrough the "laughter curtain." Studies done six years ago at the Jet Propulsion Laboratory showed that trips to the stars in reasonable times are feasible with the knowledge we have today using staged fission or fusion propulsion systems, both of which are under development. A tremendously large body of data connected with magnetoaerodynamics even suggests we might be able to build something very much like the reported UFO's—and also solve many of the problems of high speed flight and produce the electromagnetic effects so frequently associated with UFO sightings. "It's impossible" is said instead of "We don't know how."

Literally hundreds of reports from all over the world also testify to the existence of humanoid creatures associated with UFO's on the ground. Once again ridicule has kept the facts from being known. More than 200 landings have been documented for 1954 alone.

There are good pictures of UFO's from all over the world—most of which have also not received the publicity that they deserve.

A good example of the ridiculousness of the professional skeptics' attitude is the statement that "life as we know it cannot exist on any other body in the solar system." It sounds sensible until we note that we expect to send men to the moon and to Mars. The primary attribute of an advanced intelligent civilization is its ability to create its own environment almost everywhere, such as the bottom of the ocean, in outer space, and on the surface of airless, waterless bodies such as the moon and Mars. *For those who believe that the Mariner IV pictures of Mars proved that there isn't life there, it should be pointed out that of 10,000 pictures taken of the earth from a satellite with cameras of the same resolving power as those used on Mariner IV, only 1 (one) gave any indication of life on earth* [my emphasis, L.S.].

Max Plank once said that new truths come to be accepted not because their opponents come to believe in them but because their opponents die and a new generation grows up that is accustomed to them. Perhaps this is what will happen with UFO's.[24] — SUMMARY OF "FLYING SAUCERS ARE REAL" BY STANTON T. FRIEDMAN, NUCLEAR PHYSICIST

2ⁿᵈ PREPARED STATEMENT BY STANTON T. FRIEDMAN

UFOS AND SCIENCE

[Author-editor's note: Mr. Friedman uses the word "Reference" to mean "Footnote," or sometimes "Endnote," all of which I have appended to his text. L.S.]

I am grateful to the House Committee on Science and Astronautics for inviting me to present my views on Unidentified Flying Objects.[25] These viewpoints shall be presented in the form of answers to specific questions with the references, tables and figures presented at the end of the article. A partial list of the technical organizations to which I have presented a lecture entitled "Flying Saucers are Real" is given in Appendix 1. Appendix 3 is a reprint of an article I wrote.[26] Appendix 2 is a list of patents of saucer-like vehicles. The viewpoints are mine and mine alone and are not to be construed as those of any of the organizations to which I belong or of my employer, Westinghouse Astronuclear Laboratory. The opinions are based upon ten years of study of UFOs and discussions all over the U.S. and in Canada on a private level for eight years and a public level since late 1966 both in question and answer sessions following my illustrated talks and with newspaper, radio, and television reporters with whom I have publicly discussed this subject.

1. To what conclusions have you come with regard to UFOs?

I have concluded that the earth is being visited by intelligently controlled vehicles whose origin is extraterrestrial. This doesn't mean I know where they come from, why they are here, or how they operate.

2. What basis do you have for these conclusions?

Eyewitness and photographic and radar reports from all over the earth by competent witnesses of definite objects whose characteristics such as maneuverability, high speed, and hovering, along with definite shape, texture, and surface features rule out terrestrial explanations.

3. Haven't most sightings been identified as conventional phenomena?

Yes, of course. However, it is only the unidentified objects in which I am interested and on which I base my conclusions. The job of science is to sort data and focus on that which is relevant to the search at hand. Fewer than 1 % of Americans have hemophilia or are 7 feet tall or can run a mile in under 4 minutes—we certainly don't dispute the reality of hemophilia, Wilt Chamberlain, or 4 minute miles.

4. Are there any good unknowns?

Yes, there are very many good unknowns which have been reported and investigated and undoubtedly very many more which have not been reported because of the "laughter curtain." In the most comprehensive detailed scientific investigation ever conducted on this subject, and reported in [my endnote 3], it was found that 434 out of 2199 sightings evaluated had to be classified as Unknowns. This is 19.7% or a far higher percentage than most people have associated with UFOs. The complete breakdown is shown in Table 1. Table 2 shows the breakdown of sightings by quality. Fully one third of the 9.7% of the sightings labelled as Excellent were identified as Unknowns: one fourth of the Good sightings were labelled Unknown. All it would take to prove the reality of extraterrestrial vehicles is one good sighting not hundreds.

TABLE 1.—CATEGORIZATION OF UFO SIGHTING REPORTS[1]

Category	Number	Percent
Astronomical	479	21.8
Aircraft	474	21.6
Balloon	339	15.4
Other	233	10.6
Unknown	434	19.7
Insufficient information	240	10.9
Total	2,199	100

[1] Data from reference 3.

TABLE 2.—QUALITY DISTRIBUTION OF UNKNOWNS[1]

Quality	Number	Percent of total	Unknowns	Percent of group
Excellent	213	9.7	71	33.3
Good	757	34.5	188	24.0
Doubtful	794	36.0	103	13.
Poor	435	19.8	72	16.6
Total	2,199	100.0	434	19.7

[1] Data from reference 3.

5. Aren't most of those "unknowns" really sightings for which insufficient data is available to identify an otherwise conventional object?

Absolutely not. If there was not enough information available about a sighting it was labelled " Insufficient Information " not "Unknown"—again contrary to what many people believe about UFOS.

6. Were there any differences between the Unknowns and the knowns?

A "chi square" statistical analysis was performed comparing the Unknowns in this study to all the "knowns." It was shown that the probability that the unknowns came from the same population of sighting reports as the knowns was less than 1%. This was based on apparent color, velocity, etc. Maneuverability, one of the most distinguished characteristics of UFOs, was not included in this statistical analysis.

7. Weren't most sightings of very short duration, say less than a minute?

The average duration of the sightings labelled as "Unknown" was greater than that for the knowns. More than 70% of the unknowns were under observation for more than 1 minute and more than 45% for more than 5 minutes.

8. Isn't it true that UFOs have never been sighted on radar?

No, it is not. [My endnote 3] specifically mentions radar unknowns.[27] In [my endnote 4], Edward Ruppelt, former head of the official UFO investigative effort, makes specific mention of not only "Unknowns" observed on radar but of combined visual and radar "Unknowns."[28] Hynek also mention radar and visual sightings.[29]

9. Where can I get more information about "Unknowns"?

[My endnote 6] presents an unbiased description of about 160 "Unknowns."[30] [My endnote 7] includes data on over 700 Unknowns.[31] [My endnotes 8[32] and 9[33]] contain many others.

10. Why haven't the worldwide Smithsonian Network of Satellite Tracking cameras picked up "Unknowns"?

The former head of the film evaluation group concerned with the Smithsonian sky watch said that the purpose of the search was to get data on satellite orbits. If a light source on the film could be shown not to be a satellite then no further measurements were made. 10% to 15% of the plates showed anomalous light sources which were not a satellite but were not otherwise identified.[34]

11. How about the other space surveillance radar installations?

Baker in [my endnote 11] deals with this question in detail. In summary, the systems are set up to reject signals which refer to anything other than the objects of interest—typically ballistic missiles coming from certain directions.[35]

12. Aren't the reported maneuvers of UFOs in violation of existing laws of physics?

Not at all. This argument ("It's Impossible") is used when what should really be said is we don't know how to duplicate these maneuvers. Piston aircraft can't fly faster than the speed of sound and a conventional dynamite bomb couldn't have wrecked

Hiroshima and a vacuum tube circuit can't fit on the head of a pin, but surely we don't say that supersonic flight, atom bombs and microcircuits violate the laws of nature or physics. Present aircraft can't duplicate UFO maneuvers; no laws of physics have been violated by UFOS.

13. Haven't astronomers proved that trips to other stars are impossible?

Again, the answer is no. The studies[36] that conclude that trips to other stars are impossible are based upon false or unnecessary assumptions such as, assuming, that the flight be at orbital velocity.[37] The one comprehensive study of interstellar travel conducted by a JPL group actually concerned with space hardware concluded that with present technology trips to nearby stars are feasible with round trip times being shorter than a man's lifetime and *without* violating the laws of physics. They assumed that staged vehicles would be used having either fission or fusion propulsion systems.

14. Are fission and fusion propulsion systems actually being developed?

Both fission and fusion propulsion systems for space travel are under development. I have worked on both. The NEVRA program has successfully tested a number of nuclear rocket reactors suitable for use in flight throughout the solar system.[38] Flight rated systems offering substantial advantages over chemical propulsion systems could be ready in less than a decade if the current program at Aerojet General, Westinghouse Astronuclear Laboratory, and Los Alamos Scientific Laboratory is supported. [My endnotes 15[39] and 16[40]] are good reviews of the nuclear rocket program. The fusion work is not nearly as far along but has been productive at Aerojet General Nucleonics, San Ramon, California. An older review of some of the aspects of this program is given in [my endnote 17[41]].

15. Are these the only possibilities?

Not at all. This is one of the major flaws in the "non-believers" arguments; they presume that our technology is the ultimate—a presumption made by each generation of scientists in the last 75 years and proved wrong by the next generation of engineers and applied scientists. If there is one thing to be learned from the history of science it is that there will be new and unpredictable discoveries comparable with, say, relatively, nuclear energy, the laser, solid state physics, high field superconductivity, etc. It is generally accepted that there are civilizations elsewhere which are much more advanced than are we. Look what technological progress we have made in the last 100 years. Who can guess what we will accomplish in the next thousand years or what others have accomplished in the thousand or million or billion year start they may have on us. We still don't know about gravity, for example, no less anti-gravity.

16. Could UFOs be coming here from our own solar system?

They certainly could. We have no data from any other body in the solar system which definitely rules out the existence of advanced civilizations. We frequently forget that the resolution of present photographs of the other planetary bodies is extremely poor. As a matter of fact, there does seem to be a direct correlation between the number of sighting reports per unit time and the closeness of Mars to the earth. Both have periodicities of about 26 months.[42] We make certain space shots at "favorable times." The reverse may also be true but without the restrictions on payload and trajectory placed upon us by our crude, inefficient, space propulsion systems which no thoughtful engineer considers the ultimate.

17. Didn't the Mariner IV pictures prove there isn't any life on Mars?

The Mariner pictures didn't provide proof of life on Mars but they certainly didn't rule it out and were not intended to. Studies 10 of 10,000 pictures of earth taken from orbit with cameras having resolving power equivalent to those on Mariner IV provided only one picture which could be taken to indicate that there is life on the planet called earth.[43]

18. Isn't it true that life as we know it cannot exist on any other body in the solar system?

This statement, though repeated many times, is quite obviously untrue. Consider for a moment the fact that we intend to send men to the moon and by the end of the century to Mars. We expect these men to stay for a while and to return despite the fact that Mars and them on both supposedly aren't fit for life as we know it. One characteristic of an advanced technological civilization is the ability to provide suitable conditions for life almost anywhere; including under the ocean, in the void of space and on the surface of the airless, waterless moon and Mars. More and more we are also finding that life exists under almost all circumstances.

19. If we are being visited why haven't they landed?

The fact of the matter is that there are many reports of landings. The comprehensive study by scientist J. Vallee reviews 200 landings which occurred in 1954 alone; many of them with multiple witnesses giving reports of humanoids in addition to strange craft either on or just above the ground.[44] Most scientists have unfortunately not examined this data since it was published in a UFO Journal and laughter comes easier than facing up to the evidence.

20. Has the attitude of the scientific journals and professional community been changing?

There has been a quiet yet enormous change in the attitude of the technological community. I say technological to include the applied scientists and engineers who are far more responsible for the progress of the last 30 years than the academic scientists who are prone to tell us all that is impossible. Examples of the change include the publication of articles by Science,[45][46][47][48][49] Astronautics and Aeronautics,[50] the Journal of the Astronautical Sciences,[51] the American Engineer,[52][53][54] Industrial Research,[55] Scientific Research,[56][57] Aviation Week and Space Technology.[58][59]

In addition, numerous pro-UFO talks have been presented to local and national meetings of professional groups (see Appendix 1),[60][61] and the American Association for the Advancement of Science is planning a UFO seminar for a national meeting. The AIAA has even set up a UFO Committee.

21. Have there really been any electromagnetic effects associated with UFO sightings?

Indeed such reports are numerous,[62] which includes stopping of car engines and headlights, and interference with radio and TV reception, magnetic speedometers, and watches.

22. Could these conceivably be related to a propulsion scheme?

There is an enormous amount of work available concerned with magnetoaerodynamics. I received a NASA bibliography with more than 3000 references. [My endnote 39] contains abstracts of more than 800 publications dealing with interactions between vehicles and plasmas.[63] Much of this work is classified because ICBM nose cones are surrounded by plasmas. In any event, there is a body of technology which I have studied and which leads me to believe that an entirely new approach to high speed air and space propulsion could be developed using the interactions between magnetic and electric fields with electrically conducting fluids adjacent to the vehicles to produce thrust or lift and reduce or eliminate such other hypersonic flight problems as drag, sonic boom, heating, etc. These notions are based existing technology such as that included in [the following endnotes], though one would expect that a considerable development effort would be required.[64][65][66][67][68][69][70][71][72][73]

23. Have any electromagnetic propulsion systems been operated?

So far as I know no airborne system has been operated which

depended on electromagnetic forces for propulsion. At Northwestern, turning on a magnet inside a simulated re-entry vehicle with a plasma around it resulted in a change in the color of the plasma and its location relative to the vehicle. However, an electromagnetic submarine has actually been built and successfully tested. It is described in some detail in [the following endnotes].[74] [75] [76]

24. Can an EM submarine really be related to a UFO?

Dr. Way's electromagnetic submarine which, incidentally, is silent and would be quite difficult to detect at a distance is directly analogous to the type of air-borne craft I envision except that the shape of the aircraft would most likely be lenticular and the electrically conducting seawater would be replaced with an electrically conducting plasma of ionized air.

25. Would lenticular vehicles fly? I certainly think so. We seem to believe that airplanes have the only possible shape probably because the Wright brothers plane had the same outline which in turn was like that of birds. As pointed out by Chatham in [the following endnote], flight is still only a byproduct of high forward velocity leading to the need for long runways and high speed landings and takeoff.[77] Present airplanes are quite obviously inefficient in terms of fuel consumption, payload fraction, and volume of air and airport space per passenger. After all the SST will only carry a few hundred passengers though it will occupy the space of a football field capable of holding at least ten times as many people. Fuel weight is greater than payload weight and neither is a very high fraction of system weight. It is interesting to note that most scientific progress has come from doing things differently rather than using the same technique—microcircuits aren't just smaller vacuum tubes; lasers aren't just better light bulbs. Many people are not aware that the U.S. Patent Office has granted more than ten patents for what one might honestly call flying saucer-shaped craft all of which claim great maneuverability and the ability to rise vertically. Some can supposedly hover. None of these use magnetoaerodynamic techniques. For those who are interested, the patents are listed in Appendix 3. This list hasn't been up-dated for a couple of years.

26. Have any members of your audience seen any UFOs?

I have taken to asking whether any members of my audiences have seen what they would call a UFO. Typically 3-10% are willing to raise their hands and usually there are others who approach me privately. These data, though limited, tend to support the Gallup Poll of 1966 which revealed that 5 million adult Americans claimed to have observed a UFO. Interestingly enough the official files contain fewer than 12,000 reports.

27. Were these sightings by your audience reported to investigative bodies?

In general, no. At Los Alamos Scientific Laboratory 25 of the 600 listeners indicated that they had seen something odd but only one had reported what he had seen.

28. Is there some way to get more data about UFOs besides reading reports?

There are several approaches that should be taken.

    a) Lift the "laughter curtain" so that more observers are willing to report what they see and more scientists will become involved.

    b) Using existing technology establish instrumented investigative teams and automated observation instrumentation such as that recommended by Dr. Baker before the Committee on Science and Astronautics.

    c) A world wide communication and study effort should be begun.

    d) A very large survey should be conducted to determine the characteristics of the objects that have been observed. The most comprehensive picture we have of ball lightning resulted from carefully conducted surveys by McNally[78] and Rayle.[79] UFOs in my opinion are definitely not ball lightning or other natural plasmas but are analogous to ball lightning and earthquakes in that their appearance cannot be predicted and they cannot be reproduced in the lab or in the field but they have been observed.

29. Are there any other references of interest to scientists?
Yes, [see my following endnotes].[80,81,82,83,84,85,86]

30. Haven't you biased your comments by not discussing at any length the work of Marcowitz, Menzel, and Klass?

The paper by Marcowitz[87] and the books by Menzel[88,89] and Klass[90] will undoubtedly be read by scientists of the 21st century as "classics" illustrating a non-scientific approach to UFOS by people who, for whatever reason, would not examine the data relevant to UFOs or advanced technology. Marcowitz was totally wrong about fission and fusion propulsion systems, didn't even consider electromagnetic propulsion, and was obviously unaware of current technology and the data such as I mentioned earlier about UFOS. McDonald[91] has discussed Menzel's approach in detail, but let me also point out that in [the following endnote][92] fewer than 30 sightings ever listed as "unknowns" were discussed and no mention was made of the 434 "Unknowns" of [the following endnote],[93] or even the 71 Excellent Unknowns of this study. I agree with Klass on only one item, many people have observed glowing plasmas; but I believe they were adjacent to vehicles rather than ball lightning or corona discharge. He didn't even consider this possibility despite all his talk about plasmas and despite the enormous amount of plasma-vehicle data which is available. In summary, *I feel that these three gentlemen have made strong attempts to make the data fit their hypotheses rather than trying to do the much more difficult job of creating hypotheses which fit the data* [my emphasis, L.S.].[94] — PREPARED STATEMENT BY STANTON T. FRIEDMAN, NUCLEAR PHYSICIST

FRIEDMAN'S APPENDIX 1

S. T. Friedman has talked about UFOs to these groups (partial list):
Engineering Society of Detroit.
Engineering Society of Baltimore.
Los Alamos Scientific Laboratory.
Local sections of the American Institute of Aeronautics and Astronautics in Pittsburgh, Pennsylvania; Wichita, Kansas; Cumberland, Maryland; Waco, Texas; San Antonio, Texas; Raleigh, North Carolina; New York, New York.
Local sections of the Institute of Electrical & Electronic Engineers in Pittsburgh, Pennsylvania; Wilmington, Delaware; Salisbury, Maryland; New London, Connecticut.
Professional Engineers of Western Pennsylvania.
American Nuclear Society in Pittsburgh and Las Vegas, Nevada.
Pittsburgh Chemists Club.
Computer Simulation Council of Western Pennsylvania.
Dravo Corporation Engineers Club, Pittsburgh.
Society of American Military Engineers.
Universal Cyclops Corporation Engineers Meeting.
22nd Annual Frequency Control Symposium.
Duke University.
Wesleyan University.
University of Texas.
Carnegie Mellon University.
University of Illinois, Chicago.
West Virginia University.

FRIEDMAN'S APPENDIX 2

U.S. PATENTS FOR CIRCULAR AIRCRAFT

Patent No.	By	Title	Granted
3,067,967	I. R. Barr	Flying machine	Dec. 11, 1962
2,772,057	J. C. Fischer, Jr.	Circular aircraft and control system therefor	Nov. 27, 1956
2,947,496	A. L. Leggett	Jet-propelled aircraft	Aug. 2, 1960
2,801,058	C. P. Lent	Saucer-shaped aircraft	July 30, 1957
2,876,964	H. F. Streib	Circular wing aircraft	Mar. 10, 1959
2,997,013	W. A. Rice	Propulsion system	Aug. 22, 1961
3,124,323	J. C. M. Frost	Aircraft propulsion and control	Mar. 10, 1964
2,876,965	H. F. Streib	Circular wing aircraft with universally tiltable ducted power-plant	Mar. 10, 1959
2,939,648	H. Fleissner	Rotating jet aircraft with lifting disk wing and centrifuging tanks	June 7, 1960
3,103,324	N. C. Price	High velocity, high altitude VTOL aircraft	Sept. 10, 1963

FRIEDMAN'S APPENDIX 3

FLYING SAUCERS ARE REAL

There are a few standard responses to any statement that "the earth is being visited by intelligently controlled vehicles whose origin is extraterrestrial." The simplest is ridicule, coupled with a comment that flying saucers are figments of the imagination, or optical illusions, or motes in the eye, or hoaxes, or misidentified conventional phenomena seen under unusual circumstances by untrained observers. These, however, are all Identified Flying Objects [IFOs], and not the Unidentified Flying Objects with which my statement is concerned.

The next simplest response is: "We are certainly not alone in the Universe and surely some civilizations are more advanced than ours, but interstellar travel is not feasible because of the vast distances between such civilizations and the great quantity of energy and time required for the trip." These critics ignore our lack of data on intercivilization distances the possibility of unknown (to us)flight technology, and studies in this area [see Friedman's note 1]. Another response is that the reported activity of UFOs is not rational since, if "they" were advanced enough to get here, they would surely try to communicate with us.

These responses avoid coming to grips with the reported data. Those interested in data—and there is plenty of it [see Friedman's endnotes 2-13]—are advised to consult the References and derive a hypothesis other than extraterrestrial vehicles to fit the facts, rather than to try to make the facts fit the hypothesis that "we are not being visited because" (in 25 words or less).

A particularly interesting aspect of the data from all over the world is that electromagnetic effects are frequently observed in association with the presence of UFOs, along with the fact that many observations suggest that what is being observed is a "vehicle" having a plasma region adjacent to it—"vehicle" because of its metal-like surface, large size, maneuvers indicating intelligent control, well-defined shape, surface features such as "port holes, antenna, landing gears, lights, etc.; and plasma because of bright glows rather than color, changes in the color of the glow associated with changes in velocity, luminous boundary layers, and appearance on film of regions not seen by the naked eye. The EM effects include interference with the operation of automobile engines, radios, and headlights; interference with the operation of radio and TV sets, compasses, magnetic speedometers, power systems; residual magnetism in metal objects, watches, etc. [see Friedman's endnote 13].

During the past decade a vast amount of terrestrial technology, much of it classified, has been developed concerning the interactions of airborne vehicles and plasmas. The development of lightweight, compact, high-field superconducting magnets has also led to much work on the potential benefits to be derived from placing a magnet within a high-speed vehicle to interact with a plasma around the vehicle. Such a combination might be used to reduce vehicle heating, control aerodynamic drag, exert control forces on the vehicle, provide power for its operation, open a "magnetic" communications window, and change the vehicle radar profile. In addition, the magnets might be used to provide shielding against space radiation. Numerous reports cover such applications [see Friedman's endnotes 13-17].

A review of this literature and an extrapolation of existing technology suggest that with considerable effort an entirely new EM approach to hypersonic flight might be developed which, in many respects, could duplicate UFO characteristics. In turn, this leads to the notion that observed UFO behavior is not so unreasonable as might at first appear to be the case. The measurement of EM parameters of UFOs could well provide information on both UFO characteristics and new propulsion."[95]

FRIEDMAN'S ENDNOTES ("REFERENCES")

(Items 2, 3, 4, 5, 6, 13 available from UFO Research Institute, Suite 311, 508 Grant St., Pittsburgh, Pa. 15219.)

1. Spencer, D. F. and Jaffe, L. D., "Feasibility of Interstellar Travel," *Acta Astronautica*, Vol. IX. Fasc. 2, 50–58, 1963.
2. "The UFO Evidence," National Investigations Committee on Aerial Phenomena, 1964.
3. Olsen, T., "The Reference for Outstanding UFO Sighting Reports," Oct. 1966. $5.95.
4. "Project Blue Book: Special Report No. 14," Davidson, 3rd Edition, 1966, $4.00.
5. "Humanoids—Worldwide Survey of Landings and Alleged Occupants," Special Issue of *Flying Saucer Review*, October-November, 1966. $2.60.
6. McDonald, James E., "UFOs: Greatest Scientific Problem of Our Times," Oct. 1967. $1.00.
7. Hynek, J. Allen, "UFOs Merit Scientific Study," *Science*, Oct. 21, 1966, and *Astronautics & Aeronautic*, Dec. 1966. p. 4.
8. Vallee, Jacques, "Anatomy of a Phenomenon," Regnery, 1965, $4.95. Ace Books, 1966, $.60.
9. Vallee, Jacques and Janine, "Challenge to Science—The UFO Enigma." Regnery, 1966, $5.95. Also paper, $.75.
10. Michel, A., "The Truth About Flying Saucers," Criterion, 1956. $3.95.
11. Michel A., "Flying Saucers and the Straight Line Mystery," Criterion, 1958. $4.50.
12. Ruppelt, E. J., "The Report on Unidentified Flying Objects," Doubleday, 1956. $2.95.
13. Maney, Prof. C. A., and Hall, Richard, "The Challenge of Unidentified Flying Objects," 1961. $3.50.
14. Literature Search No. 541, "Interactions of Spacecraft and Other Moving Bodies with Natural Plasmas," Dec. 1965, Jet Propulsion Laboratory, 182 pages, 829 references.
15. Jarvinen, P. O., "On the Use of Magnetohydrodynamics During High Speed Reentry," NASA-CR-206, April 1965.
16. Nowak, R., et al, "Magnetoaerodynamic Re-entry," AIAA Paper 66–161, AIAA Plasmadynamics Conference, March 2–4, 1966.
17. Kawashima, Nobuki and Mori, Sigeru, "Experimental Study of Forces on a Body in a Magnetized Plasma," *AIAA Journal*, Vol. 6, No. 1, Jan. 1968, pp. 110–113.

S. T. Friedman Westinghouse Astronuclear Laboratory
(Ed.—For additional comments on UFOs see the following *AIA* references:
The Wheel in the Middle of the Air, Solomon Golomb, *Sounding Board*, August 1966, p. 16 . . . *Letters*, Nov. 1966, p. 6, George Earley and Brent L. Marsh on Saucer Doctrine . . . UFOs—Extraterrestrial Probes? by James E. McDonald, *Sounding Board*, Aug. 1967, p. 9 . . . Munday Jr., John C., "On the UFOs," Bulletin of the Atomic Scientists, Dec. 1967.
Also see December 8, 1967, issue of *Science*, No. 3806, pp. 1265 and 1266. Letters to Ed.)

CHAPTER TEN

R. LEO SPRINKLE

PREPARED STATEMENT BY R. LEO SPRINKLE

To The Honorable J. Edward Roush, M. C., Ind., Chairman, Symposium on Unidentified Flying Objects, The Committee on Science and Astronautics (The Honorable George P. Miller, M. C. Calif., Chairman), The House of Representatives, Washington, D.C. 20515.

From R. Leo Sprinkle, Ph.D., Counselor and Associate Professor of Psychology, University of Wyoming, Laramie, Wyoming, 82070.

Re: Symposium on Unidentified Flying Objects.

INTRODUCTION

☛ Thank you for your kind invitation (July 22, 1968) to submit a statement to you and your colleagues in regard to the problem of Unidentified Flying Objects (UFOS). I recognize some of the difficulties which confront you gentlemen, particularly in relation to the amount of information which becomes available to you when you wish to arrive at informed decisions. Also, I recognize some of the difficulties which confront college professors, especially when they try to be lucid and brief!

Elsewhere, a more extensive attempt has been made to present views on the psychological implications of UFO reports. (Sprinkle, R. L., "Psychological implications in the investigations of UFO reports," in Lorenzen, L. J., and Coral E., *Flying Saucer Occupants*. N.Y.: A Signet Book, 1967, pp. 160-186.) Thus, in submitting this short statement, the attempt is made to present my personal views on the significance of UFO investigations. Hopefully, these personal views can be of some assistance to you in considering the statements being submitted to you by the distinguished scientists whom you have invited to participate in the symposium.

PERSONAL VIEWS OF UFO INVESTIGATION

I accept the hypothesis that the earth is being surveyed by spacecraft which are controlled by representatives of an alien civilization or civilizations. I believe the "spacecraft hypothesis" is the best hypothesis to account for the wide range of evidence of UFO phenomena. (For a more informed description of various UFO hypotheses, see Salisbury, F. B., "The scientist and the UFO." *BioScience*, January 1967, pp. 15-24.)

I have read thousands of reports and I have talked with hundreds of persons about their UFO observations; either I must accept the view that thousands of people have observed physical phenomena, or I must accept the view that some persons have the ability to project mental images in such a manner that other persons can observe, photograph, and obtain physical evidence of those metal images. (For a more extended discussion of these hypotheses, see Jung, C. G., *Flying Saucers*, N.Y.: Harcourt, Brace and Company, 1959, pp. 146-153.)

On two occasions, each time in the presence of a person who shares my claim, I observed unusual aerial phenomena which I could neither identify nor understand. My first observation of a "flying saucer" led me to change my position from "scoffer" to that of "skeptic." My second observation of a UFO led me to change my position from "skeptic" to "unwilling believer."

As a result of my second observation, I began to join organizations (NICAP and APRO) and conduct investigations. The Grants-in-Aids Committee of the Society for Psychological Study of Social Issues, a Division of the American Psychological Association, provided a small grant, and Richard Hall, former assistant director of NICAP, cooperated in a study of the attitudes of persons interested in UFO reports. (See enclosure.) Later, I became a consultant to APRO. Mr. and Mrs. L. J. Lorenzen, directors of APRO, encouraged my interest in learning more about the psychic aspects of UFO phenomena.

At the present time, I am conducting a non-supported investigation of some psychological attributes of persons who claim to experience psychic impressions of UFO phenomena, including their impressions of possible motives of UFO occupants. As a member of the Parapsychological Association and the American Society of Clinical Hypnosis, I am interested in the possibility that reliable observations are being made by persons who claim to see UFO occupants and, in some unusual cases, experience "mental communication" with these UFO occupants. Use of hypnotic techniques with UFO observers seems to be useful procedure, in some cases, to obtain further information about UFO observations. (See enclosure "Personal and Scientific Attitudes: A Survey of

Persons Interested in UFO Reports.")

CONCLUSIONS

If these reports by UFO observers are found to be reliable and valid, I believe we shall enter the threshold of a most exciting and challenging period in man's history. In my opinion, the attempt to achieve contact with other intelligent civilizations is a goal which is worthy of great personal and social effort.

I plan to do what I can in furthering investigations of these phenomena, in hopes that these efforts can assist the contributions of other investigators.

However, I believe that the mysteries are too deep, the investigations are too difficult, and the implications are too great for these efforts to be made on an informal basis. I believe that the establishment of an international research center is the most appropriate method to follow in reaching the goal of greater understanding of these phenomena. If this method is not feasible, then I believe that a national research center is needed for continuous, formal investigation of the physical, biological, psycho-social, and spiritual implications of UFO phenomena.

Establishment of a continuing research center could provide for those facilities, equipment, and personnel to conduct the necessary field work and theoretical investigations of UFO reports. In my opinion, the staff of such a research center should be encouraged to avail themselves of scholars and experts in various disciplines, including astronomical, mathematical, physical, chemical, biological, medical, psychological, sociological, military, technical, legal, political, theological, and parapsychological fields of knowledge.

I recognize some of the difficulties which attend such a proposal; I recognize some of the arguments which have been, are being, and will be raised against such a proposal. However, I trust that you gentlemen are aware that the present difficulties of enacting such a proposal are inconsequential when compared to the historical impact created by those persons who dare to exert that leadership which could determine the powers, purposes, and persons who control the spacecraft which we call "Unidentified Flying Objects."

Thank you for your attention to these comments. I shall be most happy to respond to any question which you may have about these or related views.

Respectfully submitted, R. LEO SPRINKLE, Ph.D.

PERSONAL & SCIENTIFIC ATTITUDES: A SURVEY OF PERSONS INTERESTED IN UFO REPORTS[1]

By R. Leo Sprinkle, University of Wyoming

SUMMARY

A questionnaire survey was conducted among 3 groups: 26 Ph.D. faculty and graduate students in a Psychology Department (Psychology); 59 graduate students enrolled in an NDEA Guidance Institute (Guidance); and 259 members of an organization which is interested in "flying saucers" or Unidentified Flying Objects (UFOS), the National Investigations Committee on Aerial Phenomena (NICAP). It was hypothesized that there would be no differences between the scores of the three groups on the *Personal Attitude Survey* (Form D, Dogmatism Scale, Rokeach, 1960) and the *Scientific Attitude Survey* (Sprinkle, 1962).

The results showed significant differences (P<.001 between the 3 groups with respect to their mean scores on both inventories, with the NICAP group scoring higher on both "dogmatic" and "scientific" inventories, followed by the Guidance group and Psychology group, respectively. Also, the survey showed differences in regard to social status and education. Psychology and Guidance subjects received an *Index of Social Status* (McGuire & White, 1955) which would classify them in the Upper Middle Class, while NICAP subjects would be classified mainly in the Upper Middle and Lower Middle Classes. The average years of education were tabulated as follows: Psychology, 18.8 years; Guidance, 17.2 years; and NICAP, 14.0 years.

The results suggest that the NICAP group is more "dogmatic" and more "scientific" than the Psychology and Guidance groups. There are two feasible interpretations of these results: 1) The *Scientific Attitude Survey* (Sprinkle,1962) is not useful in assessing "scientific" attitudes, and/or 2) the two inventories have assessed the tendency of the 3 groups to exhibit the "Yeasay-Naysay" pattern (Couch & Keniston, 1960). The latter interpretation indicates that there may be more "Yeasayers" (those with an agreeing response set or a readiness to affirm) in the NICAP group, followed by the Guidance group, and Psychology group, respectively.

---

1. This study was supported by the Grants-in-Aid Committee, the Society for the Psychological Study of Social Issues, a division of the American Psychological Association.

SPRINKLE'S REFERENCES 1

Couch, A., & Keniston, K. "Yeasayers and naysayers: Agreeing response set as a personality variable." *J. abnorm, soc. Psychol.*, 1960, 60, 151-174.

McGuire, C., & White, G. D. "The measurement of social status." Research paper in *Human development*, No. 3 (revised), Dept. of Educ. Psychol., Univ. of Texas, March 1955.

Rokeach, M. *The Open and Closed Mind.* N.Y.: Basic Books, 1960.

Sprinkle, R. L. "Scientific Attitude Survey." (Unpublished attitude inventory), 1962.

SOME USES OF HYPNOSIS IN UFO RESEARCH
By R. Leo Sprinkle, University of Wyoming

ABSTRACT

First, a brief review of UFO literature is presented. References are cited which offer hypotheses (Salisbury, 1967) to account for UFO observations, and positions (Sprinkle, 1967) taken by various investigators in regard to the significance of UFO reports.

Second, examples are described in the use of hypnotic techniques in the investigation of persons who have observed UFO phenomena. Advantages and disadvantages of hypnosis are discussed in regard to obtaining further information from UFO observers.

Third, some speculations are offered in regard to the possible relationships of paranormal or ESP processes and the observations of UFO phenomena. Cases are described which indicate possible relationships of hypnotic and psychic experiences of UFO observers.

Fourth, some suggestions are presented for further investigation of UFO phenomena through the use of hypnotic and parapsychological procedures. These procedures may be useful to assess the reliability of information from UFO observers.

In conclusion, the speaker believes that investigation of UFO reports should proceed along as many lines as there are interested investigators. Considerations of hypnotic procedures and techniques are only one aspect of these investigations, but these considerations may be helpful in obtaining and evaluating information submitted by UFO observers.[96]

SPRINKLE'S REFERENCES 2

APRO Bulletin. Aerial Phenomena Research Organization, 3910 E. Kleindale Road, Tucson, Arizona, 85716.

Flying Saucer Review. 49a Kings Grove, London, S. E. 15, England.

NICAP, *The UFO Investigator*. National Investigations Committee on Aerial Phenomena, 1536 Connecticut Avenue, N.W., Washington, D.C. 20036.

Sable, M. H. *UFO Guide: 1947-1967*. Rainbow Press Company, P.O. Box 937, Beverly Hills, California, 90213, 1967.

Salisbury, F. B. "The scientist and the UFO." *Bio-Science*, January 1967, pp.15-24.

Sprinkle, R. L. "Psychological implications in the investigation of UFO reports," in Lorenzen, L. J. and Coral E. *Flying saucer occupants*. N.Y.: A Signet Book 1967, pp. 160-186.

CHAPTER ELEVEN

PREPARED STATEMENT OF GARRY C. HENDERSON

UFO'S DEFINITELY DO NOT EXIST—DO THEY?

INTRODUCTION

☛ It is not the purpose of this essay to specifically reiterate the feelings of certain members of the scientific community regarding UFO phenomena; rather it is the objective here to briefly review the state of the problem and to analyze the means at hand of acquiring information which would be sufficiently reliable to convince the scientific community and others of the hardware existence or fallacy of the UFO.

Although the common image persists of the scientist as an infallible font of wisdom and knowledge, *the majority of reported activities of scientists relating to UFO studies has been nonprofessional by nature, i.e., prominent scientists have addressed themselves to the problem in a manner which they would certainly not approach problems within their respective fields* [my emphasis, L.S.]. Such an example is the unfortunate selection of the University of Colorado team headed by a respected scientist, with the result that the squirrel-case atmosphere usually associated with UFO interest has been augmented by built-in bias and confusion, rather than eliminated by one group of scientists' involvement. One scientist has even published an article in Science (15 September 1967) implying that competent scientists would accept magic (or "semi-magic") as an answer to the existence of UFO's, and that our limited capabilities in their current stage may be our ultimate technical heritage. Is there not the slightest chance that, even today, there actually remain a few physical phenomena which we do not understand, or of which we are not even aware—and perhaps a few we misinterpret, but for which we are shrewdly able to concoct a convenient "law" by way of appealingly sufficient (but not necessary) explanation? If "others" exist, are they limited to our level of advancement?

As noted in the Special Report of the USAF Scientific Advisory Board Ad Hoc Committee to Review Project Blue Book (March 1966) some sightings classified as "identified" resulted from too meager or too indefinite evidence to permit positive listing in that category. The keys to scientific achievement have several notches, but the material is comprised of competent open-mindedness, which appears to have been all to commonly lacking in the topic of concern. Historically, many of the most astonishing accomplishments have been performed by those who persisted even in the fog of ridicule exuded by their capable but narrow-minded colleagues. Several professional, qualified observers with proper instrumentation, planning, and time should be able to devise schemes in an unbiased manner to (1) determine what UFO's *are not*, then (2) determine what, if anything tangible, they *are*.

AVAILABLE INFORMATION

Most thoughtful persons will dismiss the theatrical claims of trips on "saucers," cavorting with little green men, and the like; however, some very plausible reports from highly trained, capable, and reliable individuals cannot be so readily discarded by anyone willing to admit that there are still a few things we do not understand. God help us if our military and commercial pilots and radar facilities so commonly mistake temperature inversions, balloons, atmospheric disturbances, the planet Venus, etc. for maneuvering vehicles. Have you ever tried to convince two veteran pilots that the object they reported sighting on a clear day with CAVU ["Ceiling And Visibility Unlimited"] conditions, free of traffic lanes, showing on their radar screen, exhibiting high maneuverability, in close proximity, etc., is meteoric debris? If so, then the wrong people are being examined.

To my knowledge, all "facts" on UFO's, here and abroad, exist in the form of visual sightings and a few, apparently unretouched photographs. Not to discredit the value of unaided observation, but with our degree of technological sophistication, these are hardly the sort of facts to justify the position of the Air Force (and a few scientists) in their proclamations of "non-existence." The public has been led to believe that everything has been done to either prove or disprove the existence of UFO's—rubbish! Available information of a truly reliable nature should tend to increase activity, not place it in neglect, or worse, in ridicule.

Classified (or "unavailable") reports, mostly by the military, rob the public and scientific parties who are interested in and willing to participate in the UFO investigations. How can we even begin to evaluate for ourselves if we must depend nearly 100 percent on information doled out by the news media alone? Many scientists communicate with each other on the subject, often at scientific meetings (aside), but this route is hardly sufficient to establish a recognized basis for realistic study of the problem.

The current USAF trend seems to be merely a statement that UFO's do not pose a threat to the security of the United States, and therefore warrant neither credence nor further concern. Similar words come from some of the few Congressmen with whom I have communicated. The discovery of Noah's Ark in Times Square would not necessarily pose a threat to national security either, but it would certainly be a find worthy of the most intensive investigation whether certain individuals accepted its existence or not.

REQUIRED INFORMATION

Is it not obvious that what we need to establish the existence or non-existence of UFO's is not merely a review of sighting incidents, but an implemented plan to acquire hard facts? Rapid, accurate reporting of sightings is obviously a valuable tool in studying UFO phenomena, but many of the most creditable observers (military personnel and airline pilots, for example) are not only hesitant to do so, they are understandably adamant when facing the alternatives of silence versus inviting ridicule, and possibly jeopardizing their positions. The obvious addition to gathering interview data is to enlist the aid of the impersonal machine. Evaluate, compile, and catalog reported data according to: time of day and year; atmospheric conditions (cloudy, humid, temperature, calm, and other easily recountable gross observations); geographic location; approximate size, shape, altitude, velocity, heading, maneuvers; and phenomena reportedly associated with the UFO presence. This can be done with existing information. Update and upgrade the files with new data by soliciting information (particularly military and commercial pilots). Then prepare a plan designed by scientists, engineers, pilots, and perhaps psychologists, on means to acquire instrument observations of UFO's hopefully coupled to visual observations.

Field-instrument packages could easily be placed in areas where UFO sightings are most concentrated, perhaps according to the time of day or year, atmospheric conditions, or some factor suspected to be related to sighting activity. Such packages might be composed largely of military "surplus" instrumentation such as an infrared scanner, an active rf unit, a wide-band electromagnetic detector, a directional radiation counter and ionization gauge, a high-speed photographic camera, a three-component magnetometer, and recording environmental devices (temperature, humidity, barometric pressure, etc.). If it became advantageous to include a higher degree of sophistication, such items as a tracking television camera, a communications telemetry system, a sensitive audio recorder with a directional antenna might be added. Deployment and maintenance of the field package could easily be performed by military, university, or industrial technicians, but all data reduction and interpretation should be done by competent scientists familiar with the respective measuring techniques.

We should anticipate gathering sufficient data leading to proof of the existence or non-existence of UFO's, and, if they are real, the size(s), shape(s), flight characteristics (speed, rates of turn and climb, preferred direction of travel, etc.), possible means of propulsion and navigation, perhaps the establishment of communications, and eventually their origin.

Questions of expense and management responsibilities immediately come to mind, but I think the government would be surprised how many qualified scientists, engineers, and technicians would be willing to participate on a low-dollar, volunteer, "as can" basis in support of such a program. At least an inexpensive newsletter could be distributed to the scientific and pilot groups for comments, as a start. Because of the history of wasted funds and unwieldy publicity associated with the UFO problem, the public may not be very receptive to such a proposal just for the pure joy of attempting to resolve the problem unless it turns out that the UFO's are irrefutably proven to be extraterrestrial in origin, thereby gaining incentive as a popular curiosity. A Working Group on UFO's could be painlessly commissioned, much as other working groups comprised of scientists and engineers.

CONCLUSIONS

If there are UFO's in existence here, and *if* they are extraterrestrial, by mere intuition I seriously doubt that they would be manned. I know of no animal to take the reported g's undergone by some UFO's. [Though note that many beings, usually humanoidal, have been seen inside airborne UFOs, and also entering and exiting landed UFOs. L.S.]. In all due fairness to those who believe otherwise, we must readily admit that only a few years ago spacecraft, airplanes, nuclear power, television, human transplants, and many other items presently taken for granted were "impossible," even deemed foolish for consideration.

Conditions in our solar system appear to limit life as we know it (the catchphrase) to Earth, but the probability of almost identical environments just within the visible universe is extremely high. Even if, for some reason obscure to me, life must exist "as we know it," there are, in my opinion, innumerable possibilities of such existence. Manned travel over the required distances would take life-support systems, fuel, and means of propulsion beyond our present ability to deliver in time for us to realize results: therefore it *must* be impossible if we can't do it!

Certain publicized activities under contract purport to be concerned with scientific and engineering studies related to UFO's (for example, Raytheon's Auto-metric Division, Stanford Research Institute, University of California, National Center for Atmospheric Research, Ford Motor Company, etc.). These may yield worthwhile results in the UFO study if the primary goal is not to pursue funded research for its own sake, as is too often the case.

There is only one concrete proposal which I would extend at this time. Either we admit (1) that past funds are wasted; (2) that our technology is not up to the job; (3) that we can afford to ignore one of the potentially most significant "phenomena" in the recorded history of the human race; (4) that we will close our minds to that part of human curiosity which seeks to extend our knowledge; and (5) that we are willing to make these decisions on the flimsiest of evidence, i.e., for-the-most-part-personal opinions, *or* we will make a long-overdue, concentrated, unemotional effort to ascertain (1) the existence or non-existence of hardware UFO's, and if they exist, (2) the origin of UFO's, (3) the means of propulsion, navigation, and associated operational characteristics of UFO's, (4) the intent of the presence of UFO's, and (5) surely a multitude of knowledge, and perhaps greatly extended capability which would result from studying a UFO craft and communicating with the occupants, if any.[97] — DR. GARRY C. HENDERSON, SENIOR RESEARCH SCIENTIST, SPACE SCIENCES, FORT WORTH, TEX.

CHAPTER TWELVE

PREPARED STATEMENT OF ROGER N. SHEPARD

SOME PSYCHOLOGICALLY ORIENTED TECHNIQUES FOR THE SCIENTIFIC INVESTIGATION OF UNIDENTIFIED AERIAL PHENOMENA

ABSTRACT

☛ Even if our interest is in the study of UFO's as some sort of extraordinary physical phenomenon (whether of natural or, possibly, of intelligent, extraterrestrial origin), our study cannot ignore the inescapable fact that nearly all of our evidence comes—not from physical measuring instruments—but from human observers. So far, however, we have consistently sold the human observer short. Indeed, in neglecting to make use of psychologically oriented techniques that would more fully enable observers to bring to bear their really rather remarkable powers of perception and recognition, we may have been forfeiting the opportunity of obtaining evidence from independent observers that would be sufficiently convergent and well-defined to clarify the true nature of the phenomena.

The Extent to Which the Apparent Unpredictability of UFO Phenomena Hinders Their Scientific Study

The scientific investigation of a set of phenomena becomes possible whenever those phenomena exhibit some discernible degree of order or pattern. Scientific study is of course greatly facilitated when, as in astronomy, the order strongly emerges in the form of a space-time pattern of the very occurrences themselves. For, only then, can we arrange to have suitably trained observers suitably "training" powerful and, hence necessarily, highly directional recording and measuring instruments on the right place at the right time.

Indeed, by contrast with astronomical phenomena, those loosely classified together as "Unidentified Flying Objects" ("UFO's") or (with perhaps somewhat less commitment implied as to their real nature) "Unidentified Aerial Phenomena" sometimes appear almost inaccessible to scientific study. Certainly phenomena that are rare and fleeting are difficult enough but, if in addition they are totally capricious and unpredictable when they do occur, then the scientific method is to no avail and we are reduced to awaiting each new happening in the same primitive state of uncomprehending docility.

The repeated successes of science, however, have encouraged us always to search for pattern and order even when none at first appears. And, although a scientific study does of course become enormously more difficult when the occurrences of the phenomena do not fall into any predictable pattern in space and time, it remains a possibility so long as some regularity exists within the phenomena themselves whenever they do happen to occur. Thus, in the field of psychopathology, even if it were the case that some psychological phenomena (such, say, as psychotic episodes) occurred wholly unpredictably—striking any person at any time, quite at random, we could still study the internal patterns of such episodes when they do strike. We might for example find that when symptom A appears, it is usually accompanied by symptom B, but seldom by symptom C, and so on. This, too, is a kind of predictability and can even lead to a degree of understanding and perhaps, eventually, to a method of treatment.

Similarly, in the case of UFO episodes, it may be possible to discover some regularities or patterns within these episodes even though any clear overall pattern in their mere occurrences (except, possibly for the tendency to unpredicted local concentrations in space and time) continues to elude us. This is not to say that efforts—such as those of Michel (1958) and of Vallee and Vallee (1966)—to detect some overall pattern should not be pursued, but only that the attempt at a scientific study need not await a positive outcome of those efforts.

The Principal Sense in Which the Problem of
UFO Phenomena is a Psychological Problem

In the meantime, a scientific study of these phenomena is not impossible—just more difficult. For, we are faced for the most part with a problem—not of making physical measurements but of interpreting verbal reports. We are faced, in short, with a problem amenable more to the methods of the psychologist than to those of the physical scientist.

It is the principal purpose of this note to propose that, despite the relatively primitive state of development of psychological science, psychological and social scientists and even, indeed, law enforcement specialists have devised some techniques that could as well be applied to further the scientific study of UFO's.

I do not mean to suggest by this that most reports of UFO's can probably be shown to arise from purely psychological aberrations such as illusions, hallucinations, delusions, after-images, and the like. On the contrary, a careful examination of most of the best-documented cases has convinced me—as at least one psychologist who has studied rather extensively into the fields of normal and psychopathological perception—that very few such cases can be explained along these lines. Indeed, I have the impression that the claims that the UFO's reported even by seemingly responsible citizens represent lapses of a basically psychopathological character have generally come from people who have neglected to study closely either into the literature on psychopathology, or into that on UFO's, or (in many cases, I fear) both.

*The Desirability of Separating Three
Psychologically Extreme Types of UFO Cases*

I have so far ignored the reports of the so-called "contactees" and related cultists who seem to form a relatively distinct class and who are generally readily identifiable without the benefit of extensive psychological training. Insofar as possible, I should also like to disregard the reports of out-and-out hoaxters. Admittedly, however, these present a somewhat more troublesome problem to which we must return later—particularly in connection with photographs and other types of alleged physical evidence.

Of course there always are ambiguous cases which are difficult to place certainly within the triangle defined by the three, psychologically distinct "corners" representing the deluded contactee, the conscious pranksters, and the involuntary but responsible witness of some real but puzzling phenomenon. However, as has often been remarked, the existence of twilight does not deter us from distinguishing between night and day—nor should it.

In fact, science generally proceeds most rapidly by focusing first on the purest and most clear-cut cases, and by leaving for later any "mop-up" operation of dealing with the remaining cases that are to varying degrees complicated, mixed, messy, borderline, or obscure. Thus the social scientist who is primarily interested in studying the formation and perpetuation of delusional belief systems will do well to focus precisely on the members of the "contactee" cults, the archetypal examples of which are situated in the south-western United States and are heavily constituted by persons (often predominantly women rather past middle age) who have relatively little formal education together with a history of professed beliefs in the mystical, the spiritual, or the occult. Likewise, a clinical psychologist interested in the ways in which socially responsible adult behavior emerges or fails to emerge out of the play and testing behavior of childhood may learn something from an intensive study of any of the almost canonical cases of adolescent boys who, often in pairs and in accordance with an almost tiringly regular pattern, attain at least transitory notoriety by submitting their photograph of a "flying saucer"—replete with dome, antenna, and, perhaps, portholes and fins.

Just so, if we are, as here, primarily concerned with the possibility of unexplained but objective phenomena taking place within our atmosphere, then we should eschew not only the two pure sorts of cases just mentioned, but also the various more-or-less obscure or ill-defined cases falling somewhere within the "triangle." (Indeed, to throw all such intermediate cases together, without adequate regard for the reliability or credibility of each report—as some investigators have tended to do for the purposes of compiling over-all statistics concerning "UFO activity"—can, I think, lead to a largely uninterpretable picture. For, there is then no way of assessing or parceling out the "noise" contributed by the contactee, the prankster and, of course, the many well-meaning citizens who, under unusual circumstances, will continue to misidentify familiar phenomena.) Rather, we stand to learn most from an intensive study into those numerous cases represented by the remaining "corner" of the triangle in which converging evidence from apparently involuntary, independent, and responsible witnesses strongly points to the occurrence of an objective phenomenon of an unexplained character.

The Potential Contribution of Psychological Techniques to the Study Even of Purely Physical Phenomena

Even though our primary interest is, then, on the unexplained objective and presumably physical phenomena that may give rise to such UFO reports, our problem remains as much a psychological as a physical one. For, the vast bulk of the data upon which we must base our scientific investigation comes—not from physical recording or measuring devices—but solely from one or more human observers. Moreover these are observers who were not, evidently, selected for their powers of observation or description and who have good reason to be reluctant as well—particularly in view of the likelihood of ridicule, often encouraged, curiously, by the very investigators who profess to be seeking the truth (cf.,

Fuller, 1966, pp. 211-220; Weitzel, 1967).

It is here, surely, that we have been most glaringly remiss in our attempts to put the investigation of UFO's on a scientific footing. We have, simply, failed to make anywhere near full use of the one recording and measuring instrument at our disposal; namely, our unwitting human witness.

Now it is true that one of the more exotic psychological techniques, hypnotic regression, has already been attempted with interesting—if considerably less than conclusive results in at least one UFO case of a rather sensational nature (Fuller, 1967). However, although astonishing claims have sometimes been made for the kind of detail that can be recovered under hypnosis (e.g., by McCulloch in von Foerster, 1952, p. 100), the results of controlled experiments on accuracy of recall have generally been less impressive (Reiff & Scheerer, 1959). More reliable, in my opinion, are some techniques based on certain psychological facts of a more mundane but better understood character.

The Power of Methods Guided by Recognition Rather Than Description in the Reconstruction of a Fleeting Event

It is, I suppose, a fact familiar to us all that we can take in and remember much more information than we can readily communicate to others. Contrast, for example, how easily we recognize the face of a friend in a crowd with how difficult it is to describe that face so that any other person could then do it for us. Quite generally, our powers of recognition exceed our powers of description (and, indeed, surpass anything that we have yet been able to accomplish by physical instrument or machine). In an experiment on recognition memory, I once presented human subjects with over 600 different pictures, one right after the other, and then found that they could immediately distinguish between those "old" pictures and otherwise completely comparable "new" pictures with median accuracy of over 98% (Shepard, 1967). Even when the test was not given until a week later, the discriminations between "old" and "new" pictures were still 92% correct. Moreover, the advantage of recognition over verbal description should become especially pronounced when the object or event to be remember is unfamiliar and, so, not uniquely or succinctly "captured" by readily available terms or labels.

Over and over again, witnesses of "UFO's" have provided descriptions that, while they strongly suggest that a clear view was obtained of some well-defined but extraordinary object or phenomenon, leave the investigator frustratingly in the dark as to its precise appearance or behavior. A closely viewed, "spinning metallic" object is said, for example, to have been "mushroom-shaped" or to have resembled "an inverted top." But what does this mean? What sort of mushroom? With or without the stalk? And what on earth (!) is referred to, precisely, by "a 30 foot inverted top?"

Some psychologists have been expressly studying the ways in which people come to describe nearly nondescript objects to others (e.g., Krauss & Weinheimer, 1964, 1966). Often a person will feel that the ambiguous term he comes up with (such, for example, as "an inverted top") does quite well. Possibly this is because he is picturing some particular interpretation (e.g., a particular toy that he played with as a child). For the listener who does not have that particular picture in mind, however, the description may prove either meaningless or, worse, completely misleading (cf., Glucksberg, Krauss, & Weisberg, 1966). An indication of the same sort of problem is the tendency of witnesses to say things like "it looked about the size of a football." Further circumstances make clear that they must have been referring to its apparent visual size rather than its real, physical size (which could, after all, hardly be estimated without also knowing its real, physical distance). More pertinently here, it appears that they were really talking more about its shape than its size. Possibly, the presence, so to speak, of a very vivid image in the mind of the witness causes him to lose sight of the total inadequacy of his verbal encoding of that image.

This problem is already implicitly recognized in certain situations of more obviously pressing practical concern. Investigators in cases of homicide do not rest content with the weak and fuzzy descriptions typically offered by a witness but, in addition, may employ a skilled artist (such as Richard Kenehan of the New York Police Department) to work with the witness in an attempt to reconstruct a usable likeness of the murderer's face. The witness may be asked to select eyes, eyebrows, nose, mouth, or ears from series that are systematically graded in size and shape. The witness can then help to adjust their positions on an outlined face. This will generally provide a sufficiently concrete stimulus to enable the witness, finally, to become reasonably explicit about further refinements concerning hair, complexions, lines, scars, asymmetries, and so on. In a number of cases (such as the recent one in which Richard Speck was charged with the murder of several nursing students in Chicago) a likeness constructed in this way from a single surviving witness has proved remarkably accurate and, in more than one instance, has actually led to the apprehension of the criminal (cf., Schumach, 1958).

This provides perhaps the most directly pertinent substantiation for the one central point that I want to leave with those concerned with the investigation of UFO's. Briefly, it is this: Even when an event occurs without warning, leaves little time for careful observation and, indeed, occasions extreme fear or anxiety, the average witness often retains an accurate, almost photographic record of the event—a record, moreover, that can be largely recovered from him even though he lacks the words to describe it himself. Possibly then, in allowing our investigations to depend solely upon our informant's inadequate, his misleading and, yes, his sometimes even ludicrous choice of words, we have done both him—and ourselves—a telling disservice.

The Desirability of Establishing a System Permitting Convergence of Independent Reconstructions

Admittedly, in the case of UFO's, the value of information provided by a single, isolated witness—however detailed that information may be is, by itself, always quite small (except, of course, for the witness himself!). For, from the standpoint of any other person, there is always at least the possibility of hallucination, delusion or, more likely, just plain fabrication. This is amply pointed up by the relative lack of evidential value of the many quite detailed photographs purported to have been taken of UFOS by solitary witnesses.

It is only when there turns up an otherwise inexplicably close correspondence between the information furnished by two or more independent witnesses that the evidence becomes at all compelling at the public or scientific level. But we do not provide for even the possibility of a close correspondence unless we elicit sufficiently detailed information. Thus, when one person reports a "glowing mushroom-shaped object" while another, remote witness refers to the passage of an "inverted top," we have little basis for evaluating the likelihood that they have both observed a physical object—let alone the same physical object.

Now it is true that certain rather suggestive regularities have already emerged in the more or less spontaneous reports of observers. Among the very most common, for example, are the frequent references to disk-like shapes, to extraordinary velocities, to abrupt simultaneous changes of color and direction and, perhaps most strikingly, to the so-called "pendulum" or "falling-leaf" type of descent. Still, much more detailed information concerning such things as shape could presumably be extracted by techniques (akin to those already used in criminal cases) designed to take fuller advantage of the witnesses' usually untapped but vastly more discriminating powers of recognition.

The point, here, is that such more detailed information is needed not merely for its own sake. It is needed, even before that, because the establishment of the very validity of the information in question hinges upon the demonstration of the kind of point-for-point correspondence between reports that becomes possible only when those reports are sufficiently detailed. If two, unrelated witnesses both claim to have seen a disk-shaped object at about the same time and place this is not sufficiently compelling. (Evidently! For it has already happened many times.) But, if artists working with the two witnesses, independently, construct pictures of what appears to be the very same object or, alternatively, if the two witnesses independently point to the very same drawing or photograph in an array of 50 or more different pictures of such objects, then the coincidence becomes more interesting. (And, of course, if the pictures reconstructed or singled out in this way just once turned out to coincide, also, with an actual photograph taken at the time, we should at last have opened the door for the more precise measurements of physical science—including the sophisticated and powerful photogrammetric methods being developed for the analysis and interpretation of lunar photographs.)

The establishment of a pre-tested and standardized procedure for reconstructing information by the sort of psychologically oriented techniques envisaged here, moreover, would be incomparably cheaper than the implementation of other more physically oriented schemes that have sometimes been proposed—such as the construction of a far-flung network of automatic radar-and-camera stations. For, instead of having simultaneously to cover all possible sites in advance, we could simply move in to recover the desired information after an incident is first reported.

There is, however, one unavoidable aspect of the psychologically oriented type of approach proposed here that I, anyway, regard as quite regrettable. To the extent that any detailed pictures reconstructed by these techniques are made publically available, we cannot guarantee that pictures obtained from subsequent witnesses will be suitably independent for our purposes. Consequently, rather tight security precautions would have to be imposed on the more detailed reconstructions, if the purely scientific purposes of the investigation are not to be compromised.

The Use of Concrete Stimuli to Provide a Basis for the Independent, Recognition Guided Process of Reconstruction

The need for ensuring independence of information supplied by different witnesses is in fact so great that it is doubtful whether much reliance could safely be placed on different pictures reconstructed with the help of the same artist. Despite the best intentions of the artist, he might unwittingly guide different witnesses along somewhat similar channels by means of subtle, perhaps unconscious cues (cf., Rosenthal, 1966). Moreover, even if a different artist could be supplied for each witness, we would still be left with the problem of evaluating the likelihood that any two pictures constructed in this way could have turned out as similarly as they did by chance alone.

For these reasons and for reasons of feasibility, convenience, and economy, it would be preferable to develop a standardized set of materials containing suitably representative and graded series of shapes to which each witness could independently respond. Some such materials would be needed, in any case, in order to provide stimuli suitable for tapping the witnesses' powers of recognition. Possibly even separate arrays should be constructed for distinguishable parts such as "domes" or other projections (just as separate series of eyes, or mouths may be helpful to the witness in criminal cases). By means of suitable standardization and control in the preparation and presentation of such materials, then, we could be reasonably sure that the responses of one witness are not unduly influencing the choices of another. Moreover, since each witness would make his choices from a pre-tested, fixed set of alternatives of known size, we would be in a favorable position to assess the probability that any coincidences of choice might have occurred merely by chance.

All things considered, the best procedure might be to divide the questioning of a witness into the following three distinct phases; first, the recording of the witness describing what he saw as completely as he can, in his own words, and without any cues (whether verbal or pictorial) that might bias him in one direction or another; second, the recording of his responses to the standardized, pictorial materials; and third (if the case seems to warrant it) the full reconstruction of a new picture with the help of a suitably trained artist. Such anew picture, if sufficiently novel or well-defined, might then be incorporated in future revisions of the materials used for the second phase of the interview.

The effectiveness of the proposed procedure would depend very heavily upon the amount of thought, care, skill and, above all, pretesting that went into the preparation of the materials. The arrays of alternative shapes should of course include all types of shapes that have been clearly described, sketched, or (allegedly) actually photographed by some previous witness of at least reasonable reliability. One helpful attempt at systematizing the kinds of shapes that have been reported has in fact already been published (see Hall, 1964, p. 144). However, more extensive and refined work would be necessary in order to cover the great variety of reported shapes, and to do this in a sufficiently concrete and realistic manner to promote recognition and, possibly, further specification by witnesses.

The Use of Photographs of Alleged UFOs as a Source of Concrete Test Materials

Since photographs represent an especially tempting vehicle for the hoaxter and, in addition, are easily faked, they are individually of little value as evidence except in the rare cases in which there were independent, corroborating eyewitnesses. Photographs purporting to be of UFOs are, however, surprisingly numerous. (I myself have assembled well over 150 distinct such photographs merely from published reports.) Moreover, since at least one of these photographs might be authentic and since we have no sure way of knowing in advance which one it might be, we can not afford to eliminate any distinct type that hasn't already been proved to be a fraud. In the meantime, moreover, the other, spurious photographs can serve (in somewhat the same way as the non-suspects in a police "line-up") as a means of assessing consistency of choice or, contrary-wise, mere guessing on the part of different witnesses.

The accompanying figure reproduces drawings in which I have tried to portray, as accurately as I could, representative objects from 63 of these photographs. The greater contrast of the drawings renders them more readily duplicatable than the original photographs. Moreover, I was also able in this way to reduce them all to uniform size and to eliminate background details which, although useful for estimating size or gauging authenticity, are only distracting for present purposes. Some of the photographs are from well documented or widely publicized cases while others are of more obscure or dubious origin. At least two cases have subsequently been admitted to be hoaxes, while circumstances surrounding some of the others make them difficult to dismiss in this way.

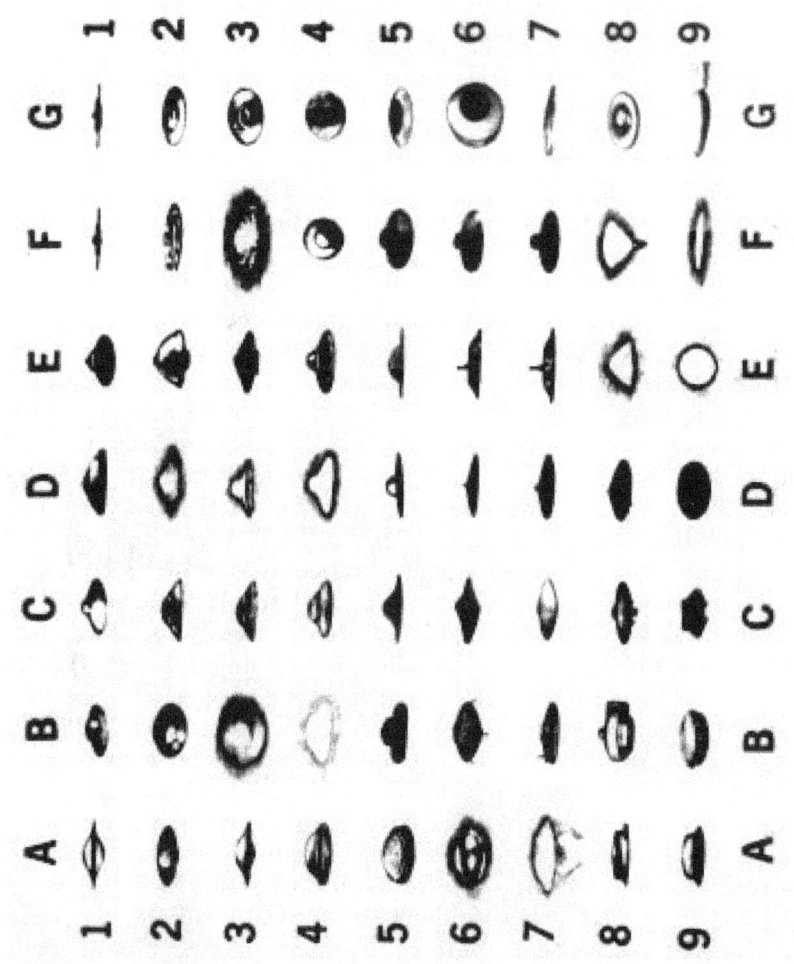

The array is intended only to convey some idea of the variety of shapes that have appeared, it does not give an adequate impression of the relative frequencies with which the different shapes have appeared. In fact, the images most commonly appearing in my total sample show either a small point, formless blob, or fuzzy ellipse of light in a night sky, or else a dark, more or less distinct ellipse (like that shown in D9) against a lighter sky. With very few exceptions, such as the rocket- or "cigar"-shaped object with "exhaust trail" (G9), which allegedly was photographed over Peru in 1952, the more well-defined objects appear to be some variant of the "saucer" or "domed disk." [Shepard's caption.]

Nevertheless, the single most striking thing about these pictures—far from being any general uniformity in their appearance—is their largely irreconcilable diversity. Whether or not this diversity is interpreted as detracting from their value as evidence, it surely cannot be taken as contributing to that value. It does, however, serve our immediate, rather different purpose of providing an initial sample from which to extrapolate and interpolate an eventual graded array of the sort that we seek for purposes of testing witnesses.

Some Incidental Comments on the Status of Photographs of Alleged UFOs as a Source of Evidence Themselves

Before leaving the photographs themselves, however, it should be noted that there are a few instances of rather suggestive similarities between these photographs. A frequently cited case is the striking resemblance between E6, reportedly taken by a farmer, Paul Trent, as he was returning home with his wife near McMinnville, Oregon, in June 1950, and E7, allegedly taken by a French military pilot near Rouen, France in March 1954.

Another example, involving several different sightings, is shown in the upper right. G1, G2, and G3 present three successive views of the same object purportedly taken by a photographer, Ed Keffel, of the Brazilian publication *O Cruziero* while he was accompanied by a journalist, Joao Martins, near Rio de Janeiro in May, 1952. The edge-on view, G1, is almost indistinguishable from another photograph, represented in F1, allegedly taken from an Argentine pursuit plane in late 1954. The "top" view, G2, seemingly resembles F2 (the left edge of which was cut off by the boundary of the original photograph), which a 15-year-old boy, Michael Savage, claims to have taken near San Bernardino, California in July, 1956. It also somewhat resembles the lighted object, F3, allegedly appearing in a color photograph taken by Joseph Sigel near Waikiki, Hawaii in June, 1959. And, finally, the "bottom" view, G3, presents the same general sort of configuration as that shown in G4, which is based upon a photograph purportedly taken by Yukuse Matsumura outside his residence in Yokohama, Japan in January 1957 (although the relative dimensions of the features appear slightly different in these last two photographs).

There are several other instances in which photographs taken by apparently unrelated individuals might be of the same object. Another view (not included in the figure) showing more of the "bottom" of the object displayed in E4 resembles the object shown in G5 and, even more closely, an object apparently hovering over a seaplane in still another photograph (also not included) of unknown origin. In a number of instances (e.g., F5, 6, and 7 or D6, 7, and 8) the degree of correspondence is more difficult to assess owing to the relatively poorer definition of the images.

Of course even very close similarities do not in themselves guarantee authenticity. Consider, for example, C2 and C3 which are strikingly similar despite the fact that the object in C2 appears over a mountain near Riverside, California in the original photograph reportedly taken by a 21-year-old man and two friends in 1951, whereas the object in C3 appears over a flock of grazing sheep in the photograph submitted by an Australian rancher in 1954. But, since the object in C2 has subsequently been admitted to be none other than a 1937 Ford hubcap, the object in C3 is presumably the same. (Another photograph later confessed to be fraudulent is represented in B6, and somewhat suspicious circumstances also surround several other photographs, including those represented in A9, B1, B8, B9, C1, and G8.)

Perhaps the safest attitude to adopt is that recommended by the National Investigations Committee on Aerial Phenomena (cf., Hall, 1964, p. 86); namely, that the evidential value at least of still photographs depends entirely upon the surrounding circumstances. An isolated photograph about which little is known, no matter how impressive it may appear in itself, is essentially worthless—except, possibly, in cases in which sophisticated photogrammetric analysis yields further detail tending to confirm the verbal account of the photographer. (I have heard that this as happened in at least one case; viz., that of the controversial series of Polaroid photographs—one of which is represented in A8—taken by a California state highway employee, Rex Heflin, near Santa Ana in August 1965.) The photographs most worthy of further intensive investigation would seem to be those for which there also were reported to be many eyewitnesses as well as other, corroborating photographs—as in the celebrated case of the "Saturn-shaped" object (A1) that was said to have been photographed from an oceanographic vessel near the Brazilian island of Trindade in January 1958 (see Lorenzen, 1966, pp. 145-153, 164–174).

The Assessment of the Representativeness
of a Set of Recognition Test-Materials

Even though an extensive effort is made to represent every sort of shape that has been reliably described, sketched, or photographed, the possibility will remain that the collection of proposed test materials will not be sufficiently representative.

Certain types of completely fraudulent shapes may unnecessarily inflate the already unwieldy collection and, more seriously, some significant types may still be missing. There are, however, ways of assessing the representativeness of any proposed collection of shapes. One is, simply, to have people describe these shapes and then to look for any pronounced departures of the relative frequencies of the various descriptive terms used from the corresponding relative frequencies in reports issuing from actual sightings of UFOs. My research assistant, Miss Shelley Meltzer, carried out an exploratory attempt at this sort of thing that may help to illustrate some of the relevant considerations.

From our total sample of photographs, 75 that seemed suitably representative were selected for this preliminary study. These included most of the 63 already portrayed in the accompanying figure, but those that were known or strongly suspected to be fraudulent were eliminated and a number of others of less sharply defined shape were added (since many reports indicate that the shape was not clearly visible). Each of 19 subjects, mostly students at Harvard University, then looked through one of three subsets of 25 of these photographs and, for each, attempted to describe the pictured object in their own words. (Immediately following that, each subject then looked through another subset of 25 and, this time, indicated the appropriateness for each photograph of each term in a fixed set that we had listed in advance on a standardized rating sheet. However this part of the experiment will not be considered in any detail here.) Of most immediate interest are the descriptive labels spontaneously produced in the 19 x 25 or 475 subject-photograph encounters.

These could now be compared with the descriptive labels appearing in a sample of 206 different representative reports of actual UFO sightings that Miss Meltzer had already extracted (for a different purpose) from a number of sources (mostly Edwards, 1966; Hall, 1964; Michel, 1967; Olsen, 1966; Ruppelt, 1956; and Vallee,1965). The accompanying table lists those descriptive terms that pertain to visual appearance but, for purposes of comparison with the mostly black-and-white photographs, excludes the many references to (chromatic) color. With one exception (#33), only terms that appeared at least twice in the sample of 206 actual reports are included, and these arranged in order of decreasing frequency of occurrence in that sample.

TABLE I

Label describing visual appearance (excluding chromatic colors)	(a) Number of occurrences in 206 actual UFO cases	(b) Number of occurrences in 475 descriptions of photographs	Appreciable discrepancies
1. Disk shaped	27	42	
2. Circular	24	24	
3. Round	22	25	
4. Metallic	19	41	
5. Domed top	15	21	
6. Starlike (point of light)	14	6	—
7. Cigar shaped	13	1	— —
8. Spherical	12	9	
9. Ball shaped	11	0	—
10. Fiery appearance	11	0	—
11. Trail of vapor or smoke	10	13	
12. Portholes or windows	10	0	—
13. Pattern of lights	9	0	—
14. White filaments emitted	9	0	—
15. Oval	6	19	+
16. Flat	6	8	
17. Elliptical	5	9	
18. Dumbbell shaped	5	0	
19. Football shaped	4	1	
20. White	4	0	
21. Saucer shaped	3	12	+
22. Egg shaped	3	5	
23. Diamond shaped	3	0	
24. Silvery	3	0	
25. Saturn shaped	2	7	
26. Top shaped	2	5	
27. Conical	2	3	
28. Washtub shaped	2	0	
29. Two washbowls rim-to-rim	2	0	
30. Two plates rim-to-rim	2	0	
31. Long tail	2	0	
32. Emitting flame	2	0	
33. Hat shaped	1	35	+ +
Total	265	251	

The two columns of numbers, then, present the resulting frequencies of occurrence (a) in the 206 actual UFO reports and (b) in the 475 opportunities for these same descriptive terms to arise in the experiment with the photographs. Direct numerical comparisons are somewhat hazardous owing to the different circumstances in which the two sets of descriptive terms arose. In terms merely of opportunities, the numbers in the second column should be about twice as large as those in the first. However, the totals for the two columns are nearly equal and, so, the real encounters evidently were relatively more productive of descriptive terms on the average. Numerically small departures or departures in which the second number is somewhere between the size of the first number and twice that size are probably not very significant therefore. The remaining positive and negative discrepancies of appreciable size are indicated by the plus and minus signs in the right-most column. Some of these are probably explainable in terms of the two-dimensional, achromatic, and stationary character of the photographs (e.g., # 10 & 13), or in terms of differences in vocabulary to be expected between the unselected witnesses and the college-educated subjects of the experiment (e.g., #9 & 15). Other discrepancies, however, suggest either that some shapes, such as the so-called "cigar" (#7), were not adequately represented in the sample of photographs, or that some shapes, such as those most frequently said to resemble a "hat" (#33), are especially likely to have been of fraudulent origin. (Among the objects included in the above figure that were often said to be hat-like are C3, which we already noted is almost certainly a 1937 Ford hubcap; E4, which doesn't seem to fit very well with the usual descriptions of UFOs; and G2, which, although it too doesn't coincide with at least my notion of a "flying saucer," does however correspond rather closely with several other photographs.) [Shepard's caption.]

The Use of the Computer in the Design of Recognition Test-Materials and in the Analysis of Results

Our sample of only 206 actual UFO cases is really too small and haphazard for the purpose of ensuring that all types of reported shapes are adequately represented in any proposed recognition array. Many descriptive terms that have repeatedly been used (such as "doughnut," "ring," "mushroom," "flattened ball," "double-convex lense," "bullet," "blimp," and "submarine") didn't happen to appear more than once in our particular sample). Ideally, for this work, one would like access to a centralized library of all reasonably documented cases—suitably coded for retrieval via computer. Indeed, at present the scientific study of the UFO problem is greatly hampered by the circumstance that the thousands of reported sightings have not been adequately coded or systematized in any uniform way and are, in fact, still scattered among such diverse and often mutually hostile organizations as the U.S. Air Force, NICAP, APRO, and the University of Colorado Project—not to mention a number of more-or-less private files assembled by individual investigators both here and abroad.

Recent developments in computer technology—particularly in computer graphics could be utilized, also, in the construction of arrays of shapes for a recognition test. Thus for any specified shape, the computer (together with suitable graphical output equipment) could automatically generate alternative pictures of the same object as viewed from any desired angle (e.g., Noll, 1965; Zajac, 1964); and could even generate other test shapes intermediate between that shape and some other specified shape. (In fact, as on-line graphical facilities become more widely available, even a relatively unartistic witness, seated in front of a suitable display device, should be able to reconstruct his own object by techniques of these general sorts.)

For the present, however, perhaps the most promising use of the computer in this connection would be in finding an optimum arrangement of the alternative test shapes in the recognition array. This is a matter of real concern owing to the large number of shapes that should be included. (Even the 63 exhibited in the above figure fall far short of covering all the varieties that have been sketched or described.) If the alternatives could somehow be arranged so that similar shapes are close together, then the witness could quickly narrow down to the most relevant region of the array in order to make his final, most refined discriminations.

In order to do this we would first need to obtain some measure of the perceived similarity between any two shapes. One could of course obtain a direct, subjective judgment of similarity from experimental subjects. However, it might be more convenient to obtain a derived measure of similarity based upon the frequency with which different subjects will sort the two shapes into the same pile, or upon the overlap in their application of the same descriptive terms to the two shapes in the kind of task described in the preceding section (cf., Rosenberg, Nelson, & Vivekananthan, in press).

Once we have any such measure of similarity for every pair, we can apply powerful new computer-based methods for mapping the objects into a two dimensional arrangement in such a way that their similarities are preserved, in so far as possible, in the spatial proximities among them (Kruskal, 1964, Shepard, 1962; Shepard & Carroll, 1966). Moreover, these same methods could yield a quantitative metric of similarity that would then enable us to specify just how similar an object identified by one witness is to the object identified by another witness. Indeed, they could even tell us something about the basic dimensions along which UFO phenomena differ or, with the help of recently perfected methods for "hierarchical clustering" (Johnson, 1967), they could provide an indication of the basically different classes into which these phenomena undoubtedly fall.

Possibly, some of these classes of unidentified aerial phenomena will turn out to be of purely natural origin. I once even ventured to suggest this for certain puzzling types of cases myself (Shepard, 1967b)—though, admittedly, attempts to develop such explanations in terms of known principles of atmospheric physics, generally, have run into competent and serious criticism (McDonald, 1968). Still, even if some of the phenomena are of natural origin, a more complete and accurate characterization of their appearance and behavior should be of some interest to the physical scientist—indeed, all the more so to the extent that they appear to conflict with known physical principles.

In any case, it appears that techniques now exist that could provide the basis for a psychologically oriented, but genuinely scientific investigation into unidentified aerial phenomena, whatever their nature may ultimately prove to be.

SHEPARD'S ACKNOWLEDGMENTS

The development of some of the techniques described here and the preliminary experimental tests of these techniques on UFO materials were carried out as one part of a more general profect on psychological scaling and data analysis supported by Grant No.

GS-1302 from the National Science Foundation to Harvard University. The author is indebted to the Foundation for its support and, also, to Miss Shelley Meltzer for her extensive assistance in the project.[98] — DR. ROGER N. SHEPARD IS PROFESSOR OF PSYCHOLOGY AT STANFORD UNIVERSITY

SHEPARD'S REFERENCES

Edwards, F. *Flying saucers—serious business.* New York: Lyle Stuart, 1966, (Also Bantam paperback S3378).

Fuller, J. G. *Incident at Exeter.* New York: G. P. Putnam's Sons, 1966, (Also Berkeley Medallion Paperback, TM 757, 375).

Fuller, J. G. *The interrupted journey.* New York: Dial Press, 1967, (Also Dell Paperback 4068).

Glucksberg, S., Krauss, R. M., & Weisberg, R. Referential communication in nursery school children: Method and some preliminary findings. *Journal of Experimental Child Psychology,* 1966, 3, 333–342.

Hall, R. H. (ED.) *The UFO evidence.* National Investigations Committee on Aerial Phenomena, 1536 Connecticut Ave., N.W., Washington, D.C., 1964.

Johnson, S. C. Hierarchical clustering schemes. *Psychometrika,* 1967, 32, 241–254.

Krauss, R. M., & Weinheimer, C. Changes in reference phrases as a function of frequency of usage in social interaction: a preliminary study. *Psychonomic Science,* 1964, 1, 113–114.

Krauss, R. M., & Weinheimer, S. Concurrent feedback, confirmation, and the encoding of referents in verbal communication. *Journal of Personality and Social Psychology,* 1966, 4, 343–346.

Kruskal, J. B. Multidimensional scaling by optimizing goodness of fit to a nonmetric hypothesis. *Psychometrika,* 1964, 29, 1–27.

Lorenzen, C. E. *Flying saucers: the startling evidence of the invasion from outer space.* New York: Signet, 1966, (Signet Paperback T3058).

McDonald, J. E. UFOs—atmospheric or extraterrestrial? Talk presented to the Chicago Chapter of the American Meteorological Society, May 31, 1968. (See, also, his contribution to this volume.)

Michel, A. *The truth about flying saucers.* New York: S. G. Phillips, 1956. (Also Pyramid Paperback T–1647)

Michel, A. *Flying saucers and the straight-line mystery.* New York: Criterion Books, 1958.

Noll, A. M. Stereographic projections by digital computer. *Computers and Automation,* 1965, 14, No. 5.

Olsen, T. M. (Ed.) *The reference for outstanding UFO sighting reports.* UFOIRC, Inc., Dept. SM 518, P.O. Box 57, Riderwood, Maryland, 21139, 1966.

Reiff, R., and Scheerer, M. *Memory and hypnotic age regression.* New York: International Univ. Press, 1959.

Rosenberg, S., Nelson, C., and Vivekananthan, P. W. A multidimensional approach to the structure of personality impressions. *Journal of Personality and Social Psychology,* 1968 (in press).

Rosenthal, R. *Experimenter effects in behavioral research.* New York: Meredith Publishing Co., 1966.

Ruppelt, E. J. *The report on unidentified flying objects.* New York: Doubleday, 1956. (Also Ace Books paperback G–537)

Schumach, M. Palette-packing cop. *New York Times Magazine,* August 24, 1958.

Shepard, R. N. The analysis of proximities: Multidimensional scaling with an unknown distance function. I. *Psychometrika,* 1962, 27, 125–140. II. *Psychometrika,* 1962, 27, 219–246.

Shepard, R. N. Recognition memory for words, sentences and pictures. *Journal of Verbal Learning and Verbal Behavior,* 1967, 6, 156–163. (a)

Shepard, R. N. Tornadoes: Puzzling phenomena and photographs. (Letter to Editor). *Science,* 1967, 155, 27–28. (b)

Shepard, R. N., and Carroll, J. D. Parametric representation of nonlinear data structures. In P. R. Krishnaiah (Ed.), *Multivariate analysis: Proceedings of an international symposium.* New York: Academic Press, 1966. Pp. 561–592.

Vallee, J. *Anatomy of a phenomenon.* New York: Ace Books, 1965 (Paperback).

Vallee, J., and Vallee, J. *Challenge to science: The UFO enigma.* Chicago: Regnery, 1966.

Von Foerster, H. (Ed.) *Cybernetics: Transactions of the eighth conference.* New York: Josiah Macy, Jr., Foundation, 1952.

Weitzel, W. . . . Into the middle of hell. *Flying saucers: UFO reports #3.* New York: Dell Publishing Co., 1967. Pp. 38–49.

Zajac, E. E. Programmed pictorial displays. *Proceedings of the 1964 symposium on digital computing.* Bell Telephone Laboratories, 1964. Pp. 33–44.

CHAPTER THIRTEEN

STATEMENT OF FRANK B. SALISBURY

The following reprint from *Bio-Science*, volume 17, No. 1, 1967, pages 15-24, was submitted by Dr. Frank B. Salisbury, head, plant science department, Utah State University, as summarizing his views on UFO's.

THE SCIENTIST & THE UFO

☛ A phenomenon is abroad in the land. Since shortly after the beginning of recorded history, but particularly during the past two decades, many people have reported visual observations of phenomena which they interpret as objects so intricate in their structure and proficient in their maneuvers that they far surpass the current human technology. The apparent objects are usually in the sky, but in a few cases they are on the ground or landing or taking off from the ground. Although they may not be flying and they may not be objects, they are called unidentified flying objects: UFO's for short.

What is the significance of these strange, typically aerial phenomena? There are many extremely important implications in the area of psychology. Perhaps the most obvious is the possibility that the UFO's may be purely psychological phenomena such as hallucinations. Of much greater importance, however,

An artist's conception of the Boiani, New Guinea, June 26-27, 1959, sighting. Object is drawn from sketches made by the witnesses at the time. Note the other two objects hovering at a greater distance. Supplied by the Aerial Phenomena Research Organization. [Salisbury's caption.]

could be the psychological questions of interpretation. These are valid regardless of what elicits the response in the witness—a real spaceship from Mars or a spotlight shining on a gossamer cloud.

The number of witnesses to these phenomena has increased tremendously in recent years (probably a sizable fraction of 1% of the world's population has been involved in "good" sightings), therefore the phenomenon is of obvious sociological importance. It could influence the relationships between nations or programs of space exploration. It might even, given the proper circumstances, develop into a panic of severe proportions. There is ample justification from the sociological standpoint for a detailed study of the UFO phenomenon.

My interest developed from the field of exobiology. If the UFO's are extraterrestrial spaceships guided by intelligent beings (as many of their witnesses insist), then they are of the most pressing interest to the exobiologist. Current speculation about life on Mars (Jackson and Moore, 1965; Salisbury, 1962, 1966) would be naive indeed if such were the case. Although they would have virtually no significance to exobiology if they are not extraterrestrial, the possibility that they might be seems great enough to merit at least a preliminary investigation.

We might well consider the UFO's from the standpoint of the philosophy of scientific method. Even if the scientific community at large were sincerely interested in the study of the phenomena, it would encounter many difficulties in knowing what approach to take. UFO sightings are events which usually cannot be repeated. The astronomer may also witness such events, e.g., the flares on Mars (Salisbury, 1962; Ley Willy and Wernher-Von Braun, 1960), but at least he is a trained observer, and none of his colleagues are likely to doubt his word. In the case of the UFO's, although many observers may be highly trained in certain aspects of contemporary modern life, few, if any, could claim much competence as carefully schooled UFO observers! Frequently, they are not trained to differentiate between observation and interpretation, and often there is a strong tendency for all but close friends to doubt their word. Here, then, is a phenomenon of nature which could, and should, be of extreme interest to the scientist. But it is a difficult one for even him to study. How do we study events which cannot be repeated and which are recorded only through the minds of observers who can scarcely resist the temptation to enlarge their stories and to intermingle the facts with their own interpretations and psychological responses?

About all we can do at present is to evaluate the reports,

although sufficient desire might make more than this possible (in the Exeter sighting described below, observers could have actually waited, fully equipped with high-speed cameras and other devices, for the return of the objects). Professor J. Allen Hynek (1966), the Director of the Observatory at Northwestern University, and for the past 18 years consultant to the Air Force in their study of UFO sightings, has often stated that to make progress we must accept the fact that the *UFO's do exist as reports* [my emphasis, L.S.]. The Air Force and several private groups have accumulated bulging files of these reports, containing everything from detailed interviews to the remnants of pancakes submitted by a witness who claimed he had received them from a space man! These reports and the many which will be obtained in the future (using hopefully better means of information gathering) are the data with which we must work, and the only data so far available. What can we do with them?

One obvious approach is to propose as many possible interpretations as can be devised and then to evaluate the data in terms of these hypotheses. The process will be a circular one, in which hypotheses are formulated on the basis of the data, and the data are then reexamined in terms of the hypotheses. In the following paragraphs, five hypotheses are discussed and then a few representative sightings are considered. The subject has been reviewed by several authors in book form, often competently, but virtually always with some degree of prejudice (for: Hall,1964; Keyhoe, 1960; Lorenzen, 1962,1966; Michel, 1958; Vallée, 1965; Vallée and Vallée, 1966—against: Menzel and Boyd, 1963).

I. Extraterrestrial Spaceships or Other Machines

Although earlier observers usually interpreted the UFO's in terms of miraculous religious events, most UFO observers during the past 19 years have suggested that the objects which they observed were extraterrestrial spaceships.

Can we eliminate the spaceship hypothesis in any rigorous scientific manner? Logically one might think of two approaches: either we must show in each and every instance ever reported that the object was not an extraterrestrial spaceship, or we must show by some sort of scientific logic that it is impossible for extraterrestrial beings to visit us.

Obviously, we cannot show in every case that a purported UFO was not an extraterrestrial spaceship. The data may not be available, and the events cannot be repeated. Furthermore, in several instances, very detailed data do exist in relation to a sighting, and yet *it cannot be rigorously stated that the UFO was not an extraterrestrial machine* [my emphasis, L.S.].

Nevertheless, this approach has been followed in an attempt to eliminate this hypothesis, notably by Professor Donald Menzel, Director of the Harvard Observatory (Menzel and Boyd, 1963) and by the United States Air Force. Menzel is aware of the logical limitations, but he takes a statistical approach. He reasons that since many sightings can be positively eliminated as extraterrestrial spaceships, those which cannot could be if only more data were available. This is an excellent example of the inductive form of reasoning which has been so productive in science. Can we confidently apply it in relation to the UFO phenomena? To do so, the cases for which ample data exist and which prove not to be spaceships must be representative of the class as a whole. To many of us this seems unlikely, since other cases fortified with considerable data cannot be eliminated as extraterrestrial machines, and in many ways they appear to have little in common with the cases which can. On purely formal grounds, then, we cannot be absolutely convinced by Menzel's approach.

It is also logically unreasonable to say with absolute certainty that it is impossible for extraterrestrial beings to visit us. Although we know a great deal about the universe, we do not yet know enough to make such an all-inclusive negative statement [my emphasis, L.S.]. Nevertheless, many of the arguments are highly compelling, and two are especially worthy of our attention.

The first argument is that the UFO's contravene the laws of nature, or more properly, that they are contrary to our experience. It is first assumed that they could not originate within our solar system because only the earth harbors intelligent life, and then it is reasoned that because of the extreme distances between stars they could not be visitors from some other planetary system. My initial contact with the UFO problem came because of my doubts in relation to the first assumption (Salisbury, 1962; 1964; 1966).

Certainly we have no conclusive or even compelling evidence that Mars might support an intelligent civilization. We do, however, have a number of observations which seem to be in agreement with this assumption. The network of lines referred to as the canals still defies explanation in terms of nonintelligent phenomena, although such an explanation may well be apparent when we obtain more data about Mars. The satellites of Mars, with their almost perfectly circular, equatorial orbits and their small size have certain of the characteristics of artificial satellites. Brilliant flares of light occasionally seen on the surface of Mars are too short in duration and too bluish-white in spectral quality to be similar to

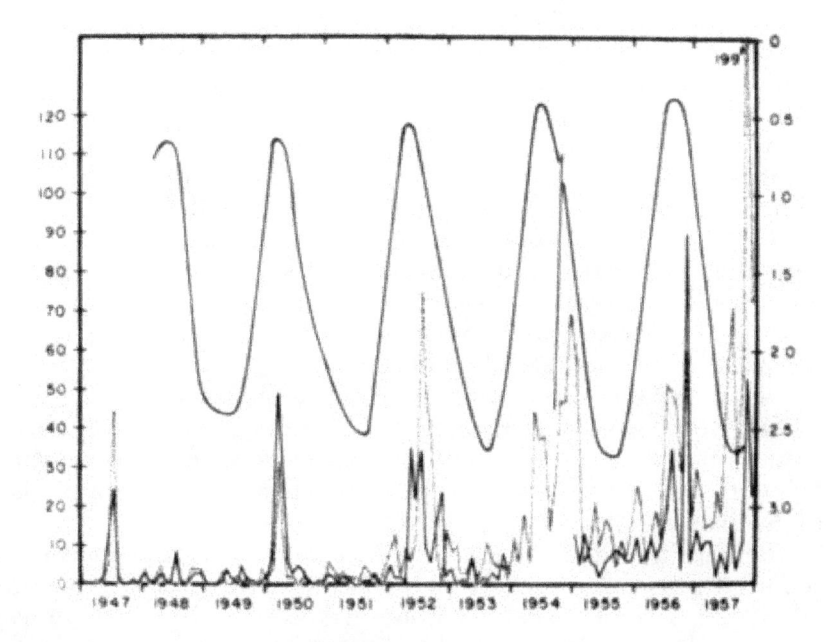

Fig. 1. UFO sightings and the oppositions of Mars. The dotted line represents "reliable" sightings in the files of the Aerial Phenomena Research Organization, 3190 E. Kleindale Road, Tucson, Arizona. They were supplied by Coral E. Lorenzen. The solid line represents sightings assembled from published reports by M. G. Quincy and presented by Jacques and Janine Vallée (1962). Quincy showed no records for 1964, and these were obtained from Michel (1958). Obviously there is no correlation between UFO sightings and the distance to Mars in 1947 and 1957, but the Vallées calculated correlation co-efficients for the other years and found that the observed correlation would be expected to occur due to chance alone less than one time in a thousand trials (they transposed all sightings two months to account for the lag apparent in 1952, 1954, and 1956). It is interesting to note that the two sighting curves agree so closely, although they obviously include many different data. A third curve, prepared by E. Buelta in Barcelona, Spain, also agrees closely. [Salisbury's caption.]

our volcanoes, yet they are too long in duration to be readily explainable as meteorite impacts. An occasional associated white cloud would seem to eliminate them as reflections. It is even possible, if one is willing to stretch the imagination a bit, to find evidences for intelligence in the Mariner photographs of Mars. These ideas have recently been discussed in considerably more detail elsewhere (Salisbury, 1966).[99]

There was an interesting correlation from 1948 to 1957 in the number of UFO sightings per unit time and the closeness to the planet Mars (Fig. 1). This was shown by Vallée and Vallée (1962; 1966) to be expected on statistical grounds less than one time in a thousand. Both Venus and Jupiter are far more prominent in the skies than Mars (both have often been misinterpreted as UFO's), and yet no such correlation exists with their apparent brightness in the skies and the number of UFO sightings.

Assuming that there is no intelligence on Mars and that the UFO's would have to cross interstellar space, can we really state with confidence that this is an impossibility? Do we know so much? Of course we do not. We are even searching for possible solutions to the problem of interstellar travel.

Perhaps the most compelling "impossibility" argument is the reported physical activity of certain UFO's (Menzel and Boyd, 1963). In several "good" sightings (those which, for reasons discussed below, do not readily fit any of the remaining four hypotheses), UFO's have appeared to accelerate at tremendous rates or even make right-angle turns while traveling at speeds of several hundred or thousand miles per hour. Although they move in the atmosphere at velocities which surely exceed that of sound, no sonic booms are heard (they are often essentially silent) nor do they appear to burn up with frictional heat. The skeptic says: "Granted that we have a lot left to learn about our universe, we surely don't expect the fundamental laws to be rejected. That we may refine them as Einstein did, it is true, but inertia is inertia, and a right-angle turn at several thousand miles per hour is a simple physical impossibility."

This may be the most compelling argument against the spaceship hypothesis, but there are two counter-arguments. First, one can simply reject the above statement. I do not see how Newton's laws could be so flagrantly violated, but others (Lorenzen, 1962: Michel, 1958; Vallée and Vallée, 1966) have come up with various suggestions. Perhaps inertia is the gravitational interaction between an object and all other objects in the universe. If this gravitational attraction could some way be severed (some mysterious antigravity shield surrounding the spacecraft, for example!), then right-angle turns at high speeds might be feasible. Would the surrounding antigravity field also nullify the sound barrier problem? Some think so. I haven't the faintest idea, but *we could be wrong about what is impossible* [my emphasis, L.S.]. Second, one might remember that not all UFO's perform "impossible" feats. The topic is sufficiently interesting if only one UFO proves to be a spaceship from Mars!

Another argument against the space-ship idea concerns the lack of formal contact with the UFO occupants. Since visiting spaceships ought to be piloted by some sort of intelligent beings, wouldn't it be reasonable to expect that they would desire contact with other intelligent beings, namely us? Or why hasn't a flying saucer landed on the United Nations Building to establish formal diplomatic relations?

This argument assumes that we can understand the motives of an extraterrestrial being. Of course we cannot. How could we know the minds of such beings? To inductively extrapolate from our own current sociological approaches to those of other intelligent entities would be to commit the logical sin of extrapolation in a most flagrant manner. It is easy to imagine several reasons why the extraterrestrials might not want to contact us. Did they plant us here as a colony many thousands of years ago and are carefully observing our evolutionary development? Do they envy us for our natural resources and want to conquer us, although present logistic problems make such an effort impossible? Are they waiting for us to straighten out our wars and race problems? Are they simply uninterested in us as contemporaries, preferring to observe us as specimens? Entomologists study the honeybees very carefully but make no diplomatic contact with the queen!

Imagine the Aborigines of Central Australia, who are still in the stone age and who have not even developed the bow and arrow. They have had no contact at all with modern civilization. What happens when a jet plane flies overhead and one of them observes it? When he tells of the huge, shiny bird that didn't flap its wings, had no feet, made an ear-splitting roar, and even had smoke coming out of its tail, surely his fellows assume that he is crazy. Or if the phenomenon becomes so common that it must be accepted as real, they could hardly be expected to deduce from it the conditions of our modern civilization, let alone our motives. "Why," they might ask, "don't the intelligent beings who guide this mighty bird land and trade bone nose-pieces with us?" Actually, many of the Aborigines, even those who have come in contact with civilized men, still interpret the airplane in a religious context, as witness the establishment of the cargo cults among these peoples (Worsley, 1959).

We cannot, then, eliminate the spaceship hypothesis, although some of the arguments against it are quite impressive. *We should, in deference to the scientific method, examine with a completely open mind any evidence which might be marshalled in favor of the hypothesis* [my emphasis, L.S.]. Let us consider the four alternatives to it.

II. Conventional Phenomena Misinterpreted

Given certain special circumstances, nearly anyone can be confused and amazed by the appearance of some conventional object which under other circumstances might cause no bewilderment whatsoever. What psychological factors lead to such misinterpretations? In various instances, reported UFO's have clearly been demonstrated to be balloons, kites, birds, conventional aircraft, artificial satellites, planets and stars, meteors, clouds, natural electrical effects such as ball lightning (Klass 1966), and optical effects such as reflections, mirages, sundogs, and defractions caused by inversion layers in the atmosphere (see Menzel and Boyd, 1963, and Air Force files). Let us consider the level of certainty in classifying a given sighting here.

Often, the sighting may be placed here with absolute certainty. A balloon reported as a UFO was never out of sight of its launchers. A perplexing light in the sky takes form as an airplane as its gets closer.

My children woke me at 6:00 a.m. in Tübingen, Germany, saying that they were watching a hovering UFO over the city. I grabbed my binoculars and watched the brilliant light move rather rapidly both toward us and away from us and even from side to side. After about a minute, I decided to make my observations more precise, backed up against a doorway, and aligned the object with a spot on the window frame. Upon doing this, it stopped moving, and we were soon able to identify it as Venus, then the morning star. Its lateral motions were apparently illusions due to our own movements, and its rapid approach and retreat were due to a thin, rapidly moving layer of mists which caused it to change intensity.

Within the last year I have positively identified UFO's over Fort Collins, Colorado (pointed out to me usually by phone) as a weather kite, the planet Venus, and the stars Vega, Capella, Betelgeuse, and Sirius. Some of the stars close to the horizon flashed red, green, and white, and only a star chart and much

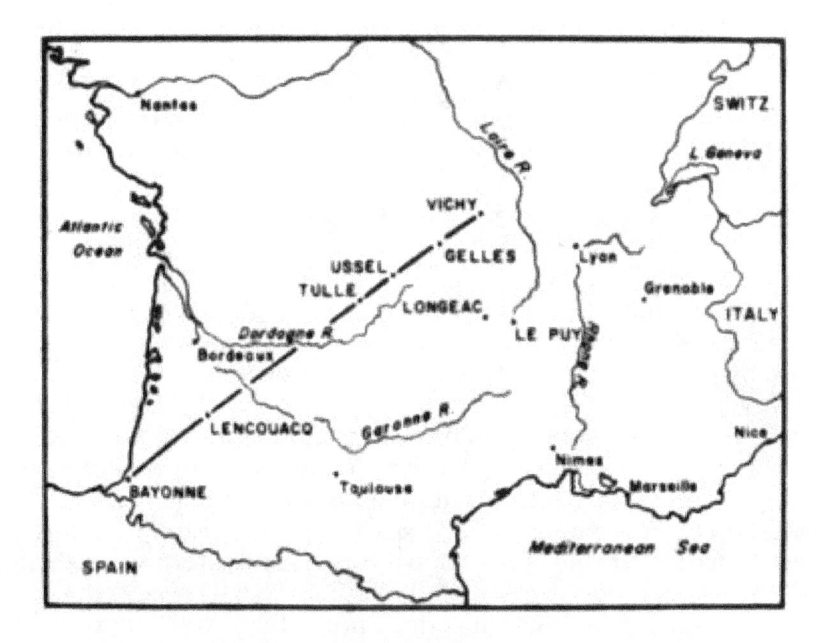

Fig. 2. Eight sightings in France for September 24, 1964 (Michel, 1958) as reported in *France-Soir, Paris-Presse,* and *LaCroix* (Sept. 26 and 28). A ninth sighting at Lantefortain-les-Baroches in northern France is not shown on the map. Sightings at LePuy and Langeac do not occur on the line, but the other six fall so close to the great circle arc indicated that no deviation can be detected on a Michelin map with a scale of 1:1,000,000. Circumstances of the six sightings on the line were very briefly as follows: Vichy, afternoon; Football players practicing in a stadium and spectators saw an elliptical, cigar-shaped object cross the sky swiftly and silently. Gelles, early night: The witnesses saw a luminous, cigar-shaped object cross the sky at fairly high speed and without noise. Ussel, about 11 p.m.: A luminous red object rose above the horizon and dived, at high speed, toward M. Cisterne, who was driving his tractor back to the barn. The object approached so closely that he jumped from the tractor and lay terrified in the field. The object hovered a few yards above the road, and in front of the tractor, remaining motionless for several minutes in complete silence. Surroundings were illuminated with a reddish light. The UFO then flew over the tractor and disappeared over the horizon in a few seconds. Two other people also saw the object, and leaves at the top of an ash tree, near where the object reportedly had hovered, were dried and curled. Tulle, 11 p.m.: M. Besse, with the aid of high-powered binoculars, watched a luminous object move rapidly in the sky, changing color from reddish to white and then to green. Lencouacq, nightfall: A single witness watched a luminous object arrive at high speed in silence, hover above meadow, and then leave again at high speed. Bayonne, afternoon: Many people watched three elliptical objects, metallic in appearance, hover in the sky, and then move away very rapidly. [Salisbury's caption.]

discussion could convince the viewer that he was not observing a spaceship.

In many other cases, data are not quite complete enough to be positive, but one can state with a high degree of certainty that a given UFO was quite likely such and such a conventional object or phenomenon.

In the most interesting cases, the sighting seems absolutely to defy explanation in these terms.

One important conclusion becomes apparent: There is a very high noise level in UFO observations. This is exactly what one might expect. People do become excited by news stories and thereby predisposed to such experiences themselves. We cannot, however, from this high noise level write off the entire phenomenon as belonging to this category of conventional objects misinterpreted. Sagan (1963) attempted to do this by pointing out the great diversity which occurs within the sightings. This might well be only the noise. Even if spaceships are visiting us, many people are still seeing conventional objects and interpreting them as spaceships.

The sightings which do not fit well into the conventional-objects-misinterpreted category have certain characteristics concerned primarily with the detail which is observed and with the nature and reliability of the witnesses. Sometimes other evidence is also available.

If only a moving light is seen at a great distance, one can hardly be tempted to run out and meet our big brothers from Mars. Even a disc or a globe with fairly sharp-appearing edges might well be an optical effect of some sort. A report is more impressive when the object is seen at close hand, especially landed on the ground. A very distinct shape with highly distinct edges, and a solid, often metallic-appearing surface is described. Windows or other markings may be apparent. Lights are frequently an associated part of the observation, and sometimes (both day and night) the brilliance was said to be so high that the observer found it difficult to continue looking at the UFO. Occasionally, one part of the UFO is described as being in motion relative to other parts. The rim of a disc may be rotating around the disc. "Occupants," both humanoid and otherwise, have been reported in conjunction with UFO's, landed and flying. The quality of a sighting is always enhanced when the time of observation is long enough for the observer to consciously consider what he is observing while he is observing it. A light that moves by in less than 5 seconds can hardly produce a very impressive account. In some cases UFO's have been observed for 1 or 2 hours or even longer.

We are primarily concerned with witnesses. Their background and training are especially important, and it is valuable when a single sighting is described by more than one witness. The likelihood of hoax is decreased if the witnesses were unknown to each other before the sighting.

In some cases an account may be supported by various forms of supplementary evidence. There are many cases in which photographs have been taken while a UFO was witnessed by several apparently competent observers. Holes have been left in the ground where a UFO had supposedly landed, or vegetation has been damaged or on fire. Occasionally (rarely), radioactivity has been detected. In one case a fence was magnetized where a UFO had passed over it. Many strange samples have been left, such as liquid residues, "angel's hair," and other materials. In no case, of course, are these things by themselves conclusive, since virtually any sort of evidence could be fraudulently produced. We remain dependent upon the reliability of the witnesses, but sometimes these secondary evidences can contribute to an evaluation of the sighting.

Many radar sightings of UFO's are on file. In a few cases, a UFO has been simultaneously observed by radar and by witnesses, both on the ground and in an aircraft. Menzel and Boyd (1963) have clearly pointed out, however, that radar evidence is far from positive proof. There are many natural atmospheric and other phenomena as well as imperfections in radar instruments which can produce so called radar angels. We must consider the argument from both sides, however. Just because radar angels are not necessarily UFO's, we are still not entitled to conclude that any unusual blip on the screen is a radar angel. We should certainly not conclude that UFO's cannot be extraterrestrial spaceships, because if they were, our radar net would pick them up. The fact of the matter is, our radar net does pick up many returns which are not identifiable in terms of known aircraft (e.g., apparent objects moving several thousand miles per hour through the atmosphere). Many of these are undoubtedly radar angels in the true sense of the word, but we can't say that some are not spaceships from Mars!

A secondary form of supporting evidence is that of pattern. While Sagan (1963) fails to see any pattern because of the noise, other investigators feel that many patterns can be established from the reports. Figure 1 (UFO's and Mars oppositions) is an example of such a pattern. Various other patterns have also appeared. Michel studied the sightings in France in 1954 and found that occasionally (Fig. 2) they appeared to fall upon great circle arcs of the earth's surface (Michel, 1958). It is extremely difficult (Menzel, 1964; Vallée, 1964) to evaluate the significance of such a pattern. In many cases, the lines could be due purely to chance. In the example illustrated, however, with six points upon a single line, one can't help but be somewhat impressed.

If all of these criteria are met for a given UFO report, then it is highly conventional object misinterpreted. The detail usually precludes this. In such a case, the UFO could be an extraterrestrial spaceship or it could fit into one of the categories discussed below.

III. Psychological Phenomena

Can the UFO's be pure figments of the mind hallucinations, dreams, and the like? Probably there are cases which this is the proper explanation, but it is a difficult one to apply to situations in which many witnesses describe with reasonable uniformity a single UFO. In such cases, the psychological explanation would have to fall back on areas such as extrasensory perception, which are really not much more respectable in modern science than spaceships from Mars. In cases in which radar observes the object at the same time that it is observed visually and/or it is photographed, we would have to postulate that one mind can project an object into the heavens in such a way that instruments such as radar and the camera detect it. This would be as exciting as spaceships!

Certainly we do not know all there is to know about the operation of the human mind, so this hypothesis cannot be completely eliminated. And even if the UFO's are spaceships, psychological factors play an important part in the phenomenon. Nevertheless, this hypothesis is not really satisfying. Probably the most detailed study of the UFO's by a psychologist was carried out by Jung (1959). He was able to document a great many extremely fascinating psychological implications of the UFO. *In his final conclusion, however, he could only state that psychological explanations were not sufficient for the phenomenon as a whole* [my emphasis, L.S.].

IV. Hoaxes or Lies

An obvious and straightforward explanation of the UFO's is that the witnesses are lying or that the object is a hoax. Yet the Air Force, always acutely aware of this possibility, explained only a very small percentage of the cases which they investigated in this way. Often it is very difficult to imagine that a hoax is involved. The witnesses give all of the outward signs of being extremely sincere: often they are emotionally upset by their recent experience. Frequently, their background and general competence seem to argue strongly against the idea of hoax. Furthermore, in sightings in which hundreds and even thousands of witnesses are involved (and a few such sightings are on record), one must reject the idea that all the witnesses were lying. If a hoax were involved, it would have to be the object itself.

Before completely eliminating this explanation, we must remember that a hoax can be amazingly effective. I saw the great Blackstone on a stage apparently pass a rapidly moving band saw blade directly through the neck of an assistant in a trance. A block of wood below the neck was sawed in half amidst much noise and flying sawdust. Yet this was admittedly a hoax. Would it be possible to some way cause an illusion in the sky which could completely fool hundreds of witnesses? I cannot absolutely say that it would not. On the other hand, in many cases, producing such an illusion would appear to be almost as great a feat as building a flying saucer itself.

One aspect of the UFO story does seem to be deeply involved in hoax. This is the so-called contactee cult. Many people now located over much of the world claim to have had direct contact with the flying-saucer people. (Adamski and Leslie, 1958; UFO International).

Perhaps the contactee is informed by mental telepathy that he should report promptly to a certain lonely spot in the desert. Upon obeying, he is met by a flying saucer whose occupants are, as a rule, beautifully humanoid and who frequently take him into their confidence by allowing him to photograph themselves and their craft, inviting him in for a look at the control panels, and perhaps taking him for a quick spin, sometimes to Mars or Venus but best of all to the mysterious planet on the other side of the sun, unobservable from mother earth.

Everything about these stories seems to cry hoax. The proof is typically a series of photographs (which could easily be fraudulent) and copious quantities of pseudoscience. Someone who had really contacted visitors from another world should surely be able to do better than that. Why should visitors from another world bother with such obscure representatives of the human race, anyway? Their message is always that man must cease his wars or be destroyed, but why should such an important message be given to someone who is bound to be considered a liar when he delivers it?

It is interesting to consider the possibility that the contactees are genuine. When considering the UFO phenomenon, all sorts of wild alternatives come to mind. If the extraterrestrials wanted to be ignored by the scientific community on earth, they could hardly choose a better and more effective way than the delivering of profound messages to the souls who presently claim contact!

V. Secret Weapons

It is possible that secret devices being tested by earthly governments are misinterpreted as extraterrestrial machines. That this explanation might account for the phenomenon as a whole is, however, quite unreasonable. To begin with, the performances of the UFO's makes our present rockets appear puny indeed. Could any modern government suppress such a capability for nearly 20 years (since 1947)?

Most convincing is the fact that the UFO phenomenon goes way back into history. UFO enthusiasts, for example, often cite the first two chapters of the Book of Ezekiel in the Old Testament as an excellent example of a flying saucer sighting, (Menzel and Boyd, 1963, indicate that it was probably a sundog, but this is a far-fetched explanation for the details reported by Ezekiel.

Vallée (1965) documents the sightings previous to 1947. He states that he has on file more than 300 UFO sightings prior to the 20th century, although he apologizes because he has never had the time to make a thorough search. He considers his cases to be only a small sample of those which might be available. They were carefully chosen for their high quality, roughly conforming to the criteria of good sightings described above. Some 60 of these 300 accounts occurred previous to 1800, and the remainder were recorded during the 19th century. The great majority of these more recent accounts were recorded in the scientific literature, particularly that of astronomy (often in the annals of the various astronomical observatories). It is important to emphasize that these are accounts which are not readily explainable as natural phenomena. Classic, for example, are the observations in Nuremberg (April 14, 1561) and in Basale (August 7, 1566) which have been analyzed in some detail by Jung (1959). Both of these sightings involved large inclined tubes in the sky from which spheres originated, an event occurring sometimes in more recent times (Vallée, 1965, cites 13 examples between 1959 and 1964). Spheres and discs appeared to fight each other in aerial dances. The inhabitants of these two relatively large cities observed this strange phenomenon for a long interval of time on the dates given.

A great attempt was made to consider the scientific accounts of the 19th century in terms of the natural universe. They were referred to as interesting cases of ball lightning or bolide meteors. Nevertheless, the descriptions are of discs and wheels and the like, and the behavior follows very closely that of the modern UFO. These "meteors" would move slowly, appear to hover, change directions, accelerate at great speeds, have an apparent diameter two or three times that of the full moon, etc. In one instance, called ball lightning, an object slowly emerged from the ocean, moved against the wind, hovered close to the ship from which it was observed, and then rushed away in the sky and disappeared in the southeast (for details, see Vallée's book, 1965). [This specific type of UFO or UAP is sometimes known as a USO, or "Unidentified Submerged Object." L.S.]

Sightings during the early part of this century were relatively few. The so-called Miracle of Fatima (Vallée,1965; Walsh, 1947), which took place on October 13, 1917, in a field at Fatima, a small village some 62 miles north of Lisbon, Portugal, is a fascinating tale, to say the least. Today it would be considered a contactee story, since three children were supposedly contacted at monthly intervals (always on the 13th of the month), beginning in May, by a beautiful, "transparent" woman dressed in white, who arrived in a globe of light. Following the first visit, other witnesses besides the children observed strange events (a buzzing noise, etc.), but only the children saw the "vision." At the time of the miracle itself, some 70,000 people were gathered in the field by Fatima to wait for the promised sign. It had been raining when suddenly the "sun" appeared through the dense cloud cover. It was a strange sun, however, looking like a flattened disc with a very definite contour, not appearing as a dazzling object, but rather having a clear, changing brightness which one could compare to a pearl. The disc began turning, rotating with increasing speed as the crowd began to cry with anguish. It then began falling toward the earth "reddish and bloody, threatening to crush everybody under its fiery wake." After an interval of dancing before the crowd, it retreated back through the clouds and disappeared forever. It would be difficult to imagine a sighting which fits the above criteria better than this one. It is also difficult to imagine that the Fatima "sun" was a secret weapon being developed by Russia or the United States!

Some Representative Sightings

Since the study of the UFO's must be based on the reports, let us consider a few sightings exemplifying various points.

1. *The Arnold Sighting, Mt. Rainier, Washington, June 24, 1947.* Although Vallée (1965) calls our attention to a fascinating wave of sightings in Scandinavia during the summer of 1946, it did not

occur to anyone at that time to consider these as extraterrestrial spaceships, but only as secret rockets being developed by Russia or the United States. The current sightings date back to that of Kenneth Arnold. Other better sightings exist for the same period, and even for several days before (as early as April), but Arnold turned his story over to the newspapers, the term "flying saucer" was coined, and the world's attention was focused on the phenomenon.

Arnold saw a formation of silvery discs flying from one peak or ridge to another around Mount Rainier in the state of Washington. By timing the elapsed period from one landmark to another, he was able to estimate their speed at not less than 1,200 miles per hour. Menzel and Boyd (1963) "explain" Arnold's sighting as a mirage brought about by inversion layers in the atmosphere which made the peaks appear to be separated from the mountains below them. Presumably, their apparent motion would be due to the motion of Arnold's airplane. A second explanation proposed by these authors is that Arnold saw the lens-shaped clouds which sometimes occur in the area. They present pictures of such clouds (which look exactly like lens-shaped clouds and not at all like the objects described by Arnold). They further cast aspersions upon Arnold's reliability as a witness by describing in some detail his subsequent actions in attempts to get publicity, etc.

Arnold is supported in his story, however, by the fact that it fits perfectly into the pattern of sightings during that period. Various authors (Hall,1964; Lorenzen, 1962) have summarized these events, and among them a recurring theme is that of formations of silvery discs. Such sightings are rare, or essentially absent, from the reports of more recent years.

It is interesting to wonder about how many apparitions of this type were observed and not reported. My wife's uncle, Mr. Earl Page, then a resident of Kennewick, Washington, had observed on July 12, 1947, a formation of six or eight silvery discs pass by his small airplane at fantastic speed. Mrs. Page and their son were present and saw the objects, which "fluttered as a group for a second or two, and then stabilized . . . alternating between these two modes." The Pages were flying north over Utah Lake. Mr. Page told his story to a few friends who laughed at him, and from then on he mentioned it to no one.

Any one of the sightings of formations of saucer-like objects during the summer of 1947 could perhaps be dis- missed from the mind. A large number of independent sightings, however, produces a pattern which is quite impressive.

2. *The Chesapeake Bay Case, July 14, 1952.* This is one of the best documented sightings on record, involving extremely high speeds and a sharp change of direction (Fig. 3). First Officer William B. Nash and 2nd Officer William H. Fortenberry were flying a commercial plane from New York to Miami, approaching Newport News, Virginia. At 8:12 (just after dark) a brilliant red glow suddenly appeared in the west. It was soon resolved as six coin-shaped objects flying in line formation. They glowed with a brilliant orange-red color on top, were estimated to be 100 feet in diameter and 15 feet thick. They moved rapidly toward the plane, at one point breaking slightly in their perfect formation as the second and third objects wavered slightly and almost overran the leader. They turned in unison on edge and reversed position in the formation, the last object moving up to the front position with the others following. They then abruptly reversed direction, moving off somewhat to the right with the original leader again in the lead position. The turn was executed almost like balls bouncing off a wall with no wavering or arc apparent. Two other objects raced out from beneath the plane and took up positions seven and eight in the formation. They decreased in brilliance just before making the far the brightest as they approached the formation; and for a brief interval or two all eight blinked out and then came back on again. They sped off, climbing to an altitude above that of the airplane, and then one by one but at random their lights blinked off and the sighting was finished. In repeating mentally their observation, the pilots estimated that it had lasted only about 12 to 15 seconds.

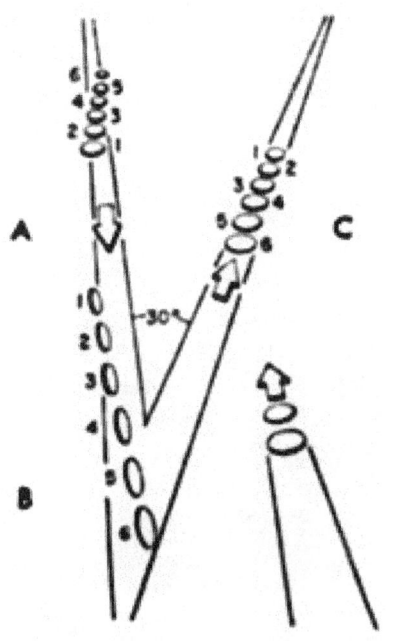

Fig. 3. The action of the Chesapeake Bay discs as reported by Nash and Fortenberry. (a) Discs at first approach. (b) They flip over and reverse order. (c) They change direction, recede, and are joined by two others (from Menzel and Boyd, 1963). [Salisbury's caption.]

Menzel and Boyd (1963), after considering many possible explanations for the sighting, concluded that the pilots must have seen the illuminated discs produced by a red searchlight shining through nearly transparent thin layers of haze. Charles Maney (1965) corresponded with Menzel for several months, considering all of the possible explanations that might come to mind. Apparently Menzel would have readily accepted several explanations if Maney had not one by one clearly demonstrated their implausibility. The pilots themselves thoroughly rejected Menzel's searchlight hypothesis, saying that they were familiar with such phenomena, and this was simply not what they observed. The details described above are certainly difficult to reconcile with a searchlight hypothesis. The extremely short duration of the sighting, however, makes one question the absolute accuracy of the account. Did some points develop a bit with discussion and remembering? Furthermore, the velocities of the UFO's calculated at between 6,000 and 12,000 mph through a dense atmosphere at 2,000 feet and including an instantaneous reversal in direction, are, to say the least, extremely difficult to fit into our present concepts of the universe. Light images could perform these maneuvers, but how could they perform some of the other maneuvers reported by the two pilots?

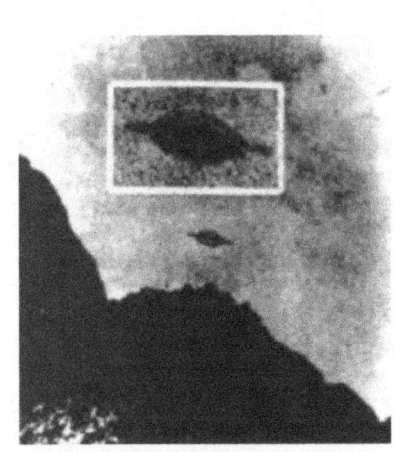

Fig. 4. The third (and best) photograph of a UFO taken by Almiro Barauna on January 16, 1958, near the island of Trindade. The insert is an enlargement of the object to the point of evident grain in the print. Supplied by The Aerial Phenomena Research Organization. [Salisbury's caption.]

This case is presented as an example of the problems met by a UFO researcher. To solve a sighting such as this to everyone's satisfaction would require turning the clock back.

3. *Trindade Island, January 16, 1958.* Figure 4 shows a photograph taken by a Mr. Almiro Barauna, a professional photographer, from the deck of the *Almirante Saldanha*, a Brazilian Navy ship. Several UFO's had been seen in the vicinity of Trindade Island (a Brazilian possession off the coast of Africa) during its reactivation as a naval base in connection with the International Geophysical Year. In the instance reported here, several sailors at

opposite ends of the ship spotted the approaching object simultaneously and began to shout the news to everyone else. Soon the approximately 100 sailors on board, including various officers, were watching the object. Mr. Barauna was preparing to take some photographs and had his camera ready. He shot six frames, of which two failed to show the object. He explained that due to the excitement he was bumped during these two and that they showed only the deck of the ship and the ocean. A darkroom was improvised below deck, the film was developed, and the minute object on it was identified by the sailors (Lorenzen, 1962).

This is an excellent sighting because of the number of witnesses involved and the excellent quality of the pictures (especially the third one, the one shown in the figure). Conventional objects can hardly explain the sighting.

Menzel and Boyd (1963) and apparently the United States Air Force consider the sighting to be a hoax. Of the available hypotheses, only this one and that of extraterrestrial machines seem to apply. The hoax explanation must also probably fail if the object was really witnessed by 100 sailors. Menzel and Boyd tell the story differently (their version is based on a report from astronomer friends of Menzel in Rio de Janeiro who did not personally investigate the incident), saying that only Barauna and two or three of his close friends claim to have seen the object. Yet newspaper reporters interviewed the sailors after the *Almirante Saldanha* landed several weeks later. I have received several reports on the sighting, including a personal conversation with Dr. Alavio Fontes, a medical doctor in Rio de Janeiro who investigated the case exhaustively.

Fig. 5. A faked photograph made by the author of two UFO's over Horsetooth Reservoir in northern Colorado. Inserts are enlargements to the point of evident grain as in Figure 4. Note that the object appears dark against the background, an effect impossible to obtain by the usual double exposure technique. The picture could have been taken and developed in the presence of witnesses—although they would not have seen the objects! [Salisbury's caption.]

These reports fully support the version that virtually all the sailors witnessed the object.

Obviously, our evaluation of the story must hinge upon this aspect. The photographs, although extremely convincing, could be fraudulent. To prove this I spent several days in an attempt to duplicate them and succeeded fairly well as indicated in Figure 5. We are still left with the question of the veracity of witnesses.

4. *St. George, Minnesota, October 21, 1965.* Driving home from a hunting trip, Mr. and Mrs. Arthur Strauch and their son Gary (age 16), and Mr. and Mrs. Donald Grew, all of Gibbon, Minnesota, sighted a hovering object and got out to observe it. Binoculars were used, and Mr. Strauch took one photograph on 804 Instamatic Kodak camera using Ektachrome X film. The photograph is shown in Figure 6 along with an artist's conception of the sighting. The object moved toward the witnesses almost directly overhead, making a high-pitched whining sound and traveling at very high speed. It disappeared in the southeast within seconds.

Fig. 6. Bottom: An artist's conception of the Saint George, Minnesota, sighting of October 21, 1965. Detail in the object represents the impression given by the witnesses. Top: A black and white reproduction of the photograph taken by Mr. Strauch. Photo and painting supplied by the Aerial Phenomena Research Organization. [Salisbury's caption.]

Much detail was observed, several witnesses were present, ample time was available, a photograph was taken, and hence this instance meets the criteria nicely.

5. *Socorro, New Mexico, April 24, 1964.* Patrolman Lonnie Zamora was following a speeder when he saw a blue flame to the southwest. He recognized the area as one which contained a dynamite shack and where teenagers sometimes tried to accelerate their cars up the steep slopes. He decided to investigate. Driving over a mesa (Fig. 7), he caught sight of something which he interpreted as an automobile standing on end with two children or small adults dressed in white clothing and standing by it. He radioed Patrolman Sam Chavez, asking for assistance, and continued down through a gully where he lost sight of the object.

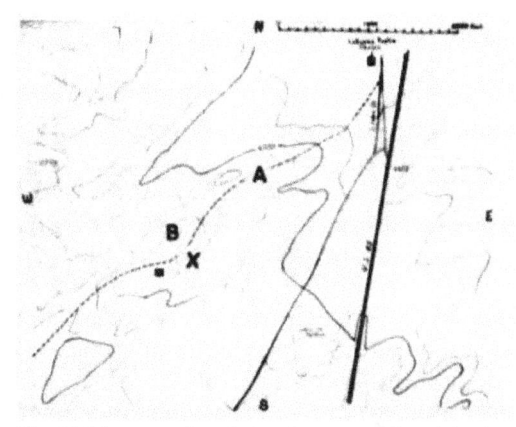

Fig. 7. The terrain of the Socorro, New Mexico, sighting, April 24, 1964. Zamora first sighted the object (at position X) from position A, interpreting it is an automobile. He parked at position B. The square indicates the dynamite shack. [Salisbury's caption.]

Coming up across the next mesa, he parked and got out of his car, moving toward the gully to see the object. It was immediately apparent that he was not observing an automobile wreck. There was a hemispherical object standing on four legs and suddenly [he heard] an ear-splitting roar. Thoroughly frightened, he turned and ran, collided with the hood of his car, and then threw himself on the ground, noticing again that the object was rising in a slanting trajectory toward the south-west. As it rose, it displayed a blue flame.

Upon investigation of the site, four distinct, rather deep impressions were found in the ground where Zamora claims to have seen the landing gear. Two smaller round depressions were in the place where a ladder was placed, leading to a marking on the object which could have been a door. Bushes below where the object had been were burning. Detailed investigations were carried out by the Air Force and by several private flying saucer investigating groups.

The sighting is a good one in terms of detail and primary evidence. A landing with observed humanoid "occupants" is also of interest. It is bad in only one respect; namely, that Zamora was the sole witness (one or two leads appeared, but other witnesses could never be located), but his apparent sincerity was impressive. Investigators studied the surrounding area for tracks of possible perpetrators of a hoax but could find none, although the ground was soft. The sighting is typical of many similar reports, particularly in France and Brazil, but occasionally also in the United States.

6. *Boiani, New Guinea, June 26, 27, 1959.* Sightings were similar on both evenings. On the evening of the 27th, Father W. B. Gill, a teacher and missionary of the Anglican Church in New Guinea,

came out of the dining hall at 6:45 p.m., looked up and saw Venus and then the large sparkling object. While he watched, some 39 others joined him (five were teachers, two were medical assistants, the rest were natives; 28 adult witnesses signed a statement). The object and two others that hovered at a greater distance are shown in the figure (see p. 15) as an artist's conception (the witnesses had no cameras but made pencil sketches during the observation). As the UFO hovered nearby, man-shaped forms appeared on the "top deck" and seemed to be working on something. Occasionally, there was a bright blue, thin beam of light which projected toward the sky. The object itself had an orangeish cast, and the "men" appeared to be dressed in silver suits of some kind. The most seen at one time were four. When one of the figures appeared to glance over the crowd, Gill waved his arm, and the figure returned the gesture. Gill and some of the natives then raised both arms, and two of the figures on the object did the same. The object came lower but did not land. The sighting lasted until 7:20 when the blue spotlight went out and the object moved into a cloud.

The witnesses, the time, and the detail make this an exceptionally good sighting, one of the best on record. The only available explanation other than the spaceship one would seem to be a complex hoax perpetrated by Gill and all of his associates.

7. *Exeter, New Hampshire, September 3, 1965.* A remarkable sighting occurred rather recently in New Hampshire and was studied and documented by several UFO investigators but particularly by Mr. John G. Fuller, a columnist for the *Saturday Review*. He has assembled his results into book form (Fuller, 1966), and a preliminary account was published in *Look Magazine* (February 22, 1966).

The sightings are remarkable not only because of their nature but in a very real sense because of Mr. Fuller's investigation. The basic sightings occurred in the early morning hours (about 2:00 a.m. to 4:00 a.m.). Patrolman Eugene Bertrand of Exeter had checked on a parked car and found a woman who told him that a huge and silent airborne object had trailed her from the town of Epping 9 miles away. The object had brilliant flashing red lights and kept within a few feet of her car. Developing tremendous speed, it had disappeared among the stars. The patrolman could not believe the story and had not even taken the woman's name.

When Bertrand checked into the police station, Norman Muscarello had just arrived and told his story. He had also seen a large dark object with brilliantly flashing lights hover above a field through which he had been walking on his way home. Patrolman Bertrand accompanied him back to the scene. Although nothing could be seen at first, horses on a nearby farm and dogs in nearby houses began making a great deal of noise, and then Muscarello screamed, "I see it, I see it!" Patrolman Bertrand turned and observed the brilliant roundish object moving toward them like a leaf fluttering from a tree. Its red lights along the sides were so brilliant that the entire area was bathed in light. It came within about 100 feet of the two witnesses, hovering with a rocking motion, absolutely silent. The lights seemed to be dimming or pulsating from left to right and then from right to left, taking about 2 seconds for each cycle. The lights were so brilliant that it was difficult to make out the shape of the object itself. It darted, turned rapidly, slowed down, and performed other such maneuvers. Patrolman David Hunt had heard the radio conversation between Bertrand and the station in Exeter and drove to the site, witnessing the object for a few minutes before it disappeared. A [Boeing] B-47 [Stratojet] flew over shortly after, providing an extreme contrast to the object which they had previously witnessed.

In Fuller's study of the case, he was able to find some 60 different people who had witnessed similar objects over a period of several days or weeks in the fall of 1965. Muscarello was so impressed by his sighting that he and his mother waited on a mountainside nearly every evening for 3 weeks following the event. On one of these evenings, they again witnessed the object. Other people in the area would park by high tension lines (in the Exeter sightings, the objects were frequently associated with power lines) and watch for the objects, occasionally being rewarded with the sight of one. This sighting is not only a good one because of the detail, the number of witnesses, and the several occasions involving comfortable intervals of time, but it adds one other extremely encouraging note. If Muscarello and other New Hampshire residents could go out and watch for the objects, occasionally being able to see them, why couldn't properly equipped scientific investigators do the same? Except for the Fatima incident, none of the other sightings have had much element of predictability. This may be simply because we have not taken the time or trouble to really look for it. Yet, it is not uncommon to find cases in which an object seen at one time returned on a later occasion (e.g., the New Guinea instance).

Serious scientific investigation of the phenomenon might be possible if it were desired by the scientific community. If a project could be set up by a number of scientists, it might be feasible to have everything in readiness for another wave of sightings such as that at Exeter or the subsequent one in the Michigan swamps. When such a wave appeared (and the proper kind of publicity might help in detecting it—although it could also contribute to the generation of a wave of fraudulent sightings!), the team of researchers might converge immediately upon the area and carry out some sort of previously planned program of investigation, busy to remain for periods of weeks to months, local people could be hired and trained in the proper techniques. Such a procedure might eventually reward us with the kind of tangible data with which science is used to dealing.[100]— DR. FRANK B. SALISBURY, HEAD OF THE PLANT SCIENCE DEPT., UTAH STATE UNIVERSITY

SALISBURY'S REFERENCES

References

Adamski, George and Leslie Desmond. 1953. *Flying Saucers Have Landed*. British Book Centre, New York. 232 pp.

Fuller, John G. 1966. *Incident at Exeter*. Putnam & Sons, New York.

Hall, Richard H. (ed.). 1964. *The UFO Evidence*. National Investigations Committee on Aerial Phenomena, Washington, D.C. 184 pp.

Hynek, J. Allen. 1966. UFO's merit scientific study. *Science*, 154: 329.

Jackson, F., and P. Moore. 1965. Possibilities of life on Mars. In *Current Aspects of Exobiology*. G. Mamikunian and M. H. Briggs (eds.). Pergamon Press, Inc., London, New York, Germany. Chapter 5.

Jung, C. G. 1959. *Flying Saucers*. Routledge & Kegan Paul Ltd., London. 184 pp.

Keyhoe, Donald E. 1960. *Flying Saucers Top Secret*. Putnam Publishing Co., Longmans, Toronto. 283 pp.

Klass, Phillip J. 1966. Many UFO's are identified as plasmas. *Aviation Week Space Technol.*, Oct. 3, p. 54.

Ley, Willy, and Wernher Von Braun. 1960. *The Exploration of Mars*. The Viking Press, New York. 176 pp.

Lorenzen, Carol E. 1962. *The Great Flying Saucer Hoax*. The William-Frederick Press, New York. 257 pp.

Lorenzen, Carol E. 1966. *Flying Saucers*. Signet Books, New York. 278 pp.

Maney, Charles A. 1965. Donald Menzel and the Newport News UFO. *Fate Magazine*, pp. 64-75 (April).

Menzel, Donald H. 1964. Global orthoteny, new pitfalls. *Flying Saucer Review*, pp. 3-4 (Sept., Oct.).

Menzel, Donald H., and Lyle G. Boyd. 1963. *The World of Flying Saucers*. Doubleday & Company, Inc., Garden City, N.Y. 302 pp.

Michel, Aimé. 1958. *Flying Saucers and the Straight-Line Mystery*. Criterion Books, New York. 285 pp.

Sagan, C. 1963. Unidentified flying objects. *The Encyclopedia Americana*.

Salisbury, F. B. 1962. Martian biology. *Science*, 136: 17-26.

Salisbury, F. B. 1964. Das Mars-

Paradoxon. *Naturwissenschaft und Medizin*, 1 (5): 36-50.

Salisbury, F. B. 1966. Possibilities of Life on Mars. Proceedings of the Conference on the Exploration of Mars and Venus, Virginia Polytechnic Institute, Blacksburg, Va., August 1965, VI: 1-16.

UFO International. Published periodically by the Amalgamated Flying Saucer Clubs of America, Inc. International Headquarters: 2004 N. Hoover St., Los Angeles, Calif.

Vallée, Jacques. 1964. The Menzel-Michel controversy, some further thoughts, *Flying Saucer Review*, pp. 4-6 (Sept., Oct.).

Vallée, J. 1965. *Anatomy of a Phenomenon*. Henry Regnery Co., Chicago, Ill., 210 pp.

Vallée, Jacques, and Janine Vallée. 1962. Mars and the flying saucers. *Flying Saucer Review*, pp. 5-11 (Sept., Oct.).

Vallée, Jacques, and Janine Vallée. 1966. *Les Phénomènes Insolites de L'Espace*. La Table Ronde, Paris. 321 pp.

Walsh, Wm. Thomas. 1947. *Our Lady of Fatima*. MacMillan Co., New York. 228 pp.

Worsley, Peter M. 1959. Cargo cults. *Scientific American*, 200: 117-128.

The End

APPENDICES

APPENDIX A

COMMITTEE ON SCIENCE & ASTRONAUTICS

George P. Miller, California, Chairman

Olin E. Teague, Texas
Joseph E. Karth, Minnesota
Ken Hechler, West Virginia
Emilio Q. Daddario, Connecticut
J. Edward Roush, Indiana
John W. Davis, Georgia
William F. Ryan, New York
Thomas N. Downing, Virginia
Joe D. Waggonner, Jr., Louisiana
Don Fuqua, Florida
George E. Brown, Jr., California
William J. Green, Pennsylvania
Earle Cabell, Texas
Jack Brinkley, Georgia
Bob Eckhardt, Texas
Robert O. Tiernan, Rhode Island
Bertram L. Podell, New York
James G. Fulton, Pennsylvania
Charles A. Mosher, Ohio
Richard L. Roudebush, Indiana
Alphonzo Bell, California
Thomas M. Pelly, Washington
Donald H. Rumsfeld, Illinois
Edward J. Gurney, Florida
John W. Wydler, New York
Guy Vander Jagt, Michigan
Larry Winn, Jr., Kansas
Jerry L. Pettis, California
D. E. (Buz) Lukens, Ohio
John E. Hunt, New Jersey

Charles F. Ducander, *Executive Director and Chief Counsel*
John A. Carstarphen, Jr., *Chief Clerk and Counsel*
Philip B. Yeager, *Counsel*
Frank R. Hammill, Jr., *Counsel*
W. H. Boone, *Chief Technical Consultant*
Richard P. Hines, *Staff Consultant*
Peter A. Gerardi, *Technical Consultant*
James E. Wilson, *Technical Consultant*
Harold A. Gould, *Technical Consultant*
Philip P. Dickinson, *Technical Consultant*
Joseph M. Felton, *Counsel*
Richard E. Beeman, *Minority Staff*
Elizabeth S. Kernan, *Scientific Research Assistant*
Frank J. Giroux, *Clerk*
Denis C. Quigley, *Publications Clerk*

APPENDIX B

BIOGRAPHY OF DR. J. ALLEN HYNEK

Born in Chicago, Ill., 1910. B.S. University of Chicago, 1931; Ph.D. (astrophysics) 1935.

Professor Astronomy, Chairman of the Department and Director of Dearborn Observatory, Northwestern University, 1960 to present [1968].

Chief of the Section, Upper Atmosphere Studies and Satellite Tracking and Associate Director, Smithsonian Astrophysical Observatory, 1956-60.

Professor, Astronomy, 1950-56, Ohio State University.

Instructor, Physics and Astronomy, Ohio State University, 1935-41; Asst. Prof. 1941-45; Associate Professor 1946-50.

Asst. Yerkes Observatory, University of Chicago, 1934.

Astronomer, Perkins Observatory, Ohio State, 1935-56.

Assistant Dean of the Graduate School 1950-53.

Supervisor of Technical Reports, Applied Physical Laboratory, Johns Hopkins University, 1942-46.

Visiting Lecturer, Harvard University, 1956–60.

Civilian with U.S. Navy 1944.

Scientific Societies: American Association for the Advancement of Science; Astronomy Society (secretary).

Specialty: Astrophysics.

Fields of Interest: Stellar spectroscopy; F type stars; stellar scintillation.[102]

APPENDIX C

BIOGRAPHY OF DR. JAMES E. MCDONALD

Born: Duluth, Minn., May 7, 1920.
Home Address: 3461 East Third St., Tucson, Ariz.
Education: University of Omaha, Omaha, Nebraska, B.A. (Chemistry) 1942.
    Massachusetts Institute of Technology, Cambridge, Mass. (Meteorology) 1945.
    Iowa State University, Ames, Ia., Ph.D. (Physics) 1951.
Professional Career: Instructor, Dept. of Physics, Iowa State University, 1946-49.
    Assistant Professor, Dept. of Physics, Iowa State University, 1950–53.
    Research physicist, Cloud Physics Project, University of Chicago, 1953-54.
    Associate Professor, Dept. of Physics, University of Arizona, 1954–56, Professor, 1956-57.
    Associate director, Institute of Atmospheric Physics, University of Arizona, 1954-57.
    Professor, Dept. of Meteorology, and Senior Physicist, Institute of Atmospheric Physics, 1958 to present.
Other activities:
    U.S. Navy, 1942-45, naval intelligence and aerology.
    Member, Panel on Weather and Climate Modification, National Academy of Sciences, 1965-present.
    Member, ESSA-Navy Project Stormfury Advisory Panel, 1966-present.
    Member, American Meteorological Society Commission on Publications, 1966-present.
    Member, Advisory panel for weather modification, National Science Foundation, 1967-present.
Professional memberships:
    American Association for the Advancement of Science, American Meteorological Society, Sigma Xi, American Geophysical Union, Royal Meteorological Society, Arizona Academy of Science, American Association of University Professors.
Personal: Married, 1945, Betsy Hunt: six children.
Fields of special interest: Atmospheric physics, physics of clouds and precipitation, meteorological optics, atmospheric electricity, weather modification, unidentified aerial phenomena.[103]

APPENDIX I

BIOGRAPHY OF DR. CARL SAGAN

Dr. Carl Sagan is Associate Professor of Astronomy in the Center for Radiophysics and Space Research at Cornell University, Ithaca, New York. He received his A.B. and B.S., an M.S. in Physics and his Ph.D. in Astronomy and Astrophysics, all from the University of Chicago. Since then he has held positions at the University of California, Berkeley; at Stanford University Medical School (as Assistant Professor of Genetics); and at Harvard University and the Smithsonian Astrophysical Observatory. Dr. Sagan's major research interests are on the physics and chemistry of planetary atmospheres and surfaces, the origin of life on earth, and exobiology. He has played a leading role in establishing, for example, that the surface of Venus is very hot and that major elevation differences exist on Mars, and has been a principal exponent of the view that organic molecules are to be found on Jupiter. Dr. Sagan has served on many advisory groups to the National Aeronautics and Space Administration and to the National Academy of Sciences, as well as such international organizations as COSPAR and the International Astronomical Union. He was a member of the Committee to review Project Blue Book for the Air Force Scientific Advisory Board. A winner of the Smith Prize at Harvard in 1964 and Condon Lecturer in the State of Oregon in 1968, Dr. Sagan is shortly to assume additional duties as Editor of the planetary sciences journal, *ICARUS*. He has been active in educational innovations, regularly teaches in the South, and is a lecturer in the astronaut training program in Houston. In addition to well over a hundred scientific papers, and several articles written for the Encyclopedias Britannica and Americana, Dr. Sagan is co-author of *The Atmospheres of Mars and Venus* (1961), *Planets* (1966), and *Intelligent Life in the Universe* (1966).[104]

> [Author-editor's note: Sagan would go on to create and host the popular 13-part 1980 TV documentary *Cosmos: A Personal Journey*, along with the bestselling companion book, *Cosmos*. L.S.]

APPENDIX E

BIOGRAPHY OF DR. ROBERT L. HALL

Born February 25, 1924, at Atlanta, Georgia. Married; 3 children.

EDUCATION
Yale University, 1941-42. B.A. 1947.
University of Stockholm, Sweden, 1947-48.
University of Minnesota, 1949–52. M.A., 1950. Ph.D., 1953.

PROFESSIONAL EXPERIENCE
1. Instructor, Extension Division, University of Stockholm, Sweden, 1948.   2. Research Assistant, University of Minnesota, 1950-52.

3. Social Psychologist in the Air Force Personnel & Training Research Center, 1952-1957. Engaged in research on performance of bomber crews, the role of the aircraft commander, and processes of evaluation of small teams.

4. Assistant Professor (1957-1960) and Associate Professor (1960–62) of Sociology. Teaching social psychology, especially the processes of mass communication and opinion change. Conducting research on social psychological aspects of higher education and effects of social interaction on the learning process.   5. Program Director for Sociology and Social Psychology, National Science Foundation, 1962-1965. Administered a program of research grants and related activities to strengthen Sociology and Social Psychology in universities in the United States and to bolster understanding in these fields through basic research.

6. Associate Professor of Sociology and Psychology (1965-66) and Professor of Sociology and Head of the Department of Sociology (since 1966), University of Illinois at Chicago Circle.

PUBLICATIONS
A number of articles in Sociological and Psychological journals and chapters in professional books. A few selected publications are listed below:

"Social influence on the Aircraft Commander's role," *American Sociological Review* 1955, 20, 292–299.

"Military Sociology," 1945-1955. Chapter in *Sociology in the United States of America*, ed. by Hans Zetterberg, Paris: UNESCO, 1966.

"Group performance under feedback that confounds responses of group members." *Sociometry*, 1957, 20, 297–305.

"The informal control of everyday behavior." Chapter in *Controlling Human Behavior*, ed. by Roy Francis, Social Science Research Center,

University of Minnesota; 1959.

"Two alternative learning in interdependent dyads." Chapter 12 in *Mathematical Methods in Small Group Processes*, ed. by Joan Criswell, H. Solomon, and P. Suppes, Stanford Univ. Press: 1962.

"The educational influence of dormitory roommates." *Sociometry*, 1963, 26, 294-318 (with Ben Willerman).

"The effects of different social feedback conditions upon performance in dyadic teams." Chapter in *Communication and Culture*, ed. by A. G. Smith, 1966, 353-364.[105]

APPENDIX F

BIOGRAPHY OF DR. JAMES A. HARDER

Born Fullerton, Calif., 1926.

B.S. California Institute of Technology, 1948; Ph.D. (fluid mechanics), 1957;

Associate Professor of Civil Engineering, University of California, Berkeley, 1962 to present;

Assistant Professor, Hydraulic Engineering, 1957-62; Resident Engineer, 1952–57;

Design Engineer, soil conservation service, U.S. Department of Agriculture, 1948-50;

U.S. Navy, 1944-45;

Scientific Societies: Fellow of the American Association for the Advancement of Science, Society of Civil Engineering;

Specialty: Engineering Science;

Fields of Interest: Hydraulic systems analysis; surface water hydrology; analog simulation.[106]

APPENDIX 6

BIOGRAPHY OF DR. ROBERT M. L. BAKER, JR.

Dr. Baker is a 36 year old scientist who received his BA with Highest Honors in Physics and Mathematics at UCLA in 1954, and was elected to Phi Beta Kappa. In 1956 he was granted a MA in Physics, and was the recipient of the UCLA Physics Prize. In 1958 Dr. Baker received a PhD in Engineering, which was the first of its kind to be granted in the nation with a specialty in Astronautics.

With respect to his academic background, Dr. Baker was on the Faculty of the Department of Astronomy at UCLA from 1959 to 1963. Since that time he has been on the Faculty of the Department of Engineering at UCLA where he currently offers courses in astronautics, fluid mechanics, and structural mechanics.

Dr. Baker is an internationally recognized expert in various fields of science and engineering. He was a research contributor to the development of preliminary orbit determination procedures utilizing radar data, astrodynamic constants, near free-molecular flow drag—all utilized in the nation's space programs. He has also developed unique theories in the area of hydrofoil marine craft design.

In private industry Dr. Baker has initiated, supervised, and conducted research programs in astronautics, physics, fluid mechanics, mathematics, and computer program design. He has contributed to problem definition and analysis of scientific and engineering problems in both industrial and military projects.

Dr. Baker's industrial career began in 1954 as a consultant to Douglas Aircraft Company. Between 1957 and 1960 he was a Senior Scientist at Aeronutronic-Philco-Ford. While in the Air Force during 1960 and 1961, he was a project officer on a number of classified Air Force projects. Between 1961 and 1964 he was the head of Lockheed's Astrodynamics Research Center, where he directed the efforts of approximately 25 scientists in various scientific areas. In 1964 Dr. Baker joined the Computer Sciences Corporation (CSC), first as Associate Manager for Research and Analysis, and later as the Senior Scientist of CSC's System Sciences subdivision. In this latter capacity he is currently involved in several Air Force, Navy, and NASA projects.

Dr. Baker represented the United States Air Force at the International Astronautical Federation meeting in Stockholm, Sweden in 1961, represented the United States at the International Union of Theoretical and Applied Mechanics European Conferences in 1962 and in 1965 and was an invitee to the Astronomical Councile of the Academy of Sciences of USSR in Moscow in 1967. He was voted an Outstanding Young Man of the Year by the Junior Chamber of Commerce in 1965. From 1963 to 1964 he was the National Chairman of the Astrodynamics

Technical Committee of the American Institute of Aeronautics and Astronautics and is currently a member of Computer Sciences Technical Committee.

Dr. Baker has been the Editor of the *Journal of the Astronautical Sciences* since 1963. He was the joint editor of the Proceedings of the 1961 International Astronautical Federation Congress and the senior author of the first textbook on astrodynamics: An Introduction to Astrodynamics published in 1960. Dr. Baker is the author of four books and over 70 technical papers (see [his] Appendix 2).

Dr. Baker's professional society memberships include the American Association for the Advancement of Science, Phi Beta Kappa, Sigma Xi, Sigma Pi Sigma, American Astronautical Society (Fellow), British Interplanetary Society (Fellow), American Institute of Aeronautics and Astronautics (Associate Fellow and member of the Computer Sciences Technical Committee), British Astronomical Society (Fellow), American Astronomical Society, American Physical Society, and Meteoritical Society.

His active security clearance is top secret.[107]

APPENDIX H

BIOGRAPHY OF DONALD H. MENZEL

Dr. Donald H. Menzel, a native of Colorado, received his Ph. D. from Princeton University in 1924. After one year as instructor at the University of Iowa, another year as Assistant Professor at Ohio State University, Dr. Menzel went to Lick Observatory in 1926, as Assistant Astronomer. While at Lick he participated in many observing programs with the large telescopic equipment. His major work, however, was in the interpretation of the spectrum of the atmosphere of the sun, from photographs taken at various total solar eclipses. He participated in the observation of two such eclipses, in the years 1930 and 1932.

In the fall of 1932, Dr. Menzel came to Harvard University, where he has been ever since, except for three years of service as a Commander in the U.S. Navy, during World War II. His studies have covered a large number of fields, from pure physics to pure astronomy. Of special concern has been the sun itself, in which field he is a recognized authority. His studies have employed a combination of observation and theory. In 1936 he was director of the Harvard-M.I.T. eclipse expedition to USSR. In 1945 he directed the joint U.S.-Canadian eclipse expedition to Saskatchewan. He has also observed the total eclipses of 1918, 1923, 1954, 1959, 1961, 1963 and 1967. In an attempt to obtain basic information outside of a total solar eclipse, Dr. Menzel developed the first coronagraph in the United States and established the station at Climax, Colorado, where it is now known as the High Altitude Observatory. Originally operated jointly by Harvard and the University of Colorado, this scientific institution is now wholly under the jurisdiction of the latter university. The observations of solar activity obtained at Climax had an immediate application to problems of solar-terrestrial relationships, especially on the propagation of radio waves. To expand the work in this field and to provide for more nearly unbroken records of solar activity, after World War II, Dr. Menzel suggested that the Air Forces establish a second solar station. After several years of site surveying, Sacramento Peak Observatory was established near Alamogordo, New Mexico on a mountain some 5000 feet above the Tularosa Basin, overlooking the White Sands Proving Ground and the Holloman Air Force Base. The large instruments, including the 16-inch coronagraph, were all designed and built under Harvard auspices, collaboratively with scientific personnel from the High Altitude Observatory. In 1956, with Air Force sponsorship, Harvard built a Solar Radio Observatory near Fort Davis, Texas, to record the radio waves of solar origin. The data from these observatories are revolutionizing our knowledge of the sun and solar activity.

In 1952 Dr. Menzel became acting Director of Harvard College Observatory and, in 1954, was advanced to the Directorship. He resigned as Director on March 31, 1966. In Harvard University he is Paine Professor of Practical Astronomy and Professor of Astrophysics. On July 1, 1966 he also accepted appointment as Research Scientist on the Smithsonian Astrophysical Observatory staff. He has lectured extensively in Spanish throughout Latin America. In 1963 he was a Visiting Professor at the University of Chile, and served as State Department Specialist for Latin America in 1964.

From 1954-56 he was President of the American Astronomical Society. He is Vice President of the American Philosophical Society, a member of the National Academy of Sciences, the American Academy of Arts and Sciences, and a large number of other professional organizations. He is a senior member of the Institute of Electrical and Electronics Engineers and a Foreign Associate of the Royal Astronomical Society. From 1948-1955 he was President of the Commission on Solar Eclipses of the International Astronomical Union, and from 1964-1967was President of the Commission on the moon. He is also a member of the International Radio Scientific Union (URSI), and the International Union for Geodesy and Geophysics. He has been Chief Scientist of GCA Corporation since 1959.He was a member of the Board of Directors of the Association of Universities for Research in Astronomy, Inc. from 1957-1966. (In 1954 he received the honorary degree of D.Sc. from the University of Denver, his Alma Mater, and their John Evans Award in 1965.)

Dr. Menzel has been a prolific writer. His books, articles, and scientific papers cover a broad field, and have been translated into many languages. He has even ventured briefly into the realm of science fiction.

His book, *Our Sun*, published by the Harvard University Press, is one of the so-called Harvard series on astronomy and a standard reference work, despite the fact that it is written in popular style for the general public. Two popular books on the subject of Flying Saucers, the second written with Lyle Boyd and published in 1963, analyze the various reports and demonstrate conclusively that these highly controversial "objects" are only various manifestation of different natural phenomena, not machines from outer space. His first book on Flying Saucers was translated into Russian. He has lectured extensively on UFO's around the world, including South America and Mexico.

Dr. Menzel's interest in promoting good writing by scientists, led him to produce *Writing a Technical Paper*, co-authored by Professor Howard Mumford Jones of the Harvard Department of English and Lyle Boyd, a science editor. He is also author of a *Field Guide to the Stars and Planets*, a popular handbook for beginning astronomers.[108]

APPENDIX I

BIOGRAPHY OF STANTON T. FRIEDMAN

Born July 29, 1934, Elizabeth, New Jersey.

B. Sc.—Physics, M. Sc.—Physics, University of Chicago 1955, 1956.

Since 1966—Westinghouse Astronuclear Laboratory, Pittsburgh; NERVA nuclear rocket Program—Fellow Scientist concerned primarily with radiation shielding experiments and nuclear instrumentation.

1963-1966—Allison Division, General Motors, Indianapolis, Indiana. Military Compact Reactor program (responsible for all shielding aspects), magnetohydrodynamics, desalination, other projects.

1959-1963—Aerojet General Nucleonics, near San Francisco. Development of various nuclear systems for space and terrestrial applications; Fusion propulsion for space, consultant on radiation shielding.

1956-1959—General Electric, Aircraft Nuclear Propulsion Department, Cincinnati. Experimental and analytical aspects of radiation shielding for nuclear aircraft.

Mr. Friedman has a relatively unique background in advanced technology, having been actively involved in the development of all of the following advanced systems: nuclear aircraft, nuclear power for space, terrestrial nuclear power, nuclear rockets, fusion rockets.

Professional affiliations include the American Physical Society, the American Nuclear Society, the American Institute of Aeronautics and Astronautics, the Aerial Phenomena Research Organization, the National Investigations Committee on Aerial Phenomena. Mr. Friedman is on the Board of Directors of the UFO Research Institute of Pittsburgh and on the Standards and Program Committees of the Shielding Division of the American Nuclear Society.

Mr. Friedman has presented papers at technical society meetings and has chaired sessions at such meetings. He has written numerous classified and unclassified reports and has published articles on UFOs as well as on radiation shielding.

Mr. Friedman has made dozens of radio and TV appearances across the United States and in Canada. These include the Joe Pyne Show (Los Angeles-radio), Long John Nebel (New York City), the J. P. McCarthy Show in Detroit, all four TV stations in Pittsburgh, and others in Raleigh, Akron, Detroit, Baltimore, Toronto, Waco, Phoenix, Calgary, Albuquerque, etc. [Note: Mr. Friedman also wrote a backcover blurb for my book, *UFOs and Aliens: The Complete Guidebook*. The author-editor, L.S.]

Mr. Friedman, his wife, and three children reside at 702 Summerlea Street in Pittsburgh, Pennsylvania.[109]

APPENDIX J

BIOGRAPHY OF R. LEO SPRINKLE

R. LEO SPRINKLE, PH. D., UNIVERSITY OF
WYOMING, LARAMIE, WYO.

Name: Ronald Leo Sprinkle.
Born: August 31, 1930, Rocky Ford, Colorado, U.S.A.
Education:
  Elementary Education: Washington School, Rocky Ford, Colorado
  High School: Rocky Ford High School, Rocky Ford, Colorado
  Academic scholarship received from the University of Colorado:
  B.A. in Psychology, Education, Sociology, and History, University of Colorado, August 1962.
  M. P. S. (Master of Personnel Service) in Counseling, University of Colorado, August 1956.
  Ph. D. in Counseling and Guidance, University of Missouri, August 1961.
Professional Experience:
  Residence hall supervisor, Men's Residence Halls, University of Colorado, 1954-956.
  Instructor-Counselor, Counseling Services, Stephens College, Columbia, Missouri, 1956-1959.
  Acting Director of Extra Class Activities, Stephens College, Columbia, Missouri, 1959-1961.
  Assistant Professor of Psychology, University of North Dakota, Grand Forks, North Dakota, 1961-1964.
  Assistant Director of Counseling Center, University of North Dakota, 1962-1963.
  Director, Counseling Center, University of North Dakota, 1963-1964.
  Associate Professor of Guidance Education, University of Wyoming, 1964-1965.
  Counselor and Assistant Professor of Psychology, University of Wyoming, 1965-67.
  Counselor and Associate Professor of Psychology, University of Wyoming, 1967.
Professional Affiliations:
  Member of the American Psychological Association, (Divisions of Counseling Psychology, and the Society for the Psychological Study of Social Issues).
  Life member of the American Personnel and Guidance Association, (Divisions of American College Personnel Association, Association of Counselor Education and Supervision,

Association for Measurement and Evaluation in Guidance, and professional member of National Vocational Guidance Association).

Licensed as Professional Psychologist in Wyoming, January 1, 1966.

Certified as Counseling Psychologist by the Board of Examiners, North Dakota Psychological Association, May 11, 1962.

Member of American Association of University Professors.

Member of Psi Chi (Psychology Honorary).

Associate Member of Parapsychological Association.

Member of Wyoming Personnel and Guidance Association.

Member of Wyoming Psychological Association.

Member of American Society of Clinical Hypnosis.

Life Member of American Association for the Advancement of Science.

Professional Organizational Activities:

State delegate to the annual meeting of the American Association of State Psychology Boards, St. Louis, Missouri, August, 1962.

Secretary, Board of Examiners, North Dakota Psychological Association (NDPA), 1962-63.

President, Board of Examiners, NDPA, 1963–64.

Member of Commission VIII, Student Health Programs, of the American College Personnel Association, APGA, 1962 to present. (Chairman of symposium sponsored by Commission VIII at the APGA Convention, Minneapolis, Minnesota, April 13, 1965.)

Publications:

"Measured vocational interests and socioeconomic background of college students." *The College of Education Record*, University of North Dakota, 1962, 47, No. 4, 54-56.

"Counselor competence and the nature of man." *The College of Education Record*, University of North Dakota, 1962, 47, No. 5, 70-73.

With Gillmor, D. "A first step in evaluation." *The Superior Student*. (InterUniversity Committee on the Superior Student, Boulder, Colorado) 1964, 6, No. 2, 30-33.

"Psychological implications in the investigation of UFO reports." In Lorenzen, L. J. and Coral E. *Flying saucer occupants*. N.Y.: A Signet Book, 1967. Pp. 160-186.

Professional Research and Writing:

"Permanence of measured vocational interests and socio-economic background": Unpublished Ph.D. dissertation, sponsored by Dr. Robert Callis, University of Missouri, 1961.

"Student health and demands for academic accomplishment: an attitude survey of students at the University of North Dakota." Unpublished manuscript presented at the American Personnel and Guidance Association (Commission VIII, Student Health Programs), Boston, Massachusetts, April 8, 1963.

"A hypothetical view of communication and human evolution."

Unpublished manuscript presented at the North Dakota-South Dakota Psychological Association Convention, May, 1963.

Received a small grant ($278.00) from the Grants-In-Aid Committee, Society for the Psychological Study of Social Issues, in support of a study to investigate the relation of personal attitudes and scientific attitudes. (Survey of persons interested in UFO reports.) (Manuscript has been rejected by the Journal of Social Issues.)[2][3]

Military History:

United States Army, Artillery, 1952-1954; graduated as Honor Student No. 1, Class No. A-5324, 7th Army NCO Academy, Munich, Germany; served as corporal in 194th F.A.Br., Wertheim, Germany.

Personal Hobbies:

Reading; composing verses and songs; observing and participating in athletics; travel; home work-shop activities.

Personal Information:

Married on June 7, 1952 to Marilyn Joan Nelson (born in Gurley, Nebraska, on April 28, 1930; and graduated from the University of Colorado with a B. Mus. Educ. in June 1953); oldest son, Nelson Rex Sprinkle, born February 20, 1958; younger son, Eric Evan Sprinkle, born on March 22, 1961; youngest son, Matthew David Sprinkle, born on May 4, 1964; daughter, Kristen Martha, born on April 16, 1967.[110]

---

2. "Psychological Problems in Gathering UFO Data"; a paper presented in a symposium sponsored by Division 21, Engineering Psychology, at the American Psychological Association convention, Washington, D.C., September 4, 1967. [Sprinkle's footnote.]

3. "Some Uses of Hypnosis in UFO Research"; a paper which will be presented at the annual meeting of the American Society of Clinical Hypnosis; Chicago, Ill.; October 10-13, 1968. [Sprinkle's footnote.]

APPENDIX K

BIOGRAPHY OF GARRY C. HENDERSON

DR. GARRY C. HENDERSON, SENIOR RESEARCH SCIENTIST, SPACE SCIENCES, FORT WORTH, TEX.

Garry C. Henderson was born in Brownwood, Tex., on October 23, 1935. He received the B.S. degree in mathematics from Sul Ross State College, Alpine, Tex., in 1960, the M.S. degree from Texas A&M University, College Station, in geophysical oceanography in 1962, and the Ph.D. degree in geophysics from Texas A&M University in 1965.

He held the post of Research Assistant in the Texas A&M Research Foundation from 1960-1963. During this time he served as a technician, operator, and data interpreter with the LaCoste-Romberg S-9 Sea-Surface Gravity Meter. From February-June 1962 he worked for Dr. G. P. Woollard aboard the NSF Polar Research Vessel *Eltanin* where he was in charge of testing the S-9 gravity meter and interpreting meter performance. He was an IBM 709 operator and senior programmer in the physical sciences for the Texas A&M Data Processing Center until the latter part of 1964. He received his Ph.D. while in the position of Chief Marine Geophysicist for Oceanonics, Inc., where he worked in techniques, instrumentation, and interpretation in the fields of gravimetry, magnetics, electrical methods, and computer operations. He joined the Applied Research Group of the Fort Worth Division of General Dynamics in the latter part of 1965. Since that time he has been engaged in studies of the methodology, instrumentation, and interpretation of geophysical investigations on lunar and planetary surfaces, particularly in the fields of gravimetry and electrical methods. He is currently Project Leader on the lunar surface gravimeter/surveying system and leader of the space sciences section of the applied Research Group.

Dr. Henderson is a member of Alpha Chi, the American Geophysical Union, the Society of Exploration Geophysicists, the American Astronautical Society, the Marine Technology Society, the Working Group on Extraterrestrial Resources, and Sciences Subcommittee Chairman of the Marine Geodesy Committee.[111]

APPENDIX L

BIOGRAPHY OF ROGER N. SHEPARD

DR. ROGER N. SHEPARD

Roger N. Shepard is professor of psychology at Stanford University. Previously he was professor and then director of the psychological laboratories at Harvard and, for eight years before that, member of technical staff and then department head at the Bell Telephone Laboratories. He obtained his Ph.D. in experimental psychology from Yale (1955), and has published some 30 technical and scientific papers on human perception and memory and on computer methods for discovering patterns in large arrays of data. Although he has had a long-standing interest in the problem of UFOs, this is his first paper on this particular subject.[112]

APPENDIX M

BIOGRAPHY OF DR. FRANK B. SALISBURY

Frank Boyer Salisbury.
Born: Provo, Utah, August 3, 1926. Married 1949; six children.
B.S. University of Utah 1951; M.A. Utah 1962; Ph.D. California Institute of Technology 1955.
Army Air Force 1945.
Field is plant physiology.
Assistant Professor of Botany, Pomona College 1954-5.
Assistant Professor of Botany, Colorado State 1955-61.
Full Professor, Colorado State 1961-66.
Professor and Head, Dept. of Plant Science, Utah State University 1966–.
Member: AAAS; Society of Plant Physiology; Ecological Society; Astronautical Society.
Interests: Physiology of flowering; space biology; physiological ecology.

NOTES

ALL FOOTNOTES, ENDNOTES, & NOTES IN GENERAL ARE MINE, UNLESS OTHERWISE INDICATED. L.S.

1. Hynek's 1972 categorization system was fully embedded in public awareness by the popular 1977 film, *Close Encounters of the Third Kind*.
2. For more on the worldwide government-UFO coverup, see my books, *UFOs and Aliens: The Complete Guidebook*, and *The Martian Anomalies: A Photographic Search for Intelligent Life on Mars*.
3. Hearings, pp. 1-2.
4. Hearings, pp. 3-17.
5. Hearings, pp. 18-32.
6. Hearings, pp. 32-85.
7. Hearings, pp. 85-92.
8. Source: Tacker, Lawrence, J., *Flying Saucers and the U.S. Air Force* (Princeton, N.J. (1960), and Library of Congress, *Facts About Unidentified Flying Objects* (Washington 1966). [Sagan's note.]
9. In Encyclopedia Americana (New York: Groller) and In Bull. Atom. Sci., 23, (6), 43, 1967. [Sagan's note.]
10. Hearings, pp. 95-98.
11. Hearings, pp. 99-112.
12. Hearings, pp. 112-125.
13. For the Utah film, see Baker and Makemson (1967); for the Montana film, see Baker (1968a). This latter reference is included in app. 4 to this paper. [Baker's note.]
14. Except in app. 3 to this report—a paper supplied by Dr. Sydney Walker III, concerning a hypothetical case. [Baker's note.]
15. Hearings, pp. 126-141.
16. At the time he had already logged some 2,200 hours as a chief photographer with the Navy. [Baker's note.]
17. The images on the "Utah" film appear to be a little brighter. However, possible variations in development techniques would not allow quantitative analysis in this regard. [Baker's note.]
18. Bits of aluminum foil dumped overboard by planes, often utilized as a countermeasure against antiaircraft radar. This material might possibly be in the form of large ribbons several feet long and several inches across. [Baker's note.]
19. The dimensions refer to wing spread. The actual exposed white area of a bird is usually less and depends upon the perspective of the observer. This difference has been roughly accounted for in the data given, however, if the body were the principal reflector the distance given should be reduced by a factor of 2 or 3. [Baker's note.]
20. Hearings, pp.137 -151.
21. Hearings, pp. 190-196.
22. Hearings, pp. 199-205.
23. Hearings, p. 205.
24. Hearings, pp. 196-197.
25. Letter, the Honorable Edward Rousch, to S. T. Friedman, July 1968. [Friedman's note.]
26. 2. Friedman, S. T., "Flying Saucers Are Real," *Astronautics and Aeronautics*, February 1968, p. 16. [Friedman's note.]
27. Davidson, L., *Flying Saucers: An Analysis of Project Blue Book Special Report No. 14*, 1966; $4. [Friedman's note.]
28. Ruppelt, E. J., *The Report on Unidentified Flying Objects*, Doubleday, $5.95, 1956; Ace, $0.50. [Friedman's note.]
29. Hynek, J. A., *Saturday Evening Post*, December 17, 1966. [Friedman's note.]
30. Olsen, T., *The Reference for Outstanding UFO Reports*, 1966; $5.95. [Friedman's note.]

31. Hall, R., "The UFO Evidence," 1964, NICAP; $5. [Friedman's note.]
32. Vallee, J., *Anatomy of a Phenomenon*, 1965; Regnery, $4.95; Ace, $0.60. [Friedman's note.]
33. Lorenzen, C. and J., *UFO's Over the Americas*, 1968; Signet, $0.75. [Friedman's note.]
34. Letter, Robin E. Sanborn (former chief, Film Evaluation Section, Smithsonian Astrophysical Observatory) to Los Angeles Subcommittee, National Investigations Committee on Aerial Phenomena, dated July 5, 1966.
35. Baker, R. M. L., Jr., "Future Experiments on Anomalistic Observational Phenomena," *Journal of the Astronautical Sciences*, Vol. XV, No. 1, January 1968. 11: 44-45. [Friedman's note.]
36. Markowitz, "The Physics and Metaphysics of Unidentified Flying Objects," *Science*, 157, pp. 1274-1279 (1967). [Freidman's note.]
37. "A Fresh Look at Flying Saucers," *Time Magazine*, August 4, 1967. [Friedman's note.]
38. Spencer, D. F. and Jaffe, L. D., "Feasibility of Interstellar Travel," *Acta Astronautics*, Vol. IX Fasc. 2, 50-58, 1963. [Friedman's note.]
39. Spence, R. W., "The Rover Nuclear Rocket Program," *Science*, 160: 3831, May 31, 1968, pp. 953-959. [Friedman's note.]
40. Schroeder, R. W., "NERVA—Entering a New Phase," *Astronautics and Aeronautics* 6: 5, May 1968, pp. 42-53. [Friedman's note.]
41. Luce, J. S., "Controlled Fusion Propulsion," *Proceedings of 3rd Symposium on Advanced Propulsion Concepts*, Vol. 1, Gordon and Breach Science Publishers, New York, 1963, pp. 343-380. [Friedman's note.]
42. Salisbury, F. B., "The Possibilities of Life on Mars," *Proceedings Conference on the Exploration of Mars and Venus*, Virginia Polytechnic Institute, August 1965. [Friedman's note.]
43. Kilston, S. N.; Drummond, R. R.; Sagan, C., "A Search for Life on Earth at Kilometer Resolution," *Icarus*, Vol. 5, January 1966, pp. 79-98. [Friedman's note.]
44. Vallee, J., "The Pattern of UFO Landings," *Flying Saucer Review*, Special Issue, "Humanoids: A Survey of Worldwide Reports of Landings of Unconventional Aerial Objects and Their Alleged Occupants," October-November 1966; $2. [Friedman's note.]
45. Hynek, J. A., "UFO's Merit Scientific Study," *Science*, October 21, 1966, and *Astronautics & Aeronautics*, December 1966, p. 4. [Freidman's note.]
46. Powers, W., "UFO in 1800: Meteor?" *Science*, 160, June 14. 1968, p. 1260. [Friedman's note.]
47. Rosa, R. J., Powers, W. T., Vallee, J., Gibbs, T. R. P., Steffey, P. C., Garcia, R. A., and Cohen, G., *Science*, Vol. 158, No. 3806, pp. 1265-1266 (1967). [Friedman's note.]
48. Page, T., "Photographic Sky Coverage for the Detection of UFO's," *Science*, 160, June 14. 1968. p. 1258. [Friedman's note.]
49. "UFO Project: Trouble on the Ground." *Science*, 161. July 26, 1968, pp. 339-342. Baker, R. M., Jr., "Observational Evidence of Anomalistic Phenomena," *J. Astronaut.*, August 1967. [Friedman's note.] August 1967.
50. Baker, R. M., Jr., "Observational Evidence of Anomalistic Phenomena," *J. Astronaut.* August 1967. [Friedman's note.]
51. Baker, R. M., Jr., "Observational Evidence of Anomalistic Phenomena," *J. Astronaut. Sci.*, XV, No. 1, January-February 1968. [Friedman's note.]
52. Morse, R. F., "UFO's and the Technological Community," *American Engineer*, 38:5. May 1968, pp. 24-28. [Friedman's note.]
53. Fowler, R. E., "Engineer Involvement in UFO Investigations," *American Engineer*, 38:5, May 1968, pp. 29-31. [Friedman's note.]
54. Moller, P. S., "Engineering Professor Teaches UFO Course at the University of California," *American Engineer*, May 1968, pp. 32-34. [Friedman's note.]
55. "UFO Study Credibility Cloud?" *Industrial Research*, June 1968. [Friedman's note.]
56. *Scientific Research*, May 13, 1968, p. 11. [Friedman's note.]
57. *Scientific Research*, May 30, 1968. [Friedman's note.]
58. Klass, P. J., *Aviation Week and Space Technology*, August 22, 1966, p. 48, see also October 10, 1966, p. 130. [Friedman's note.]
59. Klass, P. J., *Aviation Week and Space Technology*. October 3, 1966. p. 54. [Friedman's note.]
60. Morgan, D. L., Jr., "Evaluating Extreme Movements of UFO's and Postulating an Explanation of Effects of Tones on Their Maneuverability," Design Engineering Conference, ASME Meeting, New York, May 15-18, 1967, Session 10. [Friedman's note.]

61. Earley, G. W., "Unidentified Flying Objects: An Historical Perspective," Design Engineering Conference ASME Meeting. New York. May 15-18 1967. Session 10. [Friedman's note.]
62. Maney, Prof. C. A., and Hall, R., *The Challenge of Unidentified Flying Objects*, 1961, $3.50. [Friedman's note.]
63. Literature Search No. 541, "Interactions of Spacecraft and Other Moving Bodies With Natural Plasmas," December 1965, Jet Propulsion Laboratory; 182 pages, 829 references. [Friedman's note.]
64. Jarvinen, P. O., "On the Use of Magnetohydrodynamics During High Speed Reentry," NASA-CR-206, April 1965. [Friedman's note.]
65. Nowak, R., et al., "Magnetoaerodynamic Reentry," AIAA Paper 66–161, AIAA Plasma-dynamics Conference, March 2-4, 1966. [Friedman's note.]
66. Kawashima, N. and Mori, S., "Experimental Study of Forces on a Body in a Magnetized Plasma," AIAA Journal, Vol. 6, No. 1, January 1968. pp. 110-113. [Friedman's note.]
67. Ericson, W., Maciulaitis, A., and Falco, M., "Magnetoaerodynamic Drag and Flight Control," Grumann Research Department Report, RE 232J, November 1965. [Friedman's note.]
68. Smith, M. C., "Magnetohydrodynamic Re-entry Control," January 1965, Rand Corporation Memo, RM-4380-NASA. [Friedman's note.]
69. Cambel, A. B., "The Phenomenological Aspects of Magnetogasdynamic Re-entry." Presented at the 10th Midwestern Mechanics Conference, Colorado State University, August 1967. [Friedman's note.]
70. Porter, R. W., and Cambel, A. B., "Magnetic Coupling in Flight Magnetoaerodynamics," *AIAA Journal*, Vol. 5, No. 4, April 1967, pp. 803-805. [Friedman's note.]
71. Kranc, S., Porter, R. W., and Cambel, A. B., "Electrodeless Magnetogasdynamic Power during Entry," *Journal of Spacecraft and Rockets*, Vol. 4, No. 5, June 1967, pp. 813-815. [Friedman's note.]
72. Seemann, G. R., Cambel, A. B., "Observations Concerning Magnetoaerodynamic Drag and Shock Standoff Distance," *Proceedings of the National Academy of Sciences*, Vol. 55, No. 3, pp. 457-465, March 1966. [Friedman's note.]
73. Porter, R. W., Cambel, A. B., "Comment on Magnetohydrodynamic-Hypersonic Viscous and Inviscid Flow near the Stagnation Point of a Blunt Body," *AIAA Journal*, May 1966, 952-953. [Friedman's note.]
74. Way, S., "Propulsion of Submarines by Lorentz Forces in the Surrounding Sea," ASME paper 64-WA/ENER-7, Winter Meeting, New York City, November 29, 1964. [Friedman's note.]
75. Way, S., "Electromagnetic Propulsion for Cargo Submarines." Paper 67-363, AIAA/SNAME Advanced Marine Vehicles Meeting, Norfolk, Virginia, May 22-24, 1967. [Friedman's note.]
76. Way, S., Devlin, C., "Prospects for the Electromagnetic Submarine," Paper 67-432, AIAA 3rd Propulsion Joint Specialist Conference, Washington, D.C., July 7-21 , 1967. [Friedman's note.]
77. Chatham, G. C., "Towards Aircraft of the 1980's," *Astronautics and Aeronautics*, July 1968, pp. 26-38. [Friedman's note.]
78. McNally. J. R., "Preliminary Report on Ball Lightning," 2nd Annual Meeting of Division of Plasma Physics, American Physical Society, Gotlinburg, Tennessee, November 1960. [Friedman's note.]
79. Rayle, W. D., "Ball Lightning Characteristics." NASA-TN-D-3188, January 1966. [Friedman's note.]
80. Berliner, D., "The UFO From the Designer's Viewpoint," *Air Progress*, October 1967. [Friedman's note.]
81. Salisbury, F. B., "The Scientist and the UFO," *Bio Science*, January 1967, pp. 15–24. [Friedman's note.]
82. Zigel, F., "Unidentifiable Flying Objects," *Soviet Life*, February 1988. pp. 27-29. [Friedman's note.]
83. "Saucers, Hoax or Hazards," *Engineering Opportunities*, September 1967, pp. 17–24; 44-50. [Friedman's note.]
84. Kachur, V., "Space Scientists and the UFO Phenomenon: An Informal Survey," *Bio-space Associates Report No. 672*, August 1967. [Friedman's note.]
85. Hynek, J. A., "How to Photograph a UFO," *Popular Photography*, 62.3, March 1968, p. 69. [Friedman's note.]
86. McDonald, J. E., "UFOs: Greatest Scientific Problem of Our Times," October 1967, available from UFORI Suite 311, 508, Grant Street, Pittsburgh, Pa., 15219, $ 1.00. [Friedman's note.]
87. Markowitz, "The Physics and Metaphysics of Unidentified Flying Objects," *Science*, 157, pp. 1274-1279 (1967). [Friedman's note.]
88. Menzel, D. H., *Flying Saucers*, Harvard, 1953. [Friedman's note.]
89. Menzel, D. H., and Boyd, L., *The World of Flying Saucers*, Doubleday, 1963. [Friedman's note.]
90. Klass, P. J., *UFOs Identified*, 1968. [Friedman's note.]

91. McDonald, J. E., "UFOs: Greatest Scientific Problem of Our Times," October 1967, available from UFORI Suite 311, 508, Grant Street, Pittsburgh, Pa., 15219, $ 1.00. [Friedman's note.]
92. Menzel, D. H., and Boyd, L., *The World of Flying Saucers*, Doubleday, 1963. [Friedman's note.]
93. Davidson, L., "Flying Saucers An Analysis of Project Blue Book Special Report No. 14," 1966; $4.
94. Hearings, pp. 214-219.
95. Hearings, p. 222.
96. Hearings, pp. 207-210.
97. Hearings, pp. 210-213.
98. Hearings, pp. 223-235.
99. See my book, *The Martian Anomalies: A Photographic Search for Intelligent Life on Mars*. L.S.
100. Hearings, pp. 236-245.
101. Hearings, p. II.
102. Hearings, pp. 2-3.
103. Hearings, p. 18.
104. Hearings, pp. 85-86.
105. Hearings, p. 99.
106. Hearings, pp. 112-113.
107. Hearings, pp. 125-126.
108. Hearings, pp. 198-199.
109. Hearings, pp. 213-214.
110. Hearings, pp. 206-207.
111. Hearings, pp. 210-211.
112. Hearings, p. 223.

BIBLIOGRAPHY
And Suggested Reading

Hearings Before the Committee on Science and Aeronautics, U.S. House of Representatives, 90th Congress, 2nd Session, July 29, 1968. Washington, D.C.: U.S. Government Printing Office, 1968.

Seabrook, Lochlainn. *The Goddess Dictionary of Words and Phrases: Introducing a New Core Vocabulary for the Women's Spirituality Movement.* Springhill, TN: Sea Raven Press, 1997.

———. *Britannia Rules: Goddess-Worship in Ancient Anglo-Celtic Society—An Academic Look at the United Kingdom's Matricentric Spiritual Past.* Springhill, TN: Sea Raven Press, 1999.

———. *The Book of Kelle: An Introduction to Goddess-Worship and the Great Celtic Mother-Goddess Kelle, Original Blessed Lady of Ireland.* Springhill, TN: Sea Raven Press, 1999.

———. *UFOs and Aliens: The Complete Guidebook.* Springhill, TN: Sea Raven Press, 2005.

———. *Carnton Plantation Ghost Stories: True Tales of the Unexplained from Tennessee's Most Haunted Civil War House!* Springhill, TN: Sea Raven Press, 2005.

———. *Christmas Before Christianity: How the Birthday of the "Sun" Became the Birthday of the "Son."* Springhill, TN: Sea Raven Press, 2009.

———. *Everything You Were Taught About the Civil War is Wrong, Ask a Southerner!* Springhill, TN: Sea Raven Press, 2010.

———. *Christ Is All and In All: Rediscovering Your Divine Nature and the Kingdom Within.* Springhill, TN: Sea Raven Press, 2014.

———. *Jesus and the Gospel of Q: Christ's Pre-Christian Teachings As Recorded in the New Testament.* Springhill, TN: Sea Raven Press, 2014.

———. *Jesus and the Law of Attraction: The Bible-Based Guide to Creating Perfect Health, Wealth, and Happiness Following Christ's Simple Formula.* Springhill, TN: Sea Raven Press, 2016.

———. *Seabrook's Bible Dictionary of Traditional and Mystical Christian Doctrines.* Springhill, TN: Sea Raven Press, 2016.

———. *The Bible and the Law of Attraction: 99 Teachings of Jesus, the Apostles, and the Prophets.* Springhill, TN: Sea Raven Press, 2020.

———. *The Martian Anomalies: A Photographic Search for Intelligent Life on Mars.* Cody, WY: Sea Raven Press, 2022.

INDEX

INCLUDES TOPICS, PEOPLE, KEYWORDS, SPELLING VARIATIONS, & KEY PHRASES

AAAS, 365
AAL, 79
aborigine, 24
aborigines, Australian, 319
abrupt disappearance, 78
abscissa, 225, 226
acceleration, 82, 97, 211, 213, 222, 226
acceleration and jerk estimation, 222
Adamski, George, 325
Adickes, Robert, 85
Admiralty Bay, Antarctica, 119
advanced extraterrestrial societies, 220
advanced technology, 34, 47, 51, 63, 165, 277, 355
advertising planes, 257
aerial dances, 326
aerial feats, 199, 200
Aerial Phenomena Research Organization, 42, 61, 288, 313, 317, 330, 332, 355
aerial phenomenon, 27, 58, 67
aerodynamic drag, 234, 235, 280
aerodynamic lift forces, 201
Aerojet General Nucleonics, 273
aerology, 343
aeronautical device, 83
aeronautical engineer, 89
Aeronutronic-Philco-Ford, 351
aerospace program, 52
aerospace programs, 60, 197
Africa, 130, 247, 330
after-image, 258, 259
after-images, 258, 259, 297
Aiken, SC, 94
Air Defense Command, 215
Air Force, 18, 21-23, 28, 32-35, 39, 42, 44, 49-51, 53-56, 67, 79-82, 100, 130, 151, 156, 168, 170, 175, 191, 192, 196, 208, 210, 217, 222, 223, 225, 232, 234, 236, 239, 241, 242, 246, 247, 254-257, 259, 262-264, 268, 290, 310, 315, 316, 320, 324, 331, 333, 345, 347, 351, 353, 365

Air Force Filter Center, 191
Air Force ground personnel, 79
Air Force Intelligence, 67
Air Force investigators, 82
Air Force lieutenant, 32
Air Force Office of Scientific Research, 170
Air Force Personnel & Training Research Center, 347
Air Force Scientific Advisory Board, 170, 345
Air Forces, 28, 29, 353
air show, 49, 89
air traffic, 31, 93
air-borne objects, 132
airbases, 191
airborne devices, 77, 127
airborne particles, 116
airborne radar, 130
airborne radar contact, 123
airborne radar sets, 124
airborne-particle hypothesis, 117
airborne-radar, 132
aircraft, 27, 29-31, 35, 41, 49, 60, 66, 69, 71, 76, 77, 79-82, 86, 90, 91, 93, 94, 96, 101, 104, 107, 108, 110, 111, 113, 123-125, 128-134, 137, 139, 142, 150, 151, 153, 169, 185, 199, 200, 202, 210, 211, 215, 219, 225, 227, 228, 232, 234, 242, 256, 272, 273, 276, 320, 323, 347, 351, 355
aircraft control and warning station, 133
aircraft control operators, 31
aircraft dispatcher, 30
aircraft origin, 80
aircraft radar, 134
aircraft strobe light, 41
aircraft strobe lights, 66
aircraft study, 90
aircrew, 81
airline duty, 83
airline pilots, 46, 50, 76, 81, 131, 255, 291
airline regulations, 86

airline-crew sightings, 79
airline-pilot sightings, 81, 132
airliner, 77, 191, 199, 200, 211
airliner crew, 77
airlines pilot, 109
airmen, 30, 31, 122, 123, 128, 129
airpath, 84, 124
airplane fuselage reflections, 210
airplane reflections, 212, 227, 256
airport manager, 89
airports, 13
airspace, 127
Alabama, 77
Alamogordo Skyhook releases, 88
Alamogordo, NM, 353
Albuquerque, NM, 94
alien technology, 62
Allen, George, 85
alloying elements, 244
alloys, 218, 244
Almirante Saldanha (Brazilian navy ship), 330, 331
Alpha Chi, 361
Alpine, TX, 361
altitude, 22, 77, 78, 82, 83, 86, 87, 89-91, 94, 97, 101, 104, 108, 113, 121, 122, 124, 128, 130, 136, 146, 149, 169, 199, 200, 226, 227, 232, 242, 243, 291, 329, 353
aluminum, 92, 227, 244, 245
amateur astronomers, 109
amateur movie cameras, 210
amber, 30, 86, 87
amber color, 87
ambient wind, 126
ambient wind shears, 126
ambiguity, 178, 183, 186, 189, 193, 194
ambiguous data, 224
ambiguous situation, 178, 182, 189
ambulance, 44
American Academy of Arts and Sciences, 354
American Association for the Advancement of Science, 275, 341, 343, 349, 352, 358
American Association of State Psychology Boards, 358
American Association of University Professors, 343, 358
American Astronautical Society, 235-237, 352, 361
American astronomers, 114
American Astronomical Society, 24, 109, 352, 354

American Geophysical Union, 343, 361
American Institute of Aeronautics and Astronautics, 278, 352, 355
American Meteorological Society, 343
American Nautical Almanac, 232, 260
American Newspaper Publishers Association, 48, 99
American Nuclear Society, 278, 355
American Personnel and Guidance Association, 357, 358
American Philosophical Society, 354
American Physical Society, 24, 277, 352, 355
American Psychological Association, 284, 286, 357, 359
American Rocket Society, 234, 235
American Society of Clinical Hypnosis, 284, 358, 359
American Society of Newspaper Editors, 84, 156
American UFO reports, 112
Americans, 2, 130, 247, 268, 270, 276
analog simulation, 349
analysis of results, 310
Anderson, John A., 94, 95
Andrews Air Force radars, 50
Andromeda, 162
anecdotal data, 41, 45, 69
anecdotal nature, 23, 37, 248
Angel Falls, Venezuela, 211
angel type refractive anomalies, 134
angels, 134, 215, 323
angel's hair, 323
angles of elevation, 87, 153
Anglican Church, 333
angular elevation, 66, 112, 113, 118, 149
angular elevations, 102, 131
angular estimation, 45, 68
angular size, 66, 89, 90, 94, 113, 122, 123, 151
animal-reactions, 97
animals, 24, 38
Annual Meeting of the Meteoritical Society, 233
anomalistic alarms, 215
anomalistic data, 213, 216, 219, 221, 222
anomalistic motion picture data, 212
anomalistic observational phenomena, 209, 213, 216, 218, 219, 221-223, 232, 239, 272
anomalistic phenomena, 209, 210, 212, 213, 215, 217, 218,

220-222, 224, 232, 238, 275
anomalistic radio signals, 220
anomalous luminous phenomena, 213
anomalous radar propagation, 131
anomalous trails, 213
ANPA, 99
Antarctic Ocean, 119
Antarctica, 119, 407
antenna, 144, 280, 292, 298
anthropocentrism, 171
anthropology, 99
anthropomorphic, 184
antigravity, 319
antigravity field, 319
antigravity shield, 319
anxiety, 178, 179, 181-186, 189, 193, 194, 301
AOP, 209
apartment windows, reflections from, 257
apathy, 36
Appalachia, 407
apparitions, 253, 256, 257, 259, 328
appendages, 46
Applied Physical Laboratory, 341
applied research, 197, 361
APRO, 42, 61-63, 66, 74, 95, 97, 144, 155, 157, 284, 288, 310
APRO Bulletin, 74, 95, 155, 157, 288
Arctic regions, 151
Arecibo radio telescope, 161
Argentina, 174, 231
Arizona, 11, 18, 41, 59, 60, 137, 148, 154, 212, 233, 288, 317, 343
Arizona Academy of Science, 343
armchair-psychologizing, 63
armchair-researching, 64
Arnold Sighting, 327
Arnold, Kenneth, 27, 76, 91, 149, 328
Arrey, NM, 117
ARS, 131, 132, 235-237
ARTC, 131, 132
ARTC scope, 131
artificial earth satellites, 169
artificial plasmoids, 63
artificial satellites, 236, 316, 320
artist, 300, 303, 407, 408
artists, 302, 407
Askania camera, 138
Askania cameras, 139
Askania operators, 139
ASNE, 84, 156
assessing testimony, 190
assimilation, 181

Associated Press, 84
Associated Press wire stories, 84
Association of Universities for Research in Astronomy, 354
astigmatic image, 118
astrodynamic constants, 236, 351
astrodynamical laws, 214, 215
astrodynamics, 211, 232, 233, 235-238, 351, 352
Astrodynamics Technical Committee, 352
astronaut training, 345
Astronautical Society, 235-237, 352, 361, 365
astronautics, 11, 17, 41, 59, 153, 154, 235-237, 265, 270, 273, 275-278, 283, 339, 351, 352, 355
astronomer sightings, 114
astronomical basis, 27, 28
astronomical causes, 22, 28
astronomical community, 170
astronomical explanation, 28, 79, 86, 104
astronomical objects, 170
astronomical optical sensors, 213
astronomical phenomena, 53, 110, 170, 295
astronomical plate measuring, 225
astronomical tracking systems, 72
astronomy, 11, 18, 19, 21, 25, 38, 108, 125, 139, 159, 163, 167, 172, 173, 204, 213, 216, 219, 243, 295, 326, 341, 345, 351, 353, 354, 407
Astronomy Society, 341
astronuclear plant, 11, 267
astrophysics, 341, 345, 354
asymmetric current surges, 147
ATIC, 211, 225
Atkins, Chet, 407
Atlanta, GA, 77, 347
Atlantic Ocean, 122
atmosphere, 25, 62, 76, 110, 118, 131, 148, 149, 151, 169, 178, 183, 188, 193, 203, 215, 216, 232, 234, 256, 262, 289, 298, 318, 320, 323, 328, 330, 341, 353
atmospheric conditions, 124, 133, 291
atmospheric effects, 131, 138
atmospheric electricity, 148, 152, 343
atmospheric explanations, 43, 50
atmospheric mirages, 212, 215, 229
atmospheric optics, 112, 264

atmospheric physics, 18, 41, 42, 59, 60, 62, 147, 148, 154, 166, 218, 264, 311, 343
atmospheric plasmas, 147
atmospheric processes, 51
atmospheric refraction, 124
atmospheric scintillation, 118
atmospheric subsidence, 150
atmospheric temperature, 153
atmospheric turbulence effects, 118
atmospheric-electrical effects, 63
atmospheric-electrical plasma, 120
atmospheric-electrical plasmas, 51
atom bombs, 273
atomic bomb, 205
atomic fusion, 205
atoms, 207
attorneys, 179, 190, 195
audio recorder, 292
auditory observation, 190
Aurora Borealis, 37, 169, 257
Australia, 43, 46, 54, 129, 139, 143, 157, 172, 245, 319
Australian UFO cases, 142
Australian UFO witnesses, 130
authoritative information, 189
automobile, 23, 69, 180, 280, 332, 333
automobiles, 62, 181
Aviation Week, 51, 147, 275
Aviation Week magazine, 147
aviator, 256
azimuth, 30, 78, 116, 226, 227, 232
azimuth-change, 78
B-29, 78, 79, 134
B-47, 335
Bachmeier, James F., 80, 81
Back and Kerckhoff study, 181
back-illumination, 121
baffling phenomena, 42, 101
Baker, Frank E., 138
Baker, Robert M. L., Jr., 11, 18, 135, 209, 216, 229, 241, 243, 244, 277
Baker-Nunn cameras, 213
ball lightning, 51, 64, 87, 120, 142, 147, 152, 188, 215, 217-219, 222, 257, 264, 277, 278, 320, 326, 327
ball lightning hypothesis, 51
ball-shaped craft, 168
ballistic missile, 214
balloon hypothesis, 90, 91, 121, 123
balloon-drift, 104
balloon-tracking missions, 81
ballooning spiders, 228

balloons, 27, 29, 30, 33, 35, 49, 60, 81, 87, 88, 90, 91, 101, 104, 117, 120-122, 124, 126, 138, 169, 212, 228, 256-258, 262, 263, 290, 320
banking, in the sun, 256
Barauna, Almiro, 330, 331
Bardot, Brigitte, 267
barium, 206, 244, 245
Barnes, Harry G., 132
barnyard animals, 24
barometric pressure, 291
Basale, Sierra Leone, 326
Bayonne, France, 321
Baytown, TX, 106
BBC, 46, 84
Beallsville, OH, 145
beam axis, 153
Beatty, James, 191
bee-like motion, 88
Beeman, Richard E., 339
bees, 111
belief formation, 185, 194
belief systems, 177
beliefs, 26, 178, 183, 185-187, 189, 194, 245, 247, 298
Bell and Howell Automaster, 224
Bell Telephone Laboratories, 363
Bell, Alphonzo, 52, 339
BEMEWS radars, 214
Berry, Amelia, 96
Bertrand, Eugene, 334, 335
beryllium, 206
Besse, M., 321
best evidence, 29
Betelgeuse, 320
Beverly, MA, 108
biconvex object, 91
binocular magnification, 112
binoculars, 49, 89, 91, 92, 96, 109, 114, 121, 137, 320, 321, 332
bird-brains, 32
birds, 3, 90, 153, 169, 210, 212, 222, 227-229, 232, 257, 276, 320, 409
birds in flight, 210, 228
Bismarck, ND, 134
Bittick, James D., 138
bizarre events, 219
Black Widow, 57, 128
blackout, 57, 58, 146
Blackstone (magician), 325
blimp-shaped craft, 310
blimp-sized object, 144
blimps, 257
blink-interval, 125

blinking, 44, 67, 69, 71, 125, 126
blinking out, 329
blip, 133, 323
blips, 31, 130, 131
blobs, 132, 212
Blohl, captain, 199
blood cells, 259
blue, 30, 31, 114, 119, 202, 203, 212, 225, 256-258, 263, 268, 272, 278, 289, 332-334, 345
blue flame, 332, 333
blue spotlight, 334
bluish glow, 78, 142
bluish-white flash, 102
BOAC, 51, 84
BOAC Stratocruiser, 51
board of inquiry, 37-39, 187
bobbing, 93, 96, 104
bobbing motion, 99
Boddy, Clayton J., 88
bodily damage, 141
Boeing, 89, 238
Boeing 377 Monarch Stratocruiser, 84
Boeing B-29 Superfortress, 78-80, 134
Boeing B-47 Stratojet, 335
Boeing-Stearman Kaydet, 89
Boiani, New Guinea, 98, 333
Boise, ID, 76
bolide meteors, 326
Bolling, Edith, 407
bomber crews, 347
Boone, Pat, 407
Boone, W. H., 205, 242, 243, 339
boron fiber, 207
boron fiber reinforced composites, 207
Boston airport control tower, 126
Boston, MA, 126, 198
botany, 365
Boyd (first officer), 85
Boyd, Lyle, 354
Braniff Airlines, 80
Brazil, 174, 206, 333
Brazilian Agriculture Ministry, 206
Brazilian meteorologist, 119
Brazilians, 244
bright light, 30, 258, 261
bright lights, 44
bright stars, 169
bright-moving points of light, 228
brightness, 30, 212, 228, 229, 256, 318, 327
brilliant multicolored light, 133
Brinkley, Jack, 339
British Astronomical Society, 352
British Interplanetary Society, 233, 352
broad-spectrum electromagnetic noise, 146
Brown, George E., Jr., 339
Brown, S. H., 116
Buchanan, Patrick J., 407
buffoonery, 33
bullet-shaped craft, 129, 310
bullet-shaped object, 79
bun-shaped objects, 144
burn-injuries, and UFO contact, 146
burns, 102, 141, 144, 145
Burns, Tom, 102
Butler, Robert G., 91, 92
buzzing noise, 327
C-47, 79
CAA, 50, 76, 79, 131, 132
CAA control tower, 79
CAA radar operators, 131, 132
CAA radars, 50
Cabell, Earle, 339
California, 11, 18, 156, 160, 161, 169, 191, 192, 198, 201, 202, 206, 208, 211, 212, 228, 235, 255, 273, 275, 288, 293, 306, 307, 339, 345, 349, 365
California herring gull, 228
California Institute of Technology, 169, 255, 349, 365
Callis, Robert, 358
Cambridge, MA, 343
Campbell, Joseph, 407
Canada, 31, 156, 160, 217, 270, 355
Canadian Aeronautics and Space Institute, 152, 156
Canberra, Australia, 129
candles, 121, 262, 263
Cape Canaveral, FL, 144
Cape Province, South Africa, 117
Capella, 31, 32, 133, 260, 320
Capetown, South Africa, 130
Capitol Airlines, 132
car ignition failure, 57
car motor, 23
car stopping, 57, 58
car-buzzing, by UFOs, 143
car-stopping cases, 140, 142
car-stopping phenomenon, 143
carbon rod, 206
cargo cults, 320
cars, 43, 101, 104, 140, 142, 200, 332
Carson, Charles A., 198-201
Carson, Martha, 407
Carstarphen, John A., Jr., 339
case-credentials, 72

Cash, Johnny, 407
Cass, E. W., 103
Caucasus, 113
celestial mechanics, 29, 213, 237
celestial objects, 22
celestial phenomena, 168
Center for Radiophysics and Space Research, 18, 159, 345
Central Intelligence Agency, 169
chaff, 227
chaff, defined, 108
chaff-drop, 108
Chamberlain, Wilt, 270
chandelle maneuver, 124, 125, 128
changes of color, 301
changing colors, 31
Chapuis, J. L., 114
Charles Hayden Planetarium, 198
Chavez, Sam, 332
chemical processes, 171
chemical propulsion systems, 273
chemicals, 267
chemistry, 163, 257, 343, 345
Chesapeake Bay, 86, 329
Chesapeake Bay Case, 329
chi square statistical analysis, 272
Chicago, IL, 85, 195, 341
child's balloons, 256
Chile, 354
Chiles, Clarence S., 77, 78, 149
Chiles-Whitted sighting, 79
Chiles-Whitted Eastern Airlines case, 77
Chop, Al, 225
Christmas Before Christianity (Seabrook), 409
CIA, 59, 254, 255
cigar-shaped object, 82, 108, 321
cigar-shaped objects, 43, 46, 71
cinetheodolite, 138
cinetheodolite films, 211
cinetheodolites, 212
cirrus cloud deck, 116
cirrus clouds, 256
Cisterne, M., 321
cities, 48, 62, 72, 98-100, 108, 161, 326
civic duty, 26, 246
civil engineering, 11, 18, 208, 349
civilian radar sightings, 57
Civilian Saucer Intelligence, 75, 156
Clarke, Arthur C., 47
classic UFO cases, 148
classified Air Force projects, 351
classified programs, 121
Clem, Weir, 142

climate modification, 343
Climax, CO, 353
climb-out, 77, 87, 92
clinical psychologist, 170, 298
clinical psychology, 298
close-range airborne sighting, 83
close-range sightings, 68, 71
cloud deck, 78, 116, 120, 129
cloud phenomena, 76
Cloud Physics Project, 343
cloud reflections, 150
cloud-free skies, 93
cloud-reflection phenomenon, 150
cloudbank, 111
clouds, 31, 86, 90, 93, 111, 112, 114, 120, 126, 148-150, 169, 200, 202, 227, 256, 257, 320, 327, 328, 343
clusters of objects, 227
co-pilot, 28, 76, 77, 80, 255, 257
Cochran, CAN, 160
cockpit crew, 86, 126
coin-shaped objects, 329
Cole, John F., 111, 112
collective delusion, 196
collective ego, 197
collision course, 83, 124
color, 32, 87, 90, 116, 119, 138, 169, 173, 180, 191, 211, 225, 259, 268, 272, 276, 280, 301, 306, 308, 321, 329, 409
color film, 211, 225
color fringes, 225
Colorado, 55, 56, 139, 170, 207, 212, 215, 244, 262-264, 275, 289, 310, 320, 330, 353, 357-359, 365
Colorado project, 56, 207
Colorado Springs, CO, 215
Colorado State, 275, 365
Colorado UFO Study Group, 212
coloration, 102, 113
colored lights, 30
colors, 31, 119, 407
Columbia River, 107
Columbia, PA, 123
Columbus, OH, 27
Combs, Bertram T., 407
comet impact, 217
cometary entry, 219
cometoid impact, 217
cometoids, 216
comets, 170, 218
commercial airline pilots, 131
Commission on Solar Eclipses, 354
Committee on Science and

Astronautics, 11, 17, 41, 59, 153, 154, 265, 270, 277, 283
common aircraft, 90
communication, 53, 132, 138, 163, 189, 195, 277, 347, 348, 358
communications telemetry system, 292
communications, with UFOs, 292
compasses, 280
competent witnesses, 23, 24, 268, 270
complex organizations, 188
composites, 207
computer operations, 361
Computer Sciences Corp., 18, 224
Computer Sciences Corporation, 351
Computer Sciences Technical Committee, 352
concentration of energy, 193
conductors, electrical, 147
confabulation, 68
conflicting beliefs, 185
confusion, 23, 25, 36, 188, 289
Congressmen, 48, 291
Conklin, Howard S., 132
Connecticut, 238, 278, 288, 339
consistent evidence, 53
constellations, 225, 228
contact, 30, 31, 46, 51, 53, 54, 75, 123, 124, 128, 132, 133, 144, 145, 154, 171, 172, 179-184, 190-193, 199, 201, 210, 220, 285, 316, 319, 325, 326
contact-claims, 75
contactee, 171, 297, 298, 325, 327
contactee cult, 325
contactee cults, 297
contactees, 297, 325
continuous power source, 51
control tower operator, 30
control tower operators, 28
Convair B-36 Peacemaker, 125
conventional aircraft, 27, 29, 71, 76, 82, 91, 94, 104, 111, 128, 131, 137, 227, 228, 320
conventional explanation, 57, 116, 134
conventional explanations, 101
conventional helicopters, 71
conventional interpretation, 44
conventional photographic telescopes, 213
conventional scientific terms, 22, 35
conventional-objects-misinterpreted category, 322
coordinate-reference errors, 123

copper, 244, 245
Corinne, UT, 227
cornea, 259
Cornell University, 11, 18, 159, 167, 345
Cornell-Sydney Astronomy Center, 173
Corning, CA, 53, 136, 198, 199
corona discharge, 278
coronagraph, 353
coronal discharge, 215
cosmic ray balloons, 81
cosmic-radio intelligence, 173
Cosmos (Sagan), 345
Cotten, V. B., 116
cotton-wisp explanation, 88
course-reversal, instantaneous, 81
Cozens, Charles, 144, 145
craft-like devices, 53
Craig, Roy, 212, 244
Crawford, Cindy, 407
credibility, 37, 53, 60, 64, 68, 70, 71, 179, 180, 190, 217-221, 223, 234, 239, 275, 298
credibility evaluations of witnesses, 217
credibility of testimony, 179, 190
credible witnesses, 121, 140
crescent horns, 113
crescent-like object, 113
crescent-shaped craft, 113
crimes, 220
criminal cases, 302
Cruise, Tom, 407
Cruttwell, N. E. G., 70
crystal, 93, 149, 207
crystal lattice, 207
crystal planes, 207
crystalline structure, 206
CSC, 351
CSI, 75, 156
cultism, 44, 64
cults, 189, 297, 320
cup-shaped objects, 96
Curie, Marie, 38
current densities, 205
current human technology, 313
cushion-shaped craft, 99
cylindrical object, 146
cylindrical shape, 79
Cyrus, Miley, 407
d.c. fields, 58
Daddario, Emilio Q., 339
Dairy Queen, 106
dandelion seeds, 32
dark rings, 202, 203

darkroom personnel, 139
Darr, Jerry, 102
Davidson, Leon, 54, 155
Davis, John W., 339
Davis, W. R., 126
Daylight Kodachrome, 225
DC-3, 76-80, 85, 89, 211
DC-4, 86
DC-6, 200
Dearborn Observatory, 35, 341
deception, 268
Defense Department, 52
Delaware, 122, 278
Delta Aquarid meteor stream, 79
delusions, 167, 297
Denver, CO, 263
Department of Astronomy, 18, 19, 21, 38, 159, 351
Department of Defense, 247
Department of Engineering, 11, 18, 224, 351
Department of Sociology, 11, 18, 177, 195, 347
depressions, from landed UFOs, 140
depth-perception errors, 123
desalination, 355
desert atmospheric effects, 138
detection fan penetration, 214
detection fans, 214
detection fence, 215
deviant belief systems, 196
Dewey, John, 18, 164
Dickinson, Philip P., 339
diffuse reflection, 228
Dikeman, Kelly, 106
DiPaolo, Louis, 96
diple field, 143
dipole, 204
direct contact, with UFOs, 144
direct physical contact, 145
direction-shifts, 104
dirigible, 258
disc-like craft, 77
disc-like objects, 49, 107, 108, 123
disc-shaped objects, 46, 49, 86, 88, 89, 91, 107
discs, 51, 71, 76, 86, 107, 108, 149, 326, 328-330
discus, 89
disinterest, 186, 194
disk, 32, 43, 54, 58, 69, 162, 301, 302, 305
disk-like objects, 43
disk-like shapes, 301
disks, 32, 253
Disney company, 105

distance, 30, 32, 45, 47, 68, 71, 78, 80, 83, 84, 86, 91, 92, 98, 102, 111, 112, 116-118, 121, 124, 126, 133, 138, 161, 198, 200, 201, 204, 211-213, 219, 224, 227, 255, 256, 259, 275, 276, 300, 313, 317, 322, 334
distance estimation, 68
distortion, 112, 179, 180, 190
diurnal variation, 98
diverse aerial sightings, 27
DLs, 71
documentation, 24, 26, 73
dogfights, 256
domed craft, 138
domed disk, 69
domed objects, 96
domed-disc, 83, 138
domed-disc UFO, 138
Donaghue, Jack, 82
Dorian, Richard, 82
double-convex lense-shaped craft, 310
double-exposure photography, 169
doughnut-shaped craft, 310
Douglas Aircraft Company, 210, 225, 234, 351
Douglas C-47 Skytrain, 79
Douglas DC-3 airliner, 76, 77, 79, 85, 89
Douglas DC-4 airliner, 86
Douglas DC-6 airliner, 200
Dow Chemical Company, 244
Downing, Thomas N., 54, 339
dragons, 253
dreams, 324
Ducander, Charles F., 339
ducks, 232
ducting, 50, 131, 134, 152
ducting gradient, 50
Duluth, MN, 343
DuPont company, 94
Duran, Mrs. J. R., 96
duration, 85, 127, 129, 179, 190, 272, 316, 318, 330
dust devils, 257
DuToit, Petrus, 117
Duvall, Robert, 407
Earl of Oxford, 407
early warning network, 207
earthquake records, 244
earth's atmosphere, 118, 215, 216, 234
earth's gravitational field, 201
East River, NY, 99
Eastern Airlines, 77, 149
Eaton, John, 88

echos, radar, 50
Eckhardt, Bob, 339
eclipse information, 244
Ecological Society, 365
ecologists, 244
ecology, 365
egg-shaped craft, 119, 120
Eglin AFB, 214
Eglin Air Force Base, 223
Einstein, Albert, 164, 318
Ektachrome X film, 332
electrical disturbance, 143
electrical explanation, 85
electrical generators, 205
electrical induction, 36
electrical interference effects, 192
electrical machinery, 204
electrical malfunctioning, 204
electrical methods, 361
electrical motors, 205
electrical storms, 215
electrical system, vehicles, 142
electro-magnetic effect, 146
electrode, 206
electromagnetic detector, 291
electromagnetic disturbances, 146
electromagnetic forces, 205, 276
electromagnetic phenomena, 24
electromagnetic submarine, 276
electromagnetic theory, 204
electronic contact, 30
electrostatic attraction, 205
elevation angles, 93, 148
ellipsoidal, 111, 117, 236
ellipsoidal white object, 117
elliptical objects, 321
Eltanin (NSF polar research vessel), 361
elves, 254
EM effects, 280
EM parameters, 280
Emmet, ID, 76
emotion, 178, 179
emotional factors, 166, 187, 245
empennage, 78, 80
engine-noise, 96, 113
engine-pods, 80
engineering science, 349
engineers, 25, 36, 49, 94, 147, 273, 275, 278, 291, 292, 354
engineless craft, 81
England, 46, 58, 75, 108, 155, 288
environmental contamination, 64
environmental recording devices, 291
ephemerides, 214
Epperson, Idabel, 101

Epping, NH, 334
equipment malfunction, 215
erratic pattern, 31
erratically moving phenomena, 218
escalation of hypotheses, 44
ESP, 87, 287
ESP processes, 287
European royalty, 407
European theater, 128
European UFO sightings, 75
European wave, 75
Evans Signal Laboratory, 122
Evanston, IL, 35
Everglades, the, 144
Everything You Were Taught About the Civil War is Wrong (Seabroo, 409
evidence, 23, 29, 34, 35, 53, 57, 59, 62, 64, 65, 72, 73, 89, 95, 99, 126, 139, 141, 143-147, 154, 155, 159, 161, 165, 166, 169-172, 178, 181-188, 190, 192, 194, 195, 197, 209, 216, 218, 229, 232, 238, 245, 246, 248, 254, 255, 263, 267, 272, 275, 284, 289, 293, 295, 297, 298, 301, 304, 306, 316, 320, 322, 323, 333
evolution, 163, 175, 358
evolution of intelligence, 163
Exeter, NH, 74, 334
exhaust flames, 129
exhaust trail, 129
exhaustless craft, 124
exobiologist, 314
exobiology, 314, 345
expelling neutrons, 201
experienced observers, 77, 119
experienced pilots, 78, 86, 89
experienced radar controllers, 132
experimental balloons, 88
experimental psychology, 363
explosive sounds, 140
extraordinary velocities, 301
extraterrestrial, 17, 34, 51, 52, 59, 63-65, 68, 70, 99, 142, 153, 154, 159, 164-167, 170-175, 183, 184, 188, 193, 197, 216-221, 255, 264, 267, 268, 270, 271, 279, 292, 295, 314-316, 319, 323, 324, 326, 328, 331, 361
extraterrestrial civilization, 164, 219, 220
extraterrestrial devices, 51, 59, 63-65, 183, 184, 188, 193

extraterrestrial humanoid, 268
extraterrestrial hypotheses, 17
extraterrestrial intelligent civilization, 165
extraterrestrial life, 159, 166, 167, 174, 175, 217, 221
extraterrestrial machines, 316, 326, 331
extraterrestrial motives, 319
extraterrestrial origin, 64, 99, 142, 154, 159, 166, 167, 170-172, 295
extraterrestrial origins, 166
extraterrestrial surveillance, 52
extraterrestrial vehicles, 255, 264, 271, 279
extraterrestrials, 64, 171, 319, 325
eye, 5, 88, 92, 115, 121, 123, 126, 144, 151, 161, 170, 239, 257-259, 261-263, 279, 280
eyewitnesses, 219, 239, 304, 307
Ezekiel (Bible), 326
F type stars, 341
F-61, 128
F-84, 93
F-86, 85
F-94, 57, 123, 133
FAA, 80, 132
fabrications, 63
fabricators, 45
factor of astonishment, 137
Fahrney, D. S., 89
Fairchild AFB, 125
fairies, 254
falling-leaf motion, 82, 89, 301
falling-leaf motion descent, 301
false assumptions, 268
false information, 269
false radar returns, 134
false returns, 134, 152
familiar explanation, 190
fan penetration, 214
fanciful tales, 28
Faraday effect, 203
Faraday, Michael, 36, 38
Farmington, NM, 49, 87-89
faulty knowledge, 268
FBI, 26
fear, 115, 140, 297, 301
feathers, 257
Federal Communications Commission, 58
Federal government, 61
Federal Power Commission, 58, 146
Felton, Joseph M., 339
fiery chariots, 253

fiery fragments, 206
fillings, tooth, 57
film, 2, 12, 138, 139, 210-212, 224-231, 272, 280, 331, 332, 407
film-measuring equipments, 210
fins, 298
fireball, 41, 78, 79, 106, 136, 219, 255
fireball events, 136
fireball procession, 217
fireballs, 60, 66, 79, 169, 218, 222, 233
firedrakes, 253
Fisher's Peak, CO, 96
fission, 267, 269, 273, 277
Flagstaff, AZ, 113
flames, 97, 129
flares, 60, 63, 212, 245, 314, 316
flashing lights, 53, 168, 248, 334
flashing red lights, 334
flashlight bulbs, 121
flat disks, 32
flat earth debates, 216
flattened ball-shaped craft, 310
flight characteristics, 292
flight performance, 93
flipping motion, 90
flocks of birds, 169
Florida, 30, 113, 211, 339
Florida State University, 113
Floyd, NY, 213
fluid mechanics, 349, 351
fluttering motion, 91, 99
fly-under maneuvers, 80
flying disks, 32
flying formation, 80
flying saucer, 6, 21, 27, 28, 32, 36, 155, 156, 170, 172, 175, 198, 206, 274, 276, 283, 288, 319, 325, 326, 333, 358
flying saucer cultists, 198
flying saucer reports, 170
flying saucer situation, 32
flying saucers, 27-29, 32, 120, 130, 154-156, 168, 171, 175, 220, 232-234, 253, 257, 262, 268, 272, 273, 277-279, 284, 354
Flynn, James, 144
FOBS, 217
fog, 290
Fontes, Alavio, 331
Fontes, Olavo T., 206
football-shaped object, 136
Foote, Shelby, 407
Ford Motor Company, 293

foreign atoms, 207
foreign radar sightings, 130
foreign UFO references, 75
fork-tailed gull, 228
formation, 31, 76, 77, 80, 107, 162, 163, 185, 194, 218, 230, 258, 297, 328, 329
formations, 76, 85, 118, 169, 219, 225, 328
Fort Collins, CO, 320
Fort Davis, TX, 353
Fort Snelling, 259
Fort Worth, TX, 293
Fort, Charles, 75
Fortean reports, 76
Fortenberry, William H., 86, 87, 329
forward scatter, 153
forward-angle scattering of sunlight, 116
Fouéré, René, 119
Fowler, Raymond E., 108
fox-fire, 253
FPC, 58
FPS-17, 214
FPS-79, 214
FPS-85, 214, 215, 222, 223, 242
fractional orbital bombardment system, 217
France, 43, 52, 74, 114, 142, 238, 306, 321, 323, 333
frauds, 63
fraudulent photos, 135
Freehold, NJ, 122, 123
French Academy, 29, 34, 43
French Academy of Sciences, 29
French UFO wave, 75
frequency of sightings, 46
frictional heat, 318
Friedman, Stanton T., 11, 267, 353, 355
Fry, deputy, 198, 200
Ft. Monmouth, NJ, 122, 123
Ft. Myers, FL, 144
Ft. Sumner, NM, 110
fuel consumption, 276
fuel weight, 276
Fukuoka, Japan, 128
Fuller, John G., 334
Fullerton, CA, 349, 352
Fulton, James G., 248, 249, 267, 339
fundamental theory, 204
Fuqua, Don, 339
fused silica fibers, 207
fuselage, 80, 210
fusion, 51, 205, 267, 269, 273, 277, 355

fusion propulsion systems, 269, 273, 277
fusion research, 51
fusion rockets, 355
Galbraith, Jay W., 91, 92
Gallup poll, 247, 276
gas-balloon hoaxing, 104
Gayheart, Rebecca, 407
GCA Corporation, 354
GCI radar, 133
geese, 232
Gelles, France, 321
Gemini capsule, 160
General Dynamics, 361
General Mills Skyhook balloon program, 81
General Motors, 355
general relativity, 204
general-purpose sensor system, 213
geometric characteristics, 34
geometrically regular formations, 85
geophysical oceanography, 361
geophysical phenomena, 53
geophysics, 354, 361
Georgia, 79, 95, 339, 347
Gerardi, Peter A., 339
Gettys, John R., Jr., 138
ghosts, 46, 254, 259, 263
Gibbon, Minnesota, 332
Gilham, Rene, 144
Gill, William B., 98, 245, 333
Gillmor, D., 358
Giroux, Frank J., 339
glass rods, 207
Glass, H. F., 79
glowing object, 58, 94
glowing objects, 86
glowing ports, 97
glowing wakes, 79
GOC, 139
GOC net, 139
Goddard, Robert H., 38
gold, 2, 5, 257, 407
golden color, 138
Goldstone radars, 214
Goose AFB, 85
Goshen, IN, 85
Gould, Harold A., 339
government, 3, 12, 13, 19, 39, 40, 45, 55, 56, 61, 169, 170, 173, 187, 223, 264, 265, 292, 326, 371
government transparency, 13
Graves, Robert, 407
gravimetry, 361
gravitation, 204

gravitational attraction, 318
gravitational field, 201, 204
gravitational fields, 204
gravitational forces, 205
gravitational interaction, 318
Great Galaxy Andromeda, 162
green, 31, 93, 94, 119, 169, 172, 175, 210, 233, 257, 259, 290, 320, 321, 339
Green Bank, WV, 172
green disc, 93, 94
green disc-shaped object, 93
green fireballs, 169, 233
Green, William J., 339
Green-Rouse Productions, 210
greenish glowing object, 94
Greenstein, Jesse L., 169
Grenier AFB, 127
Grew, Donald, 332
Griffith, Andy, 407
Grignard reagents, 244
ground control interceptor station, 31
ground impressions, left by UFO, 333
Ground Observer Corps, 139, 191
ground observers, 92, 180, 191
ground radar, 57, 123, 131-133
ground returns, radar, 153
ground-based radar, 130
ground-confirmation, 76
ground-radar, 132
ground-visual sightings, 132
Grusch, David, 12
guided missiles, 89, 90
gull, 228
gulls, 225, 228
gun metal colored objects, 224
gun-camera photographs, 211
Gurney, Edward J., 339
Hahn, Otto, 38
half-circular object, 117
half-moon-shaped craft, 121
Hall, R. H., 73
Hall, Richard, 211, 284
Hall, Robert L., 11, 18, 177, 187, 195, 197, 218, 245, 347, 349
hallucination, 63, 94, 262, 301
hallucinations, 81, 263, 297, 313, 324
hallucinatory phenomena, 17
halo, 82, 138
Halter, Barton, 128
Hamilton, Ontario, CAN, 144
Hammill, Frank R., Jr., 339
hangars, 32
harbor patrolmen, 107
Harborside, ME, 111
hard data, 33, 213, 219, 221

hard-core UFO reports, 192
Harder, James A., 11, 18, 44, 47, 97, 197, 208, 242-245, 248, 349
Harding, William G., 407
hardware, 17, 23, 37, 52, 273, 289, 293
Harris, Waldo J., 91, 92, 149
Hartmann, William K., 212
Harvard, 50, 99, 147, 156, 159, 170, 232, 233, 277, 308, 312, 316, 341, 345, 353, 354, 363
Harvard College Observatory, 354
Harvard Department of English, 354
Harvard Meteor Project, 170
Harvard Observatory, 50, 147, 316
Harvard University, 156, 159, 232, 233, 308, 312, 341, 345, 353, 354
Harvard University Press, 156, 232, 233, 354
hat-like craft, 309
hats, 257
Hawaii, 306
Hawker Seafury, 129
hazard-and-hostility matters, 142
haze, 87, 110, 148, 151, 256, 330
haze layers, 87, 148
hazy rim on object, 138
headlight failures, 142
headlights, 23, 54, 169, 181, 192, 275, 280
health, 3, 39, 40, 56, 239, 358, 371, 407
heating, 275, 280
Hechler, Ken, 38, 172, 339
Heflin, Rex, 307
height-finding radar, 130
helicopter, 24, 44, 96
hemispherical object, 333
Hemphill, Oliver, 128
Henderson, Garry C., 12, 289, 293
Henshaw, Gloria, 85
Herndon, VA, 50
Hess, Seymour, 113, 114
hexagonal crystals, 206
hexagonal structure, 206
hierarchical clustering, 311
high altitude balloon observers, 82
High Altitude Observatory, 353
high civilization, 46, 72, 153
high field superconductivity, 273
high flight altitude, 78
high latitude regions, 148
high magnetic fields, 58
high speed lights, 132
high technology, 47

high velocity, 50, 86, 88
high-altitude balloons, 169
high-field superconducting magnets, 280
high-purity crystal, 207
high-speed objects, 53, 130
high-speed transport, 208
high-strength materials, 205, 207
high-tension line coronal discharge, 215
highway-accident dangers, 143
Hillenkoetter, Roscoe H., 59, 154
Hines, Richard P., 339
Hiroshima, Japan, 273
hoax hypothesis, 104
hoaxers, 45
hoaxes, 45, 63, 169, 178, 262, 279, 304, 324
hoaxing, 104
hob-goblins, 254
Hobard, Tasmania, 105
hobgoblins, 263
Hoch, Dean, 96
Hoch, Mrs. Frank R., 95, 96, 137
holes in ground, from landed UFOs, 140
holes left in ground, 323
Holloman AFB, 88
Holloman Air Force Base, 353
Hollywood, CA, 101, 104, 105
Holmes, Sherlock, 253, 263
homocentric fallacy, 47, 99
horizontal trajectory, 119, 120
Horsetooth Reservoir, CO, 330
hostile, 54, 171, 184, 185, 193, 254, 310
hostile aircraft, 185
hostile weapon system, 193
hostile weapons, 193
hostility, 54, 72, 141-143, 145-147, 154
hostility, from UFOs, 141, 145
Hot Air Theory, 262
House Committee on Science and Astronautics, 17, 59, 154, 270
House of Representatives, 17, 283, 371
houses, 58, 335
Houston Post, 106
Houston, TX, 77, 345
hovering, 23, 24, 27, 30, 43, 47, 48, 53, 57, 58, 62, 71, 72, 92, 101-104, 106, 108, 111, 121, 136, 144, 145, 168, 175, 179, 245, 270, 306, 313, 320, 332, 335

hovering craft, 30, 32
Howard, James, 84, 85
huge luminous objects, 121
human eye, 261, 263
human race, 197, 219, 224, 293, 325
humanoid creatures, 269
humanoid ETs, 171
humanoid occupants, 74
humans as specimens, 319
humans colonized by ETs, 319
humidity, 152, 153, 291
humidity gradients, 152, 153
Hunt, Betsy, 343
Hunt, David, 335
Hunt, John E., 339
hydraulic systems, 349
Hynek, J. Allen, 11, 18, 27, 35, 38-41, 43, 44, 52, 58, 67, 134, 173-175, 183, 187, 242, 246-248, 263, 272, 315
hyperbolic meteor velocities, 232
hypersonic flight, 275, 280
hypnotic experiences, 287
hypnotic procedures, 287
hysteria, 32, 63, 178, 179, 181, 185, 186, 188-190, 192-195
hysterical contagion, 178-183, 185, 188, 192, 193
hysterical contagion hypothesis, 182
IBM, 361
ICARUS, 237, 274, 345
ice crystals, 148-150, 256
Idaho, 76
identified flying objects, 34, 35, 268, 279
IFO's, 34, 35, 268, 269
ignis fatuus, 253
ignition, 43, 57, 140, 142, 143, 145, 180, 181
ignition failures, 140, 142, 143
ignition system, 43
ignitions, 142
Illinois, 11, 18, 177, 195, 278, 339, 347
illuminated blimps, 257
illusion, 94, 259, 325
illusions, 25, 279, 297, 320
Imperial Valley, CA, 160
improvised news, 189, 196
inaccurate reporting, 28
Incident at Exeter (Fuller), 74
incomplete data, 221, 263
independent investigative groups, 45
independent programs, 56
independent studies, 56
independent witnesses, 190

Indiana, 144, 248, 339, 355
indifference, 186, 194
inertia, 34, 318
inexplicable aerial phenomena, 120
injury, 141, 145
injury, from UFOs, 141
innuendo, 253
insects, 134, 153, 181, 212, 215, 229, 233, 257
Instamatic Kodak camera, 332
instantaneous course-reversal, 81
Institute of Atmospheric Physics, 18, 41, 59, 154, 343
Institute of Electrical and Electronics Engineers, 354
instrument lag, 131
instrument observations, 291
instrumental data, 44, 243
instrumental observation, 190
instrumental observing techniques, 71
instrumental recordings, 37
instrumental techniques, 243
instrumental tracking, 119
insulators, reflections from, 257
intelligence, 17, 28, 44, 49, 54, 67, 75, 128, 156, 160-164, 169-173, 175, 210, 219, 220, 256, 318, 343
intelligent extraterrestrial life, 159, 217, 221
intelligent radio communications, 167
intent of UFOs, 293
intercivilization distances, 279
intermittent occultation, 126
International Astronautical Congress, 234
International Astronautical Federation Congress, 352
International Astronomical Union, 345, 354
International Radio Scientific Union, 354
international research center, 285
International Union for Geodesy and Geophysics, 354
International Union of Theoretical and Applied Mechanics, 351
interpersonal relationships, 220
interplanetary nature, 203, 263
interplanetary significance, 202
interplanetary vehicle, 256
intersociety relationships, 220
interstellar communication, 163
interstellar space, 164, 165, 171, 219, 220, 318
interstellar space flight, 165, 171

interstellar space vehicle, 171
interstellar spacecraft, 171
interstellar travel, 47, 268, 273, 279, 318
inversion layer, 110, 151
inversion layers, 150, 151, 169, 215, 320, 328
inversions, 110, 148, 150-152, 290
inverted plate, 138
investigative files, 89
ionization gauge, 291
ionized air inversion layers, 215
ions, 28
Iowa, 79, 343, 353
Iowa State University, 343
Iron Curtain, 174
iron whiskers, 207
irrational defense of beliefs, 194
irrational systems of belief, 186
irrefutable proof, 53, 59
isotopes, 267
jack-o'-lanterns, 253
Jackson, W. T., 106
Jaffee, Daniel, 103
Jagt, Guy Vander, 339
Japan, 57, 128, 174, 306
Jefferson Davis Historical Gold Medal, 407
Jefferson, Thomas, 29
jet, 78, 82, 92, 106, 113, 118, 124, 133, 150, 180, 201, 202, 215, 216, 219, 235, 269, 275, 319
jet aircraft, 150, 202, 215, 219
Jet Propulsion Laboratory, 269, 275
jet speed, 92
jet trails, 180
jets, 47, 77, 80, 83, 92, 127, 180, 191
Johannessen, Carl, 198
John Evans Award, 354
Johns Hopkins University, 341
Jones, Howard M., 354
Journal of the Astronautical Sciences, 232, 234, 237, 239, 272, 275, 352
Journal of the British Interplanetary Society, 233
Journal of the Optical Society of America, 27
JPL, 273
JPL group, 273
Juarez, Mexico, 105
Judd, Ashley, 407
Judd, Naomi, 407
Judd, Wynonna, 407
June Bug, the, 181

Jung, Carl, 171, 284, 324, 326
Jupiter, 318, 345
Kaliszewski, Joseph J., 81-83
kangaroo, 54, 246
Kansas, 79, 278, 339
Kansas City, MO, 79
Karth, Joseph E., 339
Keffel, Ed, 306
Kenehan, Richard, 300
Kennewick, WA, 328
Kentucky, 407
Keough, Riley, 407
Kernan, Elizabeth S., 339
Kerstetter, Mr., 53
key witnesses, 47, 61
Keyhoe, Donald E., 64, 65, 73, 85, 125, 154
kinematic characteristics, 34
Kislovodsk, Caucasus, 113
kite, 95, 320
kites, 257, 320
Klass, Philip J., 51, 84, 120, 147, 152, 215, 219, 277
Klemperer, W. B., 210
Kleyweg, R. H., 117
Koepke (Capt.), 86
Korea, 182
Kratovil, Charles J., 126, 127
Krives, Fred, 107
La Paz family, 111
La Paz, Lincoln, 110
laboratory analyses, 140
laboratory experiments, 21, 163
Laboratory for Theoretical Studies, 218
landed aerial objects, 145
landed UFOs, 74, 292
landing gears, 280
landing lights, 76, 256
landings, UFO, 218
Langeac, France, 321
LANS, 101-104, 156
Lantefortain-les-Baroches, France, 321
LAPD, 103
Laramie, WY, 283
Larson AFB, 93
Las Cruces, NM, 109, 110, 151
LaSalle, A. A., 100
laser, 273
Latin America, 354
Latvia, 112
Latvian astronomers, 112
laughter curtain, 269, 271, 277
law-enforcement office, 100
law-enforcement officers, 109

laws of nature, 260, 273, 316
laws of physics, 29, 272, 273
lay misinterpretations, 60, 63
laymen, 41, 60, 62, 66, 68, 149, 150, 209
leaf fluttering motion, 335
Leick, W. H., 99-101
Leick, William, 48
Lencouacq, France, 321
lens, 12, 91, 212, 224, 227, 255, 259, 261, 328
lens flare, 212
lens-shaped clouds, 328
lens-shaped object, 91
lenticular cloud formations, 169
LePuy, France, 321
Levelland, TX, 57, 142, 180, 191, 192
liars, 23, 45
librations, 236
Liddel, Urner, 28, 120, 121
Life magazine, 75, 110, 156
life on earth, 159-161, 171, 174, 269, 274, 345
light, 24, 25, 27, 30, 31, 41, 44, 53, 54, 67, 76, 78-80, 82, 92, 101-104, 110-113, 118, 119, 121-123, 125, 126, 131, 133, 134, 136, 140, 149, 161, 164, 165, 169, 186, 191, 199, 202, 203, 212, 219, 221, 227-229, 254, 257-261, 272, 276, 305, 316, 320-322, 327, 330, 334, 335
light array, 110
light bulbs, 276
light images, 330
light leaks, 229
light plane, 82
light response, 53
light speed, 164
lighting circuits, 143
lightning, 29, 51, 57, 64, 87, 120, 142, 147, 152, 188, 215, 217-219, 222, 257, 264, 277, 278, 320, 326, 327
lightning activity, 142
lights, 25, 30-32, 35, 37, 44, 46, 53, 58, 66, 71, 76, 78, 87, 110, 124, 126, 131, 132, 142, 143, 145, 161, 168, 199, 200, 248, 256, 280, 322, 329, 334, 335
Lindheimer Astronomical Research Center, 19, 21
line formation, 329
linear distance, 213

linear velocity, 211
liquid residues, 323
Lisbon, Portugal, 327
Lissy, Walter, 107
listening post projects, 223
little green men, 175, 290
living systems, 163
local newspapers, 105, 138
Loch Raven Dam, MD, 144
Lockheed, 211, 237
Lockheed F-94 Starfire, 57, 123, 133
Lockheed P-38 Lightning, 81
Lockheed T-33 training aircraft, 122
Lockheed's Astrodynamics Research Center, 351
logic, 172, 214, 215, 253, 315
logistics, 208
London, UK, 84, 160
Long Island, NY, 122, 217
long-duration sighting, 85
Longview, WA, 49, 89, 90
Look magazine, 120, 334
looming, 111, 151, 169
Lopez, Ray, 103
Lorentz Forces, 276
Lorenzen, Coral E., 283, 317
Lorenzen, L. J., 283
Los Alamos Scientific Laboratory, 273, 277, 278
Los Angeles NICAP Subcommittee, 101, 156
Los Angeles Times, 138, 157
Los Molinos, CA, 198, 200
Louisiana, 339
Loveless, Patty, 407
Lovell, Bernard, 163
low angular velocity, 88
low velocities, 68
low-altitude objects, 101
low-altitude sighting, 97
low-flying meteors, 217
low-level inversion, 131
Lowell Observatory, 113
LSD, 46
Luce, Clare B., 98
Lukens, D. E., 339
luminescent characteristics, 34
luminescent organisms, 169
luminosity, 48, 80, 98, 113, 126, 138, 231
luminous bodies, 64
luminous boundary layers, 280
luminous object, 57, 119, 142, 258, 321
luminous objects, 50, 121, 131, 144
luminous portholes, 79

luminous ribbons, 113
luminous unidentified aerial object, 119
lunar diameter, 101, 122, 151
lunar photographs, 302
Luscombe Aircraft, 83
M.I.T., 353
M-31 galaxy, 162
Mach cone, 140
machine-like configuration, 106
machine-like craft, 93
machine-like devices, 53, 74
machine-like objects, 62
macrometeorites, 222
magazines, 21, 24, 70, 152, 253, 263
magic, 47, 289
magnesium, 78, 206, 244, 245
magnesium flare, 245
magnesium flares, 245
magnetic communications window, 280
magnetic field, 147, 203, 204
magnetic fields, 58, 143, 147, 204, 205
magnetic oscillations, 143
magnetic saturation, 146
magnetic speedometers, 275, 280
magnetics, 361
magnetization curve, 143
magnetoaerodynamic techniques, 276
magnetoaerodynamics, 269, 275
magnetohydrodynamics, 275, 355
magnetometer, 44, 291
magnetometer traces, 44
magnification, 112
man-shaped forms, 334
maneuverability, 268, 270, 272, 275, 276, 290
Maney, Charles, 330
Manhattan, NY, 51, 58
manifestations, 216
manned spaceships, 254
Manning, Robert F., 85
Marine Geodesy Committee, 361
Marine Technology Society, 361
Mariner IV, 269, 274
Mariner program, 167
markings, 138, 322
Markowitz Dual-rate Moon Cameras, 213
Mars, 3, 12, 13, 99, 159-161, 175, 195, 237, 254, 260, 268, 269, 274, 314, 316-319, 322-325, 345, 371, 409
Mars oppositions, 323
Mars, life on, 175

marsh gas, 219
Martian life, 175
Martins, Joao, 306
Martinsburg, WV, 50
Marvin, Lee, 407
Maryland, 155, 278
mass hallucination, 262
mass hysteria, 63, 178, 179, 181, 185, 186, 188, 190, 192-195
mass hysteria hypothesis, 178, 188, 190
Massachusetts, 236, 343, 358
Massachusetts Institute of Technology, 343
mathematics, 351, 361
Matsumura, Yukuse, 306
mb charts, 90
McCarthy, J. P., 355
McClave, Muriel, 83
McCrosky, R. E., 213
McDonald, James E., 11, 18, 41, 52, 55-57, 154, 165, 180, 181, 187, 198, 243, 245
McGraw, Tim, 407
McKay, Charles, 167
McMinville, OR, 306
McVay, D. R., 99
ME-163, 128
mechanical forces, 205
megagauss range, 143
Melbourne, Australia, 98, 139, 245
Meltzer, Shelley, 308, 312
mental telepathy, 325
Menzel, Donald H., 11, 28, 50, 78, 84, 87, 93, 110-112, 118, 124, 133, 147, 148, 150-152, 215, 253, 264, 265, 277, 316, 353
Mercury, 117
merging, objects, 85
Merifeld, P. M., 211
Merom, IN, 144
Messerschmitt ME-163 Komet, 128
metal, 32, 108, 206, 224, 280, 284
metal-foil, 108
metallic, 51, 89, 91, 96, 107, 109, 153, 206, 256, 321, 322
metallic fragments, 206
metallic gray, 91
metallic luster, 51, 89
metallic objects, 96
metallic-looking disc, 109
meteor cameras, 34, 213
meteor observation nets, 218
meteor physics, 60
meteor streams, 79
meteor-explanations, 86

meteor-tracks, 139
meteoric debris, 290
meteoric origin, 80
meteoric phenomena, 63
meteorite, 29, 43, 217, 318
meteorite falls, 29
meteorite shower, 43
meteorites, 29, 34, 43, 218, 232, 234, 235
meteoritic debris, 218
Meteoritical Society, 233, 238, 352
meteoriticist, 110
meteoritics, 43, 218
meteorological balloons, 117
meteorological data, 114
meteorological explanation, 86
meteorological explanations, 50, 83
meteorological optic objects, 50
meteorological optical explanations, 148
meteorological optics, 50, 84, 110, 112, 147-149, 151, 343
meteorological theodolite, 116
meteorological-optical explanation, 87
meteorological-optical UFO explanations, 152
meteorological-type UFO theories, 84
meteorology, 37, 59, 60, 113, 154, 243, 343
meteors, 22, 33, 35, 60, 83, 87, 90, 113, 136, 169, 170, 217, 222, 232, 261, 320, 326
Mexico, 30, 94, 105, 110, 117, 134, 170, 332, 333, 353, 354
Meyer, Ken, 102
Miami, FL, 86, 329
Michigan, 36, 57, 61, 95, 105, 133, 336, 339
Michigan swamps, 336
microcircuits, 273
microphotograph, 211, 231
Mid-Continent Airlines, 79
military airmen, 122
military bases, 13
Military Compact Reactor program, 355
military duty, 83
Military Establishment, 217
military jets, 92
military personnel, 12, 121, 291
military pilots, 76, 87, 133
military radar sightings, 57
Milky Way Galaxy, 162
Miller, George, 18, 39, 56, 163, 172, 283, 339
Millionshchikov, Dr., 174

Mills-Cross program, 172
Millstone radars, 214
Minczewski, Walter A., 115
Mineral Production Laboratory, 206
Minnaert, M., 260
Minneapolis, MN, 81, 82
Minnesota, 82, 182, 259, 332, 339, 347, 348, 358
minor planets, 213, 234
Miracle of Fatima, 327
mirage, 78, 85, 111, 112, 124, 131, 148, 149, 151, 255, 256, 261, 328
mirage effects, 148, 149
mirage explanations, 148, 149
mirage optics, 149
mirage phenomena, 85, 111, 148
mirage-effects, 112
mirage-refraction, 111
mirages, 27, 29, 111, 148, 151, 152, 169, 212, 215, 229, 256, 320
miraging, 131, 148
miraging effects, 131
misassociated targets, 214
misconceptions, 33, 35, 43, 127
misidentification, 22, 147, 178
misidentifications of familiar phenomena, 188
misidentified balloons, 91
misinformation, 128, 153
misobservation, 28
missiles, 89, 90, 272
Missouri, 357, 358
mist, 116, 151
modern science, 37, 204, 324
molecular biology, 56
molecules, 163, 345
monitoring programs, 139
monitoring radar systems, 243
monopulse mode, 214
Montana, 32, 210-212, 227, 231
Montgomery, AL, 77
Montgomery, deputy, 198, 200
Montreal, CAN, 152
moon, 3, 12, 81, 85, 94, 113, 118, 121, 122, 129, 149, 150, 180, 191, 202, 213, 218, 224, 256, 258, 261, 269, 274, 327, 354
moondog, 256
Mooney Mark 20A, 92
moonlight, 80, 81
Moore, Charles B., Jr., 117, 118
Moquin, Larry, 103, 104
Morse code, 53
Mosby, John S., 407
Moscow Aviation Institute, 113
Moscow, Russia, 160
Moses Lake, WA, 93
Mosher, Charles A., 339
Moss committee, 54
Moss Subcommittee, 73
motes in the eye, 279
motion plot, 225
motionless, 31, 101, 102, 104, 106, 137, 200, 321
motionless craft, 104, 200
motor vehicles, 38
motorbikes, 142
Mount Rainier, 253, 328
Mount Wilson Observatory, 169
Mountain Astrophysical Station, 113
mountains, 109, 328, 407, 408
Mt. Nebo, 92
Mt. Rainier, 64, 76, 91, 168, 327
Mueller, Robert, 79
multicolored lights, 119
multicolored luminous object, 119
multicolored object, 31
multiple witnesses, 48, 179, 190, 274
multiple-faceted exploration, 207
multiple-object case, 88
multiple-witness case, 50, 77, 88, 94, 97
multiple-witness cases, 45, 51, 87, 98
multiple-witness sighting, 136
Multiple-witness UFO cases, 98
multiple-witness UFO incidents, 69
Munger, Jay, 136, 137
Munich, Germany, 359
Muscarello, Norman, 334
mushroom-shaped craft, 300, 310
mushroom-shaped object, 301
Mysterious Invaders (Seabrook), 11
mysticism, 298
mythology of false information, 269
NAMU, 90
NAS, 83, 84, 134
NASA, 40, 43, 52, 54, 56, 167, 214, 218, 247, 275, 277, 351
NASA radars, 214
Nash, William B., 86, 87, 329
Natal Mercury periodical, 117
National Academy of Sciences, 39, 56, 170, 254, 275, 343, 345, 354
National Academy of the U.S.S.R., 264
National Center for Atmospheric Research, 293
national defense mission, 54
National Investigations Committee on Aerial Phenomena, 42, 61, 272, 286, 288, 307, 355

national leaders, 186, 194
national press, 79
National Reconnaissance Office, 13
National Science Foundation, 39, 56, 167, 181, 312, 343, 347
national scientific program, 52, 59
national security, 13, 35, 39, 291
National Vocational Guidance Association, 358
national wires, 97
natural atmospheric phenomena, 147
natural electrical effects, 320
natural phenomena, 29, 41, 63, 169, 178, 210, 212, 215, 216, 229, 264, 326, 354
natural phenomena interference, 215
natural physical phenomena, 17
natural plasmoids, 63
natural selection, 171
naval authorities, 130
naval intelligence, 343
navigation, 234, 235, 292, 293
Navy cosmic ray research program, 120
Navy jets, 83
near-horizon refraction effects, 151
Nebel, Long John, 355
Nebraska, 343, 359
Nelson, Marilyn J., 359
neutralization analysis, 245
neutrinos, 201
neutron activation analysis, 206
neutrons, 201
Nevada, 278
New England, 58, 108
New Guinea, 70, 98, 99, 245, 313, 333, 335
New Hampshire, 127, 334, 335
New Jersey, 122, 156, 234, 339, 355
New Mexico, 30, 94, 110, 117, 170, 332, 333, 353
new natural phenomenon, 188
New Orleans, LA, 79
new test vehicles, 63
New York, 48, 49, 57, 58, 75, 76, 84, 86, 99, 100, 105, 126, 127, 154-157, 160, 172, 175, 191, 195, 196, 232-234, 236, 238, 273, 275, 276, 278, 300, 329, 339, 345, 355
New York blackout, 57, 58
New York City, NY, 75, 99, 105
New York, NY, 84, 86, 329
New Zealand, 43, 130
New Zealand papers, 130
Newcomb, Simon, 260

Newhouse, Delbert C., 224-226
Newport News, VA, 86, 329
newspaper, 27, 32, 48, 73, 84, 99, 100, 127, 156, 198, 270, 331
newspaper publicity, 32, 198
newspapermen, 80, 105
newspapers, 21, 64, 70, 105, 138, 263, 264, 328
Newton, Isaac, 318
Newton, NH, 54
Newton's laws, 318
Niagara Mohawk network, 58
NICAP, 42, 48, 49, 54, 61-63, 65, 66, 71, 73, 74, 99-101, 106, 109, 144, 154-156, 272, 284, 286, 288, 310
NICAP UFO Investigator, 99
night pibal runs, 126
nighttime sights, 48
nine fluttering discs, 149
ninety-degree turns, 107
Noah's Ark, 291
noctilucent clouds, 112
nocturnal light, 27
noiseless craft, 114, 201
non-believer, 268
non-inertial course-reversal, 81
non-official UFO groups, 66
non-urban reports, 108
nonanomalistic films, 212
nonastronomical, 22, 27, 28
nonastronomical cases, 28
nonastronomical reports, 22
nonquantitative reports, 24
nonsense, 22, 41, 43, 46, 59, 97, 154, 167, 255, 261
nonterrestrial alloy, 244
nonterrestrial origin, 84
nontwinkling constellations, 228
nonuniform field, 204
North America, 232
North American Continent, 161
North American F-86 Sabre, 85
North Carolina, 181, 278, 407
North Dakota, 357-359
North Dakota Psychological Association, 358
North Dakota-South Dakota Psychological Association, 359
Northeast blackout, 146
northern California, 198, 201
northern lights, 37
Northrop F-61 Black Widow, 128
Northrop P-61 Black Widow, 57
Northwestern University, 11, 18, 21, 35, 38, 213, 263, 315, 341

Northwestern University Astronomy
    Department, 213
novel weapons, 63
Nowra Naval Air Station, 129
Nowra radar, 130
Nowra, Australia, 129
NSF, 40, 54, 361
nuclear aircraft, 355
nuclear energy, 273
nuclear power, 13, 292, 355
nuclear power plants, 13
nuclear radiations, 143
nuclear rockets, 355
nuclear war, 184, 193, 194
nuclear weapons, 64
numbness, 140
Nunn-Baker satellite cameras, 139
Nuremberg, Germany, 326
O Cruziero, 306
object-descriptions, 101
oblong object, 199
observation and interpretation, 314
observation theory, 148
observations, 28-30, 44, 46, 48, 55,
    60, 63, 68, 71, 72, 74, 83, 91,
    93, 95, 96, 99, 108, 112,
    118-120, 127, 146, 148, 149,
    152, 154, 168-170, 172,
    177-181, 186, 190, 192, 193,
    198, 201, 213, 215, 216, 228,
    233, 238, 244, 248, 256, 280,
    284, 287, 291, 313, 316, 320,
    322, 326, 353
observatories, 25, 112, 169, 326, 353
observed angular velocity, 112
occult truth, 63
occultation, 126
ocean UFOs, 327
Oceanonics, Inc., 361
oddities, 34
Odessa, Ukraine, 130
Odessa, WA, 123, 125
oersteds, 143
Office of Naval Research, 56, 81, 88
Office of Scientific Intelligence, 169
official file summaries, 89
official UFO explanations, 148
Ogra, Latvia, 112
Ohio, 27, 53, 145, 339, 341, 353
Ohio State University, 27, 341, 353
Old Testament, 326
Olsen, T. M., 87
Omaha, NE, 79, 108, 343
Ontario Hydro Commission, 58
Ontario, CAN, 58, 160
Ontario, OR, 76

open ridicule, 29
open-mindedness, 290
ophthalmologist, 22
optical deceptions, 260
optical distortions, 111
optical explanation, 84, 86
optical field, 116
optical flares, 212
optical illusions, 279
optical mirages, 169
optical paths, 111
optical phenomena, 77
optical sensor design, 220
optical space surveillance, 213
optical space tracking systems, 213
optical stimuli, 263
optical theory, 148
oral reports, 33
orange, 30, 48, 78, 97, 119, 121, 329
orange-colored trail, 119
orange-red color, 329
orange-red exhaust, 78
orange-red flames, 97
orange-red object, 121
orbit determination procedures, 351
ordinary aircraft, 31, 77
Oregon, 76, 107, 198, 224, 306, 345
organic molecules, 345
origin of life, 162, 163, 171, 345
origin of UFOs, 292
oscillation, 90
oscillation frequency, 90
outages, 57, 58, 147
outlines, 106, 107, 110, 114, 177
oval-shaped object, 145
Overall, Zan, 211
overlay trace, 225, 226
Overton, James, 136, 137
O'Brien panel, 170
O'Brien, Brian, 170
O'Keefe, John, 218
P-38, 81
P-61, 57
Pacific Northwest, 227
Pacific Ocean, 79, 160
pacing, 80
Page, Earl, 328
paintings, bleeding, 262
Palomar Observatory, 169
Pan American Airlines, 86
panic, 24, 146, 184-186, 189, 193,
    194, 314
paper, 24, 26, 27, 39, 48, 49, 90,
    105, 156, 202, 211, 216, 227,
    229, 236, 257, 265, 275-277,
    287, 354, 359, 363

parachute-shaped craft, 114
parachutes, 90
paralysis, 140, 141
paranormal processes, 287
Parapsychological Association, 284, 358
parapsychological fields, 285
parapsychological procedures, 287
paraselenae, 149
parhelia, 93, 149
Paris, France, 126, 160
particles, 32, 116
Parton, Dolly, 407
Pasteur, Louis, 38
patrolman, 109, 332, 334, 335
pattern, 27, 29, 31, 42, 46, 67, 82, 99, 105, 110, 139, 160-162, 174, 223, 225, 246, 274, 286, 295, 296, 298, 323, 324, 328
patterns, 27, 30, 140, 161, 169, 186, 195, 220, 221, 244, 296, 323, 363
Patterson, Earl, 107
Patuxent River, 134
payload fraction, 276
pear-shaped object, 211
peasant society, 25
Peking, China, 160
Pelly, Thomas M., 339
pendulum motion descent, 301
Pennsylvania, 154, 249, 278, 339, 355
Penobscot Bay, ME, 111
Pentagon, 26, 67
Pentagon press conference, 67
performance characteristics, 92
Perkins Observatory, 341
permanent photographic data, 216
Pettis, Jerry L., 44, 57, 173, 339
phantom radar returns, 134
phantom returns, 133
phased-array radar, 214, 222
phased-array radars, 222
phenomenology, 62, 101, 105, 154
Phi Beta Kappa, 351, 352
Philadelphia, PA, 51, 83
philosophers, 220
photo, 53, 96, 135-137, 332
photo reconnaissance plane, 53
photogrammetric analysis, 224, 307
photogrammetric experiment, 210
photogrammetric methods, 302
photogrammetrists, 211
photographic evidence, 23
photographic observing programs, 139
photographic plates, 169

photographic sky-survey cameras, 139
photographs, 37, 53, 159, 160, 211, 213, 230, 274, 290, 297, 301, 302, 304, 306-309, 318, 323, 325, 331, 332, 353
photos, 6, 12, 72, 135-139
physical effects, 38, 57, 140, 144, 190
physical effects of UFOs, 140
physical fragments of UFO's, 206
physical injuries, serious, 144
physical scientists, 33
physics, 18, 29, 41, 42, 59, 60, 62, 147, 148, 152-154, 163-166, 218, 232, 234, 235, 239, 264, 272, 273, 277, 311, 341, 343, 345, 351, 353
physics of clouds and precipitation, 343
physiological ecology, 365
physiological effects, 141
pibal runs, 126
pibal-station locations, 90
Pierman, S. C., 132
pilot, 27, 28, 31, 32, 51, 55, 57, 76, 77, 80, 81, 83-88, 90, 91, 109, 115-117, 121-126, 128-130, 132, 133, 255-257, 292, 306
pilot balloon, 90, 115-117, 121, 122, 125, 126
pilot balloon observation, 115, 117
pilot balloon stations, 90
pilot balloon theodolites, 115
pilot-sightings, 76, 132
pilotless aircraft, 90
pilots, 28-30, 32, 46, 49-51, 55, 72, 76, 78-81, 86, 87, 89, 121, 131-133, 149, 180, 191, 243, 255, 256, 290, 291, 329, 330
piston aircraft, 272
plane of polarization, 202, 203, 248
planetary atmospheres, 113, 345
planetary sciences, 345
planetary systems, 171
planets, 22, 60, 151, 160, 162-164, 167, 169, 171, 174, 213, 232, 234, 254, 259, 267, 320, 345, 354
Plank, Max, 269
plant physiology, 365
plants, 13, 38
plasma-UFO hypothesis, 120
plasma-UFOs, 219
plasma-vehicle, 278
plasmas, 51, 147, 152, 275, 277, 278, 280
plastic bags, 262

plastic sacks, 257
plastic-bag balloons, 263
pluralistic ignorance, 109
Pluto, 110
Podell, Bertram L., 339
Podkamenaia Tunguska River Basin, 217
Point Pleasant, NJ, 122
polar orbits, 217
polarization, 202, 203, 248
polarized light, 202, 203
Polaroid glasses, 202, 203
police, 43, 49, 53, 89, 101, 103, 107-109, 136, 144, 145, 179, 180, 190-192, 201, 243, 300, 304, 334
police officers, 49, 89, 101, 103, 108, 136, 180, 190-192, 201, 243
pollution, 39, 40
Pomona College, 365
Pooh Bah, 29
port holes, 280
Port Huron, MI, 132, 133
portholes, 79, 298
Portland, OR, 107, 224
ports, 79, 97
Portugal, 327
positive identification, 28
Poughkeepsie, NY, 191
Powell, William C., 83, 84
power distribution systems, 146
power facilities, 57
power installations, 62, 71
power line, reflections from, 257
power lines, 257, 335
power network, 58
power outage, 146, 147
power systems, 280
power-system electrical engineering, 147
Powers, W. T., 213
Prairie Network for Meteor Observations, 213
pranksters, 104, 262, 297
preexisting beliefs, 189
Prehn, K. A., 107
present-day scientific knowledge, 62, 84
preservationist, 407
Presley, Elvis, 407
Presley, Lisa M., 407
press accounts, 65, 77, 84, 101, 107, 130
press ridicule, 107
prevarication, 179, 180, 190
pre-sighting interest, 69

private groups, 62, 315
private pilot, 55, 76, 91
private pilots, 81
professional astronomers, 109, 170
professional persons, 109
professional pilots, 55
Project Bluebook, 61, 66, 74, 156, 157, 258
Project Grudge, 28, 51, 152, 155
Project Grudge report, 152
Project OXMA, 172
Project Stormfury, 343
prominent executives, 109
propagation anomalies, 57, 125, 133, 148, 152, 153
propagation-anomaly hypothesis, 132
propellers, 80
propulsion, 43, 71, 83, 114, 197, 198, 201, 203-205, 235, 269, 273-277, 280, 292, 293, 355
propulsion problem, 198, 201, 203
propulsionless craft, 114
Provo, UT, 365
pseudoscience, 325
pseudoscientism, 172
Psi Chi, 358
psychiatric evaluations, 218
psychic aspects, 284
psychic experiences, 287
psychically unbalanced individuals, 26
psycho- social implications, 285
psychological aberrations, 297
psychological effects, 97
psychological explanation, 324
psychological factors, 100, 171, 320, 324
psychological impact of contact, 223
psychological phenomena, 17, 63, 296, 313, 324
psychological responses, 314
psychological science, 296
psychological warfare, 104
psychologically oriented approach, 302
psychologically oriented techniques, 295, 302
psychologist, 22, 170, 171, 177, 178, 187, 188, 245, 296-298, 324, 347, 358
psychologists, 25, 28, 64, 69, 109, 178, 179, 182, 189, 190, 192, 291, 300
psychology, 12, 166, 183, 184, 195, 219, 283, 286, 312, 313, 347, 357-359, 363
psychopathological character, 297

psychopathology, 296, 297
psychophysical factors, 68
Pt. Pleasant, NJ, 123
public disquiet, 60
public interest, 32, 60, 183, 188, 192
public notoriety, 48
public puzzlement, 60
public safety, 13
publicity seeking, 182
Puckett, Jack, 79
Puerto Rico, 161
pulpmill smoke-drift, 91
pulsars, 163, 172
pulsating light, 102
pulsating lights, 335
puzzling flying objects, 177, 187
puzzling reports, 23, 33, 34
Pyne, Joe, 355
pyrotechnics, 105
quantitative analysis, 210, 226
quantitative details, 110
quantitative observation, 53
Quebec, CAN, 84
Quigley, Denis C., 339
Quincy, M. G., 317
race problems, 319
radar angels, 323
radar base, 191, 198-201
radar beam, 152, 262
radar beam angles, 152
radar beams, 50
radar chaff, 108
radar contact, 123, 124, 132, 180, 191, 192
radar detection, 97, 129, 169
radar energy, 153
radar equipment, 127, 133
radar evidence, 323
radar fixes, 132
radar interference, 134
radar observations, 30, 72, 169
radar observing programs, 139
radar operator, 31, 123, 128, 130, 201
radar personnel, 132, 133
radar propagation anomalies, 57, 148, 152
radar propagation physics, 153
radar refractive index gradient, 50
radar returns, 57, 124, 125, 133, 134
radar saucer flap, 262
radar scope, 31, 262
radar scope operator, 31
radar stations, 129, 179, 180, 191
radar tracking data, 53
radar UFO cases, 134

radar unknown, 72
radar unknowns, 134, 272
radar-and-camera stations, 302
radar-ducting, 131
radar-propagation, 50
radar-tracked objects, 71
radar-tracking incidents, 128
radar-visual sighting, 129, 131
radar-visual sightings, 53, 95
radarman, 128, 133
radarmen, 131
radars, reflections from, 257
radarscopes, 134
radiation counter, 291
radiation exposure, 143
radiational cooling, 150
radio, 23, 32, 90, 130, 132, 146, 161, 163, 167, 172, 173, 191, 200, 219, 220, 237, 257, 261, 270, 275, 280, 335, 353-355
radio astronomy, 163, 167, 172
radio astronomy programs, 167
radio communication, 132
radio communications, 161, 167
radio control, 90
radio signals, intelligent, 219
radio telescopes, reflections from, 257
radio waves, 261, 353
radioactivity, 143, 201, 323
radioactivity, left by UFO, 323
radios, 72, 140, 280
radiosonde, 50, 122, 123, 131, 133
radiosonde balloons, 122
radiosonde data, 131
radiosonde sounding, 133
Radnor, PA, 83
railroad ties, 140
rain, 57, 142
Rakes, Frank, 136, 137
Rammelkamp, James, 211
random turbulence, 116
Raney, Harold, 126
Rapid City, SD, 134
rapid displacements, 111
rappings, 259
rationality, 185, 194, 246
Rayburn House Office Building, 11, 265
rays on object, 119, 120
Raytheon, 293
re-entry phenomena, 63
reality of UFOs, 62, 198
recency, 190
recessional motion, 112
recognition test-materials, 310
recombination, 51

reconstruction of eyewitness accounts, 303
Recordak, 225
rectangles of light, 110
rectangular-shaped craft, 99
recurrent features, 187, 188, 190
recurring patterns of behavior, 195
red, 24, 30, 31, 44, 51, 67, 78, 83, 85, 86, 88, 97, 101-104, 117, 119, 121, 179, 190-192, 199, 200, 257, 259, 320, 321, 329, 330, 334, 335
red beam, 199
Red Bluff, CA, 179, 190, 192, 199
red Christmas ball, 102
red glow, 85, 86, 329
red light, 24, 44, 67, 101-104, 191, 199
red lights, 199, 200, 334, 335
red object, 85, 88, 101, 121, 321
red-domed discs, 51
reddish craft, 119
reddish crescent-shaped craft, 113
reddish glow, 142
reddish light, 113
Redlands University, 97, 98
Redlands, CA, 47, 51, 55, 97, 105
reflected light, 136
reflection, 106, 110, 148, 150, 151, 169, 227, 228, 256, 261, 262
reflection effects, 151
reflection of sunlight, 169
reflections, 3, 86, 150, 153, 169, 210, 212, 227-229, 256, 257, 318, 320
reflections of searchlights, 169, 257
reflectivities, 110
refracted image, 118
refraction anomalies, 149, 151
refraction phenomena, 148
refractive angular image-displacements, 118
refractive anomalies, 112, 131, 134, 149
refractive index, 50, 110, 131, 151
refractive index gradient, 50, 131
Reilly, Richard, 82
reinforced composites, 207
relativistic neutrinos, 201
reliable airlines pilots, 121
reliable observers, 29, 101
religiosity, 44
religious beliefs, 26, 245
religious cults, 189
replicated light signals, 54
report-material, 62

reportorial accuracy, 73
Republic F-84 Thunderjet, 93
research programs, 56, 351
residual magnetization, 143
residues, left by landed UFOs, 140
restricted vision, 100
retina, 258, 259, 261
retinal spot, 258
Rhode Island, 339
Richmond, VA, 115
ridicule, 25, 29, 31, 32, 37, 43, 49, 55, 70, 100, 107, 141, 182, 184, 192, 246, 248, 269, 279, 290, 291, 298
ridicule lid, 43, 44, 49, 100
ring-billed gull, 228
ring-shaped craft, 310
Rio de Janeiro, Brazil, 306, 331
Risdon, Tasmania, 105
Ritchey, James M., 132
River Derwent, 105
Riverside, CA, 307
Robertson panel, 169, 170
Robertson, H. P., 169, 255
Robins AFB, 79
rocket-like object, 79
rocket-shaped craft, 79
rocking motion, 335
Rocky Ford, CO, 357
Rogers, Wilbert S., 123
Romans, ancient, 220
Rome, Italy, 98
rooflines, 103
Rosalia, WA, 125, 126
Rosi, Soe, 102
rotating rim, on UFO, 322
Roudebush, Richard L., 339
Rouen, France, 306
round lighted objects, 108
Roush, J. Edward, 11, 18, 38, 39, 41, 52, 55, 56, 159, 164, 173-175, 177, 184, 197, 205, 208-210, 216, 229, 241-248, 264, 283, 339
rows of lights, 78
Royal Astronomical Society, 354
Royal Meteorological Society, 343
Rucker, Edmund W., 407
rumor, 63, 189, 196
rumor phenomena, 63
rumors, 189
Rumsfeld, Donald H., 18, 39, 40, 52, 166, 207, 339
running lights, 126
Ruppelt, Edward J., 64, 73, 95, 114, 122, 124, 133, 155, 225, 272

Russian astronomers, 112
Russian incident, 130
Russian UFO reports, 112
Russians, 173, 174
Ryan, William F., 55-58, 339
S-9 gravity meter, 361
Sabre, 85
Sacramento Peak Observatory, 353
Sacramento, CA, 80, 161
Sagan, Carl, 11, 159, 163, 164, 168, 172-175, 187, 241-243, 245, 322, 323
SAGE, 243
Saginaw Bay, MI, 57
saints, moving statues, 262
Salisbury, Frank B., 12, 284, 313, 336, 365
Salt Lake City Weather Bureau, 93
Salt Lake City, UT, 91, 93, 149, 227
salvation, 26, 63, 245
Samford, John A., 44, 67, 84
Samuel Goldwyn Studios, 210
San Bernardino, CA, 306
San Francisco, CA, 161
Sandia Base, 94
Santa Ana, CA, 307
Santa Monica, CA, 210
Sao Paulo, Brazil, 206
Saskatchewan, CAN, 353
satellite debris, 255
satellite reentries, 35
satellite tracking cameras, 34, 272
satellite tracking stations, 34
satellites, 32, 33, 63, 113, 139, 169, 212, 214, 215, 235-237, 259, 316, 320
Saturday Review (NH), 334
Saturn-shaped object, 307
saucer flap, 262
saucer reports, 81, 170
saucer-like objects, 328
saucer-shaped clouds, 257
saucer-shaped craft, 224
saucer-shaped objects, 96
saucers, 27-29, 32, 106, 120, 130, 154-156, 168, 171, 175, 220, 224, 232-234, 253, 256, 257, 261, 262, 267-270, 272, 273, 277-279, 284, 354
Savage, Michael, 306
Savannah River, 94, 95
Savannah River AEC facility, 94
Savannah River, GA, 95
Sazanov, Anatoli, 113
Scandinavia, 327
Scandinavia, UFO flap, 327

science, 10-12, 17, 18, 21, 23, 24, 29, 31-39, 41, 43, 47, 56, 59-62, 65, 153-155, 164-167, 171, 181, 201, 204-206, 217, 219, 233-237, 239, 257, 263, 265, 270, 273, 275, 277, 283, 284, 288, 289, 296, 297, 312, 313, 316, 324, 336, 339, 341, 343, 347, 349, 351, 352, 354, 358, 365, 371, 407
science fiction, 205, 206, 354
scientific approach, 25, 267, 277
scientific bodies, 26, 39
scientific circles, 65, 73, 135
scientific clues, 197
scientific community, 8, 40, 41, 44, 45, 52, 59, 61, 65, 153, 170, 216, 289, 314, 325, 335
scientific curiosity, 171
scientific data, 24, 36, 44, 207
scientific disciplines, 217, 220
scientific hearing, 33
scientific importance, 13, 41, 59, 61, 248
scientific laboratories, 23, 40
scientific method, 7, 32, 164, 166, 221, 296, 314, 320
scientific mystery, 60
scientific obligation, 27, 29, 37, 98
scientific organizations, 61
scientific problem, 17, 59, 60, 154, 156, 277
scientific progress, 29, 276
scientific rejection, 105
scientific respectability, 37
scientific secrets, 205
scientific societies, 19, 40, 341, 349
scientific stigma, 26
scientific study, 25, 26, 33, 198, 209, 220, 275, 295, 296, 310
scientific workers, 28
scientifically significant phenomenon, 89
scientifically trained people, 34, 35
scintillation, 118, 261, 341
scope operator, 30, 31
scope-identity, 130
Scott, George C., 407
Scott, S. E., 198-200
Sea Bright, NJ, 123
Seabrook, Lochlainn, 13, 407, 409
Seabrook's Bible Dictionary (Seabrook), 409
Seacliff, NY, 58
search for life, 59, 274
searchlight, 87, 330

searchlight hypothesis, 330
searchlights, 169, 257
Seattle, WA, 32, 76
second gravitational field, 204
secret devices, 188, 326
secret test devices, 63
seed pods, 257
Seff, Philip, 48, 97, 98
seismicity data, 244
seismologists, 244
seismology, 243
Seitz, Frederick, 56
self-luminous, 111
semi-secret advanced technology, 63
sensational conclusions, 263
separate observers, 190
separate witnesses, 195
Seven Islands, CAN, 84
severe bodily damage, 141
Shamokin, PA, 53
shape, 30, 69-71, 79, 81, 85, 86, 91, 94, 98, 101, 180, 191, 202, 268, 270, 276, 280, 291, 292, 300, 301, 307, 308, 310, 322, 335
shape-changes, 85
shape-shifting, 85
shape-shifting craft, 202
Shelbyville, KY, 57
Shepard, Roger N., 12, 295, 312, 313, 363
sheriffs' deputies, 109
Sherman, Harold, 102, 103
Shibutani, Tamotsu, 189
shimmering, 229
Shklovskii, Dr., 174
short-range ARS radar, 132
Siberia, 217
Sigel, Joseph, 306
sighting, 27, 30, 31, 35, 43, 44, 48-50, 55, 57, 58, 61, 63, 64, 67, 69, 74, 76, 77, 79, 81-87, 90, 91, 93-95, 97-101, 106-120, 122-124, 126, 127, 129, 131-133, 136, 149-151, 155, 168, 191, 192, 198, 202, 245, 253, 263, 264, 271, 272, 274, 290, 291, 313, 315, 317, 320-323, 326-335
sightings, 17, 27-30, 32-34, 42, 43, 45-48, 51, 53, 56-58, 60, 61, 63, 65-68, 71, 73-77, 79, 81-83, 86, 87, 90, 95, 97, 98, 100, 105, 107-109, 112, 114, 119, 122, 123, 128, 130, 132-137, 140, 142, 147, 150, 168-170, 172, 174, 180, 181, 186, 188, 191, 192, 207, 219, 242, 248, 249, 253, 254, 257, 262, 267-272, 275, 277, 278, 289-291, 306, 308, 310, 314-318, 321-323, 325-329, 333-336
Sigma Pi Sigma, 352
Sigma Xi, 343, 352
Signal Test Processing Facility, 213
silent craft, 30, 137, 191, 199, 200, 318, 321, 334, 335
silhouette, 31, 129
silica fibers, 207
silver, 98, 111, 122, 257, 334
silver suits, 334
silver-colored object, 122
silvery craft, 96
silvery discs, 328
silvery object, 115, 202
Sioux City Weather Bureau, 81
Sioux City, IA, 79
Sirius, 124, 125, 151, 320
size, 32, 66, 68, 71, 78, 79, 81, 86, 89, 90, 92, 94, 96, 101, 103, 107, 111-113, 118, 120, 122, 123, 129, 137, 151, 180, 191, 200, 213, 218, 224, 227, 268, 280, 291, 292, 300, 303, 304, 309, 316
Skaggs, Ricky, 407
skeptical people, 26
skeptical scientists, 46
skepticism, 46, 69, 72, 154, 170
skeptics, 68, 95, 99
skin burns, 145
skin irritations, 143
skin-burns, 144
skin-reddening, 141, 144
skin-warming, 141
sky-cover data, 81
sky-scanning, 100
sky-writing aircraft, 49
Skyhook balloons, 49, 81, 91, 121
Skyhooks, 88
slip plane, 206
Smart, Richard, 33
Smith Prize, 345
Smith, Douglas, 82
Smith, Emil J., 76, 77
Smithsonian Astrophysical Observatory, 11, 213, 264, 272, 341, 345, 354
smoke, 89, 91, 319
soaring bird speeds, 229
social advancement, 220
social groups, 189
social psychological phenomenon, 194

social psychological problem, 177, 186, 194
social psychologists, 109, 178, 179, 182, 189, 190, 192
social psychology, 177
social scientists, 33, 34, 220, 296
socialist countries, 174
Society for Psychological Study of Social Issues, 284
Society for the Psychological Study of Social Issues, 286, 357, 359
Society of Civil Engineering, 349
Society of Exploration Geophysicists, 361
Society of Plant Physiology, 365
sociological importance, 314
sociologist, 22, 178, 188
sociology, 11, 18, 177, 184, 195, 347, 357
Socorro, New Mexico, 332
sodium vapor releases, 257
soft data, 216, 217, 219, 221, 222
soil, 140, 349
soil conservation, 349
solar activity, 353
solar images, 150
Solar Radio Observatory, 353
solar system, 160, 163, 171, 254, 269, 273, 274, 292, 316
solar-terrestrial relationships, 353
solid evidence, 248
solid state physics, 273
sombrero-shaped craft, 96
Sommers, Mr., 54
sonic boom, 102, 275
sonic booms, 46, 140, 318
Sons of Confederate Veterans, 407
sound barrier, 319
sound testimony, 188
soundless craft, 30, 43, 69, 71, 102, 106, 110
soundlessness, 71
South Africa, 130, 247
South America, 74, 354
South Bend, IN, 86
South Dakota, 359
Soviet Union, 172, 173, 195
space, 12, 13, 18, 26, 29, 56, 59, 75, 77, 90, 95, 103, 107, 122, 128, 141, 143, 152, 154, 156, 159, 164, 165, 170, 171, 175, 213-215, 219-223, 232-239, 241, 253-255, 257, 267, 269, 272-277, 280, 293, 295, 296, 314, 315, 318, 319, 345, 351, 354, 355, 361, 365

space animals, 188
space biology, 365
space brothers, 26, 75
Space Defense Center, 214
space exploration, 314
space flight, 165, 171, 237
space objects, 214, 215, 222, 223, 241
space program, 59
space radiation, 280
space sciences, 12, 293, 361
space surveillance, 213, 214, 223, 241, 272
space surveillance sensors, 241
space-based surveillance, 222
spacecraft, 12, 171, 204, 206, 275, 283, 285, 292, 319
spaceship, 23, 44, 67, 68, 267, 314, 315, 318-320, 322, 324, 334
spaceship hypothesis, 67, 315, 318, 320
spaceship interpretation, 68
spaceships, 63, 67, 254, 267, 314-316, 319, 322-324, 328
Spain, 202, 262, 317
Spain Flying Field, 202
sparkling object, 334
special-purpose sensor system, 213
spectrographic equipment, 222
spectrographic test, 206
specular reflection, 150
speed, 28, 30, 32, 48, 53, 71, 76, 77, 84, 86, 88, 92, 94, 100, 107, 111, 113, 114, 118, 121, 124, 125, 127-133, 147, 165, 191, 199, 200, 208, 212, 213, 220, 224, 239, 243, 255-257, 269, 270, 272, 275, 276, 280, 291, 292, 315, 321, 327, 328, 332, 334
speed of light, 164
spheres, 112, 326
spider silk, 228
spider web reflections, 228
spider webs, 229, 257
spiders, 228
spiraling meteorite decay, 217
spirits, 254
spiritual implications, 285
spiritualism, 298
Spokane, WA, 125, 126
Sprinkle, R. Leo, 12, 283, 357
spurious images, 214, 257
spurious optical reflections, 229
Sputnik 1, 169
squall-line, 78

square holes, 138
St. George, Minnesota, 332
St. Lawrence Valley, 51
Staffordshire, UK, 46
staged fission propulsion systems, 269
Stanford Research Institute, 293
Stanford University, 12, 312, 345, 363
Stanford University Medical School, 345
star formation, 162, 163
star mirage, 256
star-like light, 30
starlike object, 112
starlike objects, 31
stars, 22, 25, 90, 125, 151, 161-165, 169, 170, 257, 259-261, 263, 268, 269, 273, 316, 320, 334, 341, 354
state troopers, 109
state wires, 97
statues, moving, 262
Stearman, 89
stellar images, 118
stellar magnitude, 30
stellar scintillation, 341
stellar spectroscopy, 341
Stephens College, 357
Stephens, Alexander H., 407
Stevens (co-pilot), 76
stewardess, 28, 76, 85
Stockholm, Sweden, 347, 351
Stoke-on-Trent, UK, 46, 75
stones, 28, 29, 43
stopwatch, 116, 117
STPF, 213, 214
strange aircraft, 29, 130
strange light, 31
strange objects, 182, 267
strange shining objects, 224
stratiform clouds, 120
Strauch, Arthur, 332
stress concentration, 207
strong magnetic fields, 143, 147
strontium, 206, 244
structural features, 106
structural mechanics, 351
study of eyewitness credibility, 219
sub-cloud altitudes, 120
sub-suns, 150
submarine-shaped craft, 310
submodulations, 163
successive nightly sightings, 207
Sul Ross State College, 361
sun, 24, 29, 48, 90, 91, 93, 96, 107, 113, 117, 118, 149, 150, 157, 161, 162, 169, 202, 210, 216, 227, 228, 255, 256, 258, 325, 327, 353, 354
sun dogs, 169
sun reflections, 227, 228
sun-glint, 90, 91, 96, 100, 107, 117, 227
sun-illumination, 113
sunburn, 144
sundog, 81, 93, 149, 255, 256, 326
sundog hypothesis, 93
sundogs, 85, 93, 148-150, 320
Sunshine Sisters, the, 407
Super-Schmidt cameras, 170
superconducting magnets, 280
superconducting sources, 205
supersonic flight, 273
supersonic speeds, 140
surface features, 270, 280
surface water hydrology, 349
surge-protectors, 147
surveillance, 46, 51, 52, 54, 59, 63, 99, 135, 153, 154, 213-215, 217, 222, 223, 241, 243, 272
surveillance radars, 135, 214, 215, 241
swamp gas, 36, 61, 95, 106, 175
swans, 232
swarm of bees motion, 111
swarming bee UFOs, 112
sworn statements, 30
symmetric pattern, 110
Symposium on Unidentified Flying Objects, 11
Syracuse, NY, 58
system outages, 57, 147
systematic evidence gathering, 194
systematizing eyewitness accounts, 304
systems of belief, 177, 183, 186-189, 194, 245
T-33, 122, 123
tailless craft, 46, 76, 79, 81, 88, 92, 96, 129
Tamaroa, IL, 57
Tampa, FL, 79
Tasmania, 43, 105
Tasmanian Hydroelectric Commission, 105
Taylor, Moulton B., 49, 89-91
Teague, Olin E., 339
technical civilization, 161-163, 171
technical corporations, 109
technicians, 36, 94, 292
technological devices, 178, 188
technological forecasting, 220

technological machinery, 17
Tehama County Sheriff's Office, 198, 199, 201
telekinesis, 261
telepathy, 325
telescope, 112, 114, 161, 162, 172, 173, 261
telescopes, 25, 169, 213, 232, 257
television antennas, reflections from, 257
television disturbance, 146
television transmission, 161
temperature, 30, 148, 150, 151, 153, 169, 223, 227, 290, 291
temperature inversion layers, 169
Tennessee, 3, 277, 407
terrestrial craft, 92, 255
terrestrial foundries, 218
terrestrial life, 172
terrestrial nuclear power, 355
terrestrial salvation, 63
terrestrial sensors, 209
terrestrial standards, 193
terrestrial technologies, 143
territorialization, 245
test maneuvers, 130
Texas, 12, 105, 106, 191, 233, 238, 278, 287, 339, 353, 361
Texas A&M University, 361
texture, 92, 270
The Concise Book of Owls (Seabrook), 409
The Martian Anomalies (Seabrook), 12, 371, 409
The New Yorker, 75
The UFO Evidence (Hall), 73
The UFO Problem, 11
theodolite, 82, 115-118
theodolite field, 116
theodolite observers, 82
theory, 29, 148, 164, 195, 196, 204, 232, 238, 245, 262, 353
theory of relativity, 164
thermal updrafts, 229
thickness-to-diameter ratio, 83
Thor-Able Star launching, 211
thunderstorm, 51, 120
thunderstorm activity, 120
Tiernan, Robert O., 339
time estimation, 68
Time magazine, 47, 156, 273
tingling, 140, 141
TIROS photograph, 160
titanium, 206
Tombaugh, Clyde, 109, 110, 151
Toulouse Observatory, 114

tower operator, 30, 81
tower radar, 132
tracking television camera, 292
trajectory, 119, 120, 122, 126, 127, 139, 241, 242, 274, 333
trajectory objects, 242
transmission properties, 118
transoceanic shortwave broadcast, 130
trapping, 50, 131, 153
Tremonton, UT, 224, 227
Trent, Paul, 306
triangulated altitude, 104
triangulation shots, 138
Trindade Island, 307, 330
Trinidad, 95, 96, 137
Trinidad, CO, 95, 96, 137
troposphere, 88, 134
tropospheric discontinuities, 153
trucks, 140
true believers, 26, 34
true frequency of UFO sightings, 97
Truk Island, 79
trusted information, 189, 190, 193
tubes, 276, 326
Tübingen, Germany, 320
tubular craft, 103
Tucson airport Weather Bureau station, 121
Tucson, AZ, 61, 121, 343
Tularosa Basin, 353
Tulle, France, 321
Tunguska event, 217
turbulence, 116, 118, 150, 229
turbulent image-displacement, 125
turn and climb, 292
TV reception, 275
TV receptions, 140
TWA, 79, 85, 86, 126, 127
TWA Constellation, 126
Twain, Mark, 46, 72
twinkling dots, 225
U.S. Air Force, 21, 168, 175, 192, 196, 268, 310
U.S. Air Force, deception of, 268
U.S. Bureau of Standards, 244
U.S. Department of Agriculture, 349
U.S. Navy, 224, 341, 343, 349, 353
U.S. Patent Office, 276
U.S. Weather Bureau, 115, 152
U.S. Weather Bureau station, 115
U.S.-Canadian eclipse expedition, 353
U.S.S. Glacier, 119
UAP, 13, 327
UAPs, 7, 13
Ubatuba magnesium, 244
Ubatuba, Brazil, 206

UCLA, 11, 18, 210, 216, 224, 234, 235, 238, 351
UCLA Physics Prize, 351
UFO, 3, 5, 6, 11, 13, 21-27, 32-38, 41-47, 49, 52, 54, 55, 57-71, 73-77, 81, 83-85, 87, 89, 93, 95, 97-100, 105, 106, 108, 109, 111, 112, 114, 115, 118-122, 126-130, 134-144, 146-156, 165, 169-171, 173, 174, 179, 185, 187, 189, 191-195, 197, 198, 201, 204, 205, 207, 212, 219, 228, 229, 233, 239, 246, 249, 253, 255-259, 261-264, 267-269, 272-277, 280, 283-293, 295-299, 308-311, 313-327, 330, 334, 355, 358, 359
UFO addicts, 256
UFO annals, 134
UFO buffs, 33, 35, 253, 255, 261, 262
UFO case material, 73
UFO cases, 24, 33, 61, 65, 73, 74, 76, 89, 98, 108, 134, 142, 143, 148, 150, 179, 297, 310
UFO confusion, 25
UFO data, 23
UFO detection, 100
UFO diple, 143
UFO documentation, 73
UFO encounters, 24, 146
UFO evidence, 62, 73, 141, 155, 195, 272
UFO interference, electronic, 140
UFO investigating group, 74
UFO investigations, 100, 275, 283, 290
UFO investigative programs, 61
UFO investigator, 68, 99, 119, 156, 288
UFO investigators, 67, 97, 135, 141, 334
UFO investigatory program, 114
UFO occupants, 75, 284, 319
UFO phenomenology, 154
UFO phenomenon, 3, 5, 6, 11, 21, 22, 24-27, 32-34, 37, 38, 70, 146, 233, 256, 277, 314, 325, 326
UFO photos, 135, 136, 139
UFO problem, 11, 21, 23, 36, 37, 41, 46, 54, 59-62, 64-66, 69, 70, 73, 75, 76, 98, 105, 111, 112, 119, 122, 127, 142, 143, 148, 152-154, 170, 255, 292, 310, 316
UFO propulsion problem, 198
UFO reference material, 74
UFO reports, 22-26, 33, 34, 36-38, 60, 63, 66, 68, 70, 71, 75, 77, 87, 93, 97, 109, 112, 120, 121, 141, 147, 169-171, 185, 187, 189, 191-195, 219, 233, 261, 272, 283-288, 298, 309, 358, 359
UFO research, 26, 154, 249, 287, 355
UFO Research Institute, 154, 249, 355
UFO scoffers, 63
UFO shapes, 305
UFO sighting, 43, 55, 57, 67, 77, 100, 115, 155
UFO sightings, 42, 45, 47, 57, 75, 81, 95, 97, 98, 105, 119, 136, 140, 142, 147, 150, 169, 192, 207, 219, 257, 267, 269, 275, 291, 308, 314, 315, 317, 318, 326
UFO skies, 25
UFO studies, 68, 100, 289
UFO Symposium, Montreal, 152
UFO, definition of, 35
UFO-type sightings, 75
UFO-witness interviewing, 60
UFOs, 3, 7, 8, 12, 13, 17, 59, 60, 62, 64, 65, 67-70, 72, 74, 81, 86, 87, 97-100, 105, 106, 109, 112, 114, 115, 120, 121, 128, 134-136, 138-144, 146-149, 151-157, 188, 190-192, 194, 265, 270-274, 276-280, 283, 286, 292, 301, 304, 306, 308, 309, 355, 363, 371, 409
UFOs and Aliens: The Complete Guidebook (Seabrook), 12, 355, 371, 409
UFO's, 21, 22, 24-26, 32-35, 37-39, 41, 44-46, 50, 51, 53, 56-58, 73, 76, 87, 98, 108, 127, 135, 140, 151, 159, 166-170, 173, 181, 183, 197, 198, 201, 202, 204-207, 209, 219, 224, 227, 229, 233, 243, 247, 248, 253-260, 262-264, 267-269, 272, 275, 289-293, 295-297, 299, 301, 313-316, 318-320, 322-324, 326, 327, 330, 354
ultra-high-speed objects, 53
UN cooperation, 187
unaided observation, 190, 290

uncommon weather conditions, 169
unconventional aerial devices, 79
unconventional aerial machine, 98
unconventional aircraft, 30, 169
unconventional objects, 81, 87, 101, 109, 130
unconventional phenomena, 142
uncorrelated targets, 215, 242
undersun, 53, 150
UNESCO, 347
unexplainable aerial performance, 95
unexplainable craft, 38
unexplainable phenomenon, 114
unexplained phenomena, 142
unfulfilled religious needs, 171
unidentifiable lights, 131
unidentified aerial objects, 27, 28, 50, 88
unidentified aerial phenomena, 295, 311, 343
Unidentified Flying Objects, 11, 17, 21, 36, 59, 81, 115, 154-156, 168-172, 174, 175, 179, 181, 182, 188, 197, 203, 224, 232, 248, 249, 270, 272, 273, 275, 277, 279, 283, 286
unidentified lights, 132
unidentified radar cases, 57
unidentified targets, 30, 131
unidentifieds, 34, 35, 81
uniform magnetic field, 204
United Flight 105, 76
United Nations building, 99
United States, 2, 6, 22, 37, 41, 43, 49, 52, 127, 160, 168, 170, 174, 197, 205, 215, 217, 234, 249, 254, 291, 297, 316, 327, 328, 331, 333, 347, 351, 353, 355, 359
universal hypotheses, 27
Universal Studios, 212
universe, 34, 59, 72, 162, 170, 175, 223, 233, 254, 279, 292, 316, 318, 326, 330, 345
university, 11, 12, 18, 21, 27, 35, 38, 41, 47, 55, 56, 59, 82, 97, 98, 100, 110, 113, 137, 139, 154, 156, 159, 167, 170, 177, 182, 195, 198, 201, 208, 212, 213, 232, 233, 235, 238, 262-264, 275, 278, 283, 286, 287, 289, 292, 293, 308, 310, 312, 313, 315, 336, 341, 343, 345, 347-349, 353-355, 357-359, 361, 363, 365
University of Arizona, 11, 18, 41, 59, 137, 154, 212, 343
University of California at Berkeley, 18, 208
University of California, Berkeley, 11, 345, 349
University of Chicago, 232, 341, 343, 345, 355
University of Chile, 354
University of Colorado project, 56
University of Colorado UFO project, 139
University of Denver, 354
University of Illinois, 11, 18, 177, 195, 278, 347
University of Minnesota, 82, 182, 347, 348
University of Minnesota Airport, 82
University of Missouri, 357, 358
University of North Dakota, 357, 358
University of Omaha, 343
University of Oregon, 198
University of Redlands, 97
University of Stockholm, 347
University of Utah, 365
University of Wyoming, 12, 283, 286, 287, 357
unknown luminous objects, 131
unknown phenomena, 57
unknowns, 50, 87, 90, 130-132, 134, 258, 268, 271, 272, 278
unswept wing, 80
unswept wings, 81
unusual aerial phenomena, 27, 29, 284
unusual external light patterns, 169
unusual injuries, 144
Upington Meteorological Station, 117
Upington, South Africa, 117
upper atmosphere, 25, 232, 341
Upper Atmosphere Studies and Satellite Tracking, 341
upper winds, 90, 122
upper-air research balloon, 124
upper-level winds, 122
upper-wind data, 104, 125, 127
upper-wind measurement, 117
urban reports, 108
URSI, 354
USAF-ATIC, 225
USO, 327
Ussel, France, 321
USSR, 112, 113, 351, 353
USSR Academy of Sciences, 113
Utah, 12, 91, 92, 210, 211, 224, 227, 229, 230, 313, 328, 336, 365
Utah Central Airport, 91

Utah Lake, 328
Utah State University, 12, 313, 336, 365
vacuum tubes, 276
Vallee, Jacques, 75, 114, 317, 326
Vallee, Janine, 317
Vancouver, WA, 107
Vandenberg AFB, 211
vapor trail, 82
vapor trails, 31
vapor-like trails, 113
Vega, 320
vegetation, burned, 333
vegetation, burned by UFO, 323
vegetation, disturbed by landed UFOs, 140
velocity, 50, 86, 88, 94, 112, 116, 117, 140, 211, 222, 226, 229, 242, 268, 272, 273, 276, 280, 291
Venezuela, 211
Venus, 23, 28, 41, 66, 92, 93, 169, 237, 257, 258, 274, 290, 318, 320, 325, 334, 345
vernier drives, 117
Vichy, France, 321
Villela, Rubens J., 119, 120
Vina Plains Fire Station, 199
Vinther, Lawrence W., 79-81
Virginia, 54, 172, 274, 276, 278, 329, 339, 407
visibility, 84, 93, 227, 290
vision, 327
visions, 263
visitors from space, 29, 156
visual contact, 30, 192
visual discrimination, 68
visual observation, 129, 179, 190, 191
visual observations, 30, 152, 170, 291, 313
visual observing programs, 139
visual unknown, 72
Vitolniek, R., 112
Voyager program, 167
Waggonner, Joe D., Jr., 339
Waikiki, HI, 306
Walker, Charles, 102
Walker, Sydney, III, 210, 239
Wall Street Journal, 246
wallabies, 246
wallaby-hunting, 70
wandering lights, 32
warning operators, 31
wars, 220, 319, 325
wartime military flying duty, 77
Washington, 6, 11, 17, 42, 49, 50, 90, 93, 95, 125, 131, 132, 155, 156, 160, 168, 169, 195, 196, 224, 262, 265, 283, 288, 327, 328, 339, 357, 359, 371
Washington National Airport, 50, 95, 131, 132, 262
Washington, D.C., 6, 11, 17, 131, 155, 156, 169, 195, 196, 224, 265, 283, 288, 359, 371
waste disposal, 40
watches, 258, 275, 280
Watson, Harold E., 210
wave of sightings, 46, 327, 336
weaponry, 105
weapons, 63, 64, 193, 326
weather balloon, 93, 121-123, 126, 127, 138
weather balloons, 120-122, 138, 256
Weather Bureau, 51, 81, 93, 104, 114-116, 121, 152, 174, 262
weather conditions, 51, 133, 169, 229
weather data, 79, 142
weather kite, 320
weather maps, 84
weather modification, 343
weather observer, 115
weather observers, 28, 46, 115
weather officer, 79
weather phenomena, 133
weaving, 104
web reflections, 228
Webb, Walter N., 198
Webb, Wells A., 201-203, 208
weird effects, 263
Wells, Gregory, 145
West Point, 260
West Virginia, 54, 172, 278, 339, 407
Westinghouse astronuclear plant, 11, 267
wet ignitions, 142
whining sound, 332
white cloud-like object, 202
white craft, 116, 117, 123, 126
white dots, 210, 225
white glow, 200
white light, 31, 123, 125
white lights, 124, 199, 200
white object, 106, 117, 123, 212
White Plains, NY, 180, 191
white running lights, 126
White Sands Proving Ground, 117, 119, 353
white vapor trails, 31
white-domed discs, 51
whitish craft, 144
Whitted, John B., 77, 78, 84, 149

wild theories, 245
Williams, John M., 79, 80
Willow Grove NAS, 83, 84
Willow Grove, PA, 83
wills-o'-the-wisp, 253
Wilson, James E., 339
Wilson, Richard, 120
Wilson, Woodrow, 407
wind, 93, 94, 104, 117, 124-127, 228, 229, 257, 327
wind speeds, 124
wind, moving against the, 104
window lights, 110
window-like craft, 110
windows, 78, 106, 110, 257, 322
windows in craft, 124
winds, 90, 91, 104, 118, 122, 125, 126, 212, 227
wingless craft, 43, 45, 46, 48, 76-79, 83, 88, 92, 96, 111, 129, 200
wingless discs, 71, 76
wingless objects, 46
Winn, Larry, Jr., 339
wire editors, 48, 97
wire services, 48, 138
wishful thinking, 64
witchcraft, 167, 264
witches, 254
Witherspoon, Reese, 407
witness-interviewing, 64, 101, 104, 108, 136
witness-reluctance, 67
witness-response, 67
witness-variance, 95
witness-variances, 101
witnesses, 22-24, 33, 37, 41-45, 47-49, 53, 55, 58, 60, 61, 63-71, 79, 81, 83-85, 87, 88, 90-92, 94-103, 106-109, 114, 118, 121, 130, 136, 137, 140, 141, 144, 145, 177, 179-183, 187, 190, 192, 195, 198, 217, 239, 245-247, 260, 263, 268, 270, 274, 298-304, 306, 309, 313, 314, 321-325, 327, 331-335
wobbling motion, 107, 111
Womack, Lee Ann, 407
Woods, Jimmie, 106
Woollard, G. P., 361
Working Group on Extraterrestrial Resources, 361
world society, 38
World War II, 49, 262, 353
World War III, 217
Wright Field, 28, 30

Wright Patterson Air Force Base, 28, 210
written reports, 33, 34
Wydler, John W., 202, 339
Wyoming, 5, 6, 12, 13, 283, 286, 287, 357, 358
Wyoming Personnel and Guidance Association, 358
Wyoming Psychological Association, 358
X-ray diffraction, 206
Yale University, 238, 347, 363
Yeager, Philip B., 339
yellow, 30, 31, 110, 113, 211, 231
yellow amber, 30
yellow light, 113
yellow lights, 110
yellow object, 211
Yerkes Observatory, 341
Yokohama, Japan, 306
Yuma, AZ, 116, 202
Zacko, Joseph, 132
Zamora, Lonnie, 332
zig-zagging motion, 107
Zigel, Felix, 112, 113, 130
zinc, 206, 244, 245
zinc strontium, 244

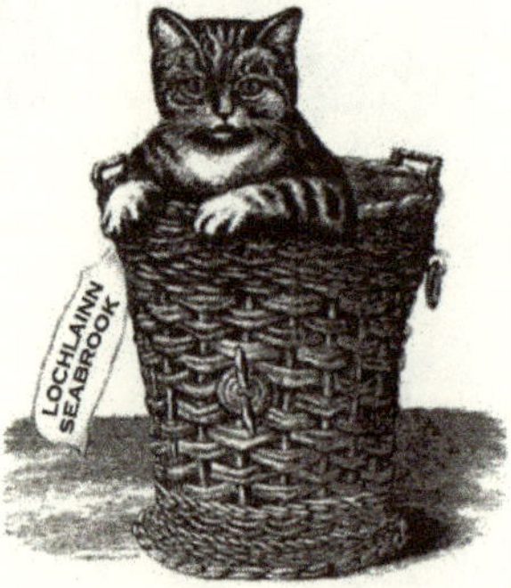
LOCHLAINN SEABROOK

MEET THE AUTHOR-EDITOR

NEO-VICTORIAN SCHOLAR LOCHLAINN SEABROOK, a descendant of the families of Alexander Hamilton Stephens, John Singleton Mosby, Edmund Winchester Rucker, and William Giles Harding, is a 7th generation Kentuckian and one of the most prolific and widely read writers in the world today. Known by literary critics as the "new Shelby Foote," the "American Robert Graves," the "Southern Joseph Campbell," and by his fans as the "Voice of the Traditional South," he is a recipient of the United Daughters of the Confederacy's prestigious Jefferson Davis Historical Gold Medal. A lifelong writer, the Sons of Confederate Veterans member has authored and edited books ranging in topics from history, politics, science, religion, spirituality, astronomy, entertainment, military, and biography, to nature, music, humor, gastronomy, etymology, onomastics, alternative health, genealogy, and the paranormal; books that his readers describe as "game changers," "transformative," and "life altering."

One of the world's most popular living historians, he is a 17th generation Southerner of Appalachian heritage who descends from dozens of patriotic Revolutionary War soldiers and Confederate soldiers from Kentucky, Tennessee, North Carolina, and Virginia. Also a history, wildlife, and nature preservationist, the well-respected polymath began life as a child prodigy, later maturing into an archetypal Renaissance Man. Besides being an accomplished and esteemed author, historian, biographer, creative, and Bible authority, the influential litterateur is also a Kentucky Colonel, eagle scout, screenwriter, nature, wildlife, and landscape photographer and videographer, artist, graphic designer, genealogist, former history museum docent, and a former ranch hand, zookeeper, and wrangler. A songwriter (of some 3,000 songs in a dozen genres), he is also a film composer, multi-instrument musician, vocalist, session player, and music producer who has worked and performed with some of Nashville's top musicians and singers.

Currently Seabrook is the multi-genre author and editor of nearly 100 adult and children's books (totaling some 30,000 pages and 15,000,000 words) that have earned him accolades from around the globe. His works, which have sold on every continent except Antarctica, have introduced hundreds of thousands to vital facts that have been left out of our mainstream books. He has been endorsed internationally by leading experts, museum curators, award-winning historians, bestselling authors, celebrities, filmmakers, noted scientists, well regarded educators, TV show hosts and producers, renowned military artists, venerable heritage organizations, and distinguished academicians of all races, creeds, and colors.

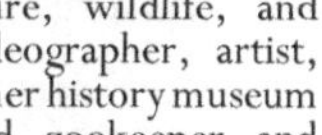

Of northern, western, and central European ancestry, he is the 6th great-grandson of the Earl of Oxford and a descendant of European royalty through his Kentucky father and West Virginia mother. His modern day cousins include: Johnny Cash, Elvis Presley, Lisa Marie Presley, Billy Ray and Miley Cyrus, Patty Loveless, Tim McGraw, Lee Ann Womack, Dolly Parton, Pat Boone, Naomi, Wynonna, and Ashley Judd, Ricky Skaggs, the Sunshine Sisters, Martha Carson, Chet Atkins, Patrick J. Buchanan, Cindy Crawford, Bertram Thomas Combs (Kentucky's 50th governor), Edith Bolling (second wife of President Woodrow Wilson), Andy Griffith, Riley Keough, George C. Scott, Robert Duvall, Reese Witherspoon, Lee Marvin, Rebecca Gayheart, and Tom Cruise.

A constitutionalist, avid outdoorsman, and gun rights advocate, Seabrook is the author of the international blockbuster, *Everything You Were Taught About the Civil War is Wrong, Ask a Southerner!* He lives with his wife and family in the magnificent Rocky Mountains, heart of the American West, where you will find him hiking, filming, and writing.

For more information on author Mr. Seabrook visit
LochlainnSeabrook.com

408 ○○ MYSTERIOUS INVADERS

Keep Your Body, Mind, & Spirit Vibrating at Their Highest Level
YOU CAN DO SO BY READING THE BOOKS OF

SEA RAVEN PRESS

There is nothing that will so perfectly keep your body, mind, and spirit in a healthy condition as to think wisely and positively. Hence you should not only read this book, but also the other books that we offer. They will quicken your physical, mental, and spiritual vibrations, enabling you to maintain a position in society as a healthy erudite person.

KEEP YOURSELF WELL-INFORMED!

The well-informed person is always at the head of the procession, while the ignorant, the lazy, and the unthoughtful hang onto the rear. If you are a Spiritual man or woman, do yourself a great favor: read Sea Raven Press books and stay well posted on the Truth. It is almost criminal for one to remain in ignorance while the opportunity to gain knowledge is open to all at a nominal price.

We invite you to visit our Webstore for a wide selection of wholesome, family-friendly, well-researched, educational books for all ages. You will be glad you did!

Artisan-Crafted Books & Merch from the Rocky Mountains

SeaRavenPress.com

LochlainnSeabrook.com
TheBestCivilWarBookEver.com
AmbianceGoneWild.com
Pond5.com/artist/LochlainnSeabrook

LOCHLAINN SEABROOK ~ 409

If you enjoyed this book you will be interested in Colonel Seabrook's popular related scientific titles:

- UFOs AND ALIENS: THE COMPLETE GUIDEBOOK
- THE MARTIAN ANOMALIES: A PHOTOGRAPHIC SEARCH FOR INTELLIGENT LIFE ON MARS
- CHRISTMAS BEFORE CHRISTIANITY: HOW THE BIRTHDAY OF THE "SUN" BECAME THE BIRTHDAY OF THE "SON"
- THE CONCISE BOOK OF OWLS: A GUIDE TO NATURE'S MOST MYSTERIOUS BIRDS
- SEABROOK'S BIBLE DICTIONARY OF TRADITIONAL & MYSTICAL CHRISTIAN DOCTRINES
- EVERYTHING YOU WERE TAUGHT ABOUT THE CIVIL WAR IS WRONG, ASK A SOUTHERNER!

Available from Sea Raven Press and wherever fine books are sold

ALL OF OUR BOOK COVERS ARE AVAILABLE AS 11" X 17" COLOR POSTERS, SUITABLE FOR FRAMING

SeaRavenPress.com